Eric Ling and Annachiara Piubello
Editors

SPACETIME
1908-2023

Selected peer-reviewed papers presented at the Third Hermann Minkowski Meeting on the Foundations of Spacetime Physics, dedicated to the 115th anniversary of Minkowski's lecture "Space and Time," 11-14 September 2023, Albena, Bulgaria

MINKOWSKI
Institute Press

Eric Ling
Department of Mathematical Sciences
University of Copenhagen

Annachiara Piubello
Institute of Mathematics
University of Potsdam

Front Cover: Most participants of the *Third Hermann Minkowski Meeting on the Foundations of Spacetime Physics* held in the famous resort Albena (near Varna) on the Bulgarian Black Sea coast from 11 to 14 September 2023. Picture taken by Vladimir Petev (Front office manager of Hotel Sandy Beach)

© Minkowski Institute Press 2024

ISBN: 978-1-998902-25-5 (softcover)

ISBN: 978-1-998902-26-2 (ebook)

Minkowski Institute Press
Montreal, Quebec, Canada
http://minkowskiinstitute.com/mip/

For information on all Minkowski Institute Press publications visit our website at
http://minkowskiinstitute.com/mip/books/

This volume is dedicated to the memory of our late colleague
Durmuş Ali Demir who passed away suddenly on February 24, 2024

Durmuş Demir presenting his paper "Poincaré Breaking in Flat Minkowski and Gauge-Restoring Emergent Gravity" at the Third Minkowski Meeting on September 11, 2023.

Contents

PREFACE

This volume contains a selection of peer-reviewed contributions to the Third Hermann Minkowski Meeting on the Foundations of Spacetime Physics, held in Albena, Bulgaria, from September 11-14, 2023. The meeting commemorated the 115th anniversary of Minkowski's groundbreaking 1908 lecture "Space and Time," which introduced the revolutionary concepts of spacetime structure and four-dimensional physics. The contributing papers are organized into four main categories, each represented by a distinct part of the volume. This classification is intended to serve as a general guide rather than an absolute framework.

In Part 1, focusing on Quantum Theories on Curved Spacetimes, we delve into the foundational aspects of gravitational and quantum phenomena. Durmuş Demir discusses an emergent gravity framework where gauge symmetries are restored via affine condensation, predicting new massive particles and suggesting a flat universe where gravity and gauge forces emerge. Pawel Gusin, Andrzej Radosz, and Romuald J. Ściborski explore possible instabilities of a quantized scalar field within a dynamically changing Schwarzschild black hole, emphasizing gravitationally driven instabilities caused by the background spacetime. Tomohiro Matsuda proposes a method for defining the vacuum on manifolds to explain local particle production phenomena such as the Schwinger, Unruh, and Hawking effects, thereby resolving issues with traditional field equation approaches.

In Part 2, dedicated to Cosmological Models, the focus shifts to innovative models explaining the evolution of the universe. Eric Ling and Annachiara Piubello provide mathematical support for inflationary cosmology within anisotropic spacetimes, showing that specific initial conditions lead to inflationary periods dominated by a cosmological constant. Seokcheon Lee explores the Milne model, a special-relativistic cosmological model equivalent to a specific case of the Robertson-Walker model with negative spatial curvature and zero energy density, demonstrating that the cosmological principle and redshift relation hold even if the speed of light varies at cosmological scales. Guido J.M. Verstraeten and Willem W. Verstraeten integrate Whitehead's philosophical concepts with Verlinde's quantum gravitation theories, proposing a Big Bounce cosmological model that challenges the traditional Big Bang origin. Sanjeeda Sultana and Surajit Chattopadhyay investigate the cosmology of a generalized holographic dark energy model using a logotropic equation of state, reconstructing the dark energy density within the universe and proving the generalized second law of thermodynamics for various cosmological horizons.

In Part 3, exploring Gravitational and Quantum Theories, researchers investigate specific gravitational anomalies and effects within different systems. Asher Yahalom examines a gravitational anomaly in wide binary star systems, proposing that it can be explained within the framework of general relativity by considering the galaxy's gravitational field. Patrick Das Gupta extends General Relativity by introducing a dynamic four-form field leading to torsion and Chern-Simons gravity, exploring its implications for dark matter and gravitational wave propagation. Piotr Ogonowski examines the relationship between the metric tensor and the field tensor, analyzing implications for the structure of spacetime and suggesting directions for further research. Orfeu Bertolami and A. E. Bernardini argue that time crystal properties naturally arise in phase-space noncommutative quantum mechanics, exemplified by the periodic oscillations of a two-dimensional quantum harmonic oscillator.

In Part 4, centered on the Ontology of Spacetime and the Nature of Time, contributions address the fundamental nature and existence of spacetime and time itself. Vesselin Petkov questions the existence of singularities in black holes as traditionally predicted by general relativity, arguing that black holes may not form as conventionally understood and highlighting discrepancies in their formation process for distant observers. Ovidiu Cristinel Stoica explores the relationship between observables and physical properties, arguing that spacetime structure is tied to the sentience of observers. Stuart Kauffman and Stephen Guerin propose a nonlocality-based model where the universe self-constructs through collective autocatalysis of entangled particles, offering explanations for cosmogenesis, cosmic inflation, and the fundamental laws of physics. Bruce M. Boman models the progression of time as an autocatalytic reaction, explaining the perpetual and asymmetric nature of time progression through kinetic reaction equations and changes in energy and entropy.

We hope this volume effectively conveys the diverse aspects of the foundations of spacetime physics, particularly those stemming from Minkowski's original vision, to the broader scientific community, showcasing innovative approaches and groundbreaking theories that push the boundaries of our understanding of the universe.

The editors

Contributors

A. E. Bernardini
Departamento de Física, Universidade Federal de São Carlos
alexeb@ufscar.br

Orfeu Bertolami
Departamento de Física e Astronomia
Faculdade de Ciências, Universidade do Porto
Centro de Física das Universidades do Minho e do Porto
orfeu.bertolami@fc.up.pt

Bruce M. Boman
Department of Mathematical Sciences
University of Delaware

Surajit Chattopadhyay
Department of Mathematics, Amity University
surajitchatto@outlook.com
schattopadhyay1@kol.amity.edu

Patrick Das Gupta
Department of Physics and Astrophysics
University of Delhi
pdasgupta@physics.du.ac.in

Durmuş Demir
Sabancı University, Faculty of Engineering and Natural Sciences
İstanbul, Türkiye

Stephen Guerin
Department of Earth and Planetary Sciences
Harvard University
stephenguerin@fas.harvard.edu

Pawel Gusin
Department of Quantum Technologies,
Wroclaw University of Science and Technology

Stuart Kauffman
University of Pennsylvania
stukauffman@gmail.com

Seokcheon Lee
Department of Physics, Institute of Basic Science, Sungkyunkwan University
skylee@skku.edu

Eric Ling
Copenhagen Centre for Geometry and Topology (GeoTop)
Department of Mathematical Sciences, University of Copenhagen
el@math.ku.dk

Tomohiro Matsuda
Laboratory of Physics, Saitama Institute of Technology, Fukaya

Piotr Ogonowski
Kozminski University
piotrogonowski@kozminski.edu.pl

Vesselin Petkov
Minkowski Institute, Montreal
vpetkov@minkowskiinstitute.org

Annachiara Piubello
Institute of Mathematics, University of Potsdam
annachiara.piubello@uni-potsdam.de

Andrzej Radosz
Department of Quantum Technologies,
Wroclaw University of Science and Technology

Romuald J. Ściborski
Department of Quantum Technologies
Wroclaw University of Science and Technology
and Jaramogi Oginga Odinga University of Science and Technology in Bondo (Kenya)

Anguel S. Stefanov
Bulgarian Academy of Sciences
angstefanov@abv.bg

Ovidiu Cristinel Stoica
Dept. of Theoretical Physics, NIPNE—HH, Bucharest
cristi.stoica@theory.nipne.ro
holotronix@gmail.com

Sanjeeda Sultana
Department of Mathematics, Amity University
sanjeedasultana98@gmail.com
sanjeeda.sultana1@s.amity.edu

Guido J.M. Verstraeten
Satakunta University of Applied Sciences, Björneborg-Pori
KAEHIE, Brussels
guido.verstraeten@kahiel.be

Willem W. Verstraeten
Royal Meteorological Institute of Belgium

Asher Yahalom
Ariel University
asya@ariel.ac.il

Part I

Quantum Theories on Curved Spacetimes

Eric Ling and Annachiara Piubello (Eds), SPACETIME 1908-2023. Selected peer-reviewed papers presented at the *Third Hermann Minkowski Meeting on the Foundations of Spacetime Physics*, 11-14 September 2023, Albena, Bulgaria (Minkowski Institute Press, Montreal 2024). ISBN 978-1-998902-25-5 (softcover), ISBN 978-1-998902-26-2 (ebook).

1 Poincaré Breaking in Flat Minkowski and Gauge-Restoring Emergent Gravity

Durmuş Demir

Abstract   In this talk, we discuss an emergent gravity completion of the Standard Model. The UV cutoff in flat spacetime gives mass to gauge fields via the loops, and we interpret the UV cutoff as the condensate of the affine curvature in view of the Poincaré-breaking nature of the UV cutoff. Affine condensation leads to the restoration of gauge symmetries and the emergence of gravity, with the prediction of new massive particles such that the new particles do not have to couple to the SM directly. In this picture, the Universe starts out flat, where the gauge forces and gravity take shape in a time set by the UV cutoff.

Keywords: flat Minkowski spacetime, emergent gravity, standard model, UV cutoff

1   Introduction

The strong, weak, and electromagnetic interactions are described by the SM as a renormalizable quantum field theory (QFT) [1]. The gravity is not included. Due to challenges in quantizing the curved metric [2] and challenges in taking QFTs into curved spacetime [3, 4, 5], the SM is unique to the flat Minkowski. As nearly-classical field theories obtained by integrating out high-energy quantum fluctuations [1, 6] in the sense of both the Wilsonian and one-particle-irreducible effective actions [7, 8], the flat spacetime effective QFTs, unlike full QFTs like the SM, can have a certain degree of affinity with curved spacetime.

As an effective QFT, the SM is endowed with a physical UV cutoff scale (not a regulator) [6]. This UV cutoff, which we denote as $\Lambda_\wp$, is the scale at which a UV completion sets in. The null LHC searches [9] indicate that $\Lambda_\wp$ could lie at any scale above a TeV. It explicitly breaks Poincaré (translation) symmetry, restricting the loop momenta ℓ^μ into the range $-\Lambda_\wp^2 \le \eta_{\mu\nu}\ell^\mu\ell^\nu \le \Lambda_\wp^2$. Then scalar mass-squareds get corrections proportional to $\Lambda_\wp^2$. Similarly, $\Lambda_\wp^4$ and $\Lambda_\wp^2$ appear in the corrected vacuum energy [6, 10]. In addition to these unnatural UV sensitivities [6], all the gauge bosons gain mass-squareds proportional to $\Lambda_\wp^2$ and these loop-induced masses clearly break gauge symmetries [11, 12]. In this sense, restoring the gauge symmetries by the Higgs mechanism [13, 14, 15] would be the most logical course of action. Standing in the way of this proposal is the stark difference between the intermediate vector boson mass (Poincaré-conserving) [16] and the loop-induced gauge boson mass (Poincaré-breaking) [17]. Indeed, in the former, the vector

Eric Ling and Annachiara Piubello (Eds), SPACETIME 1908-2023. Selected peer-reviewed papers presented at the *Third Hermann Minkowski Meeting on the Foundations of Spacetime Physics*, 11-14 September 2023, Albena, Bulgaria (Minkowski Institute Press, Montreal 2024). ISBN 978-1-998902-25-5 (softcover), ISBN 978-1-998902-26-2 (ebook).

boson mass is promoted to a scalar field, which leads to the usual Higgs mechanism [13, 14, 15]. In the latter, however, it is necessary to find a Poincaré-breaking Higgs field and employ the Higgs mechanism accordingly. In this regard, as we will discuss in this talk, we construct a mechanism in which the UV cutoff $\Lambda_\wp$ is promoted to affine curvature [18, 19] as the Higgs field such that the affine curvature condenses to the usual metrical curvature in the minimum of the metric-affine action [20, 21]. This condensation results in the defusion of the unnatural UV sensitivities, restoration of the gauge symmetries, the emergence of gravity, and the necessity of new particles beyond the SM with no necessity to couple to the SM directly. This affine condensation mechanism gives a bottom-up UV completion of the QFT with emergent gravity. The results reported in this talk have been obtained in a number of papers [17, 22, 23, 24], with the most recent one being [25].

This proceeding volume is organized as follows. In Sec. II below, we construct the effective QFT in flat Minkowski. In Sec. III, we discuss how to take a given effective QFT to the spacetime of a curved metric. In Sec. IV, we give a detailed discussion of the affine condensation mechanism by indicating explicitly how the Einstein-Hilbert term arises as a result of the condensation. In Sec. V we conclude.

2   The Effective QFT and Its UV Sensitivities

For a generic renormalizable theory of quantum fields ψ (the SM fields plus new particles), in flat spacetime at scales $\mu \ll \Lambda_\wp$, the low-energy effective action is given by [25, 17, 22]

$$S[\eta, \psi; \log \mu, \Lambda_\wp^2] = S_{tree}[\eta, \psi] + \delta S_{log}[\eta, \psi; \log \mu] + \\ + \delta S_{pow}[\eta, \psi; \log \mu, \Lambda_\wp^2] \tag{2.1}$$

after extending the dimensional regularization [26, 27] to QFTs with UV cutoff $\Lambda_\wp$ [28]. Here, the QFT is defined by the tree-level action S_{tree} in terms of the field spectrum and symmetries. The piece δS_{log} stands for the logarithmic corrections, which involve the renormalization scale μ (not the UV cutoff $\Lambda_\wp$) in parallel with S_{tree} in view of the renormalizability. The power-law correction

$$\delta S_{pow} = \int d^4x \sqrt{-\eta} \left\{ - c_O \Lambda_\wp^4 - 2M_O^2 \Lambda_\wp^2 - c_\phi \Lambda_\wp^2 \phi^\dagger \phi + \\ + c_V \Lambda_\wp^2 \text{tr} \left[V^\mu \eta_{\mu\nu} V^\nu \right] \right\} \tag{2.2}$$

contains the scalars ϕ and gauge fields V_μ such that tr[...] denotes the color trace. In this action, the loop-induced factors

$$c_O = \frac{(n_b - n_f)}{64\pi^2} \tag{2.3}$$

and

$$M_O^2 = \frac{\text{str}[M^2]}{64\pi^2} \tag{2.4}$$

arise from the flat spacetime matter loops such that n_b (n_f) stands for the total number of bosonic (fermionic) degrees of freedom in the QFT, and M^2 is the mass-squared matrix of all the QFT fields, with the supertrace $\text{str}[M^2] = \sum_s (-1)^{2s} (2s+1) M_s^2$ over the particle spin s. In action (2.2), the loop factor c_ϕ controls the quadratic corrections to scalar mass-squares [22, 23, 24].

4

3 Taking Flat Spacetime Effective QFT to Curved Spacetime

By the nature of low-energy effective actions, the matter fields ψ in (2.1) are essentially mean fields with small quantum fluctuations; namely, they are nearly-classical fields. In this regard, it should be possible to carry the effective action $S[\eta]$ to spacetime of a curved metric $g_{\mu\nu}$ smoothly as in the classical field theories. The way to do this is the general covariance map [29]

$$\eta_{\mu\nu} \to g_{\mu\nu}, \quad \partial_\mu \to \nabla_\mu \tag{3.1}$$

in which ∇_μ is the covariant derivative with respect to the Levi-Civita connection

$${}^g\Gamma^\lambda_{\mu\nu} = \frac{1}{2}g^{\lambda\rho}\left(\partial_\mu g_{\nu\rho} + \partial_\nu g_{\rho\mu} - \partial_\rho g_{\mu\nu}\right) \tag{3.2}$$

so that $\nabla_\alpha g_{\mu\nu} = 0$. However, for the metric $g_{\mu\nu}$ to be dynamical (curved), its curvature (kinetic term), such as the Ricci curvature $R_{\mu\nu}({}^g\Gamma)$, must appear in the effective QFT in curved spacetime. In classical field theories, this can be achieved simply by adding requisite curvature terms (such as the Einstein-Hilbert term) with appropriate bare constants. But, such a by-hand addition curvature is not allowed in effective QFTs. The reason is that in the effective QFTs, all masses and couplings are either generated or corrected by the loop corrections, and the addition of curvature terms contradicts the renormalized QFT. In the face of this contradiction, we are left with the sole option that curvature must arise from within the flat spacetime effective action (2.1) with the loop factors c_i (not some bare parameters). In other words, the curvature has to arise via the deformations of the existing loop corrections in δS_{pow}. In fact, deformations are seeded by the commutator $[\nabla_\mu, \nabla_\nu]$ as it leads to curvature in the form $V^\mu[\nabla_\nu, \nabla_\mu]V^\nu = V^\mu R_{\mu\nu}({}^g\Gamma)V^\nu$. The appearance of the quadratic form $V^\mu R_{\mu\nu}({}^g\Gamma)V^\nu$ implies that the object to be deformed is actually the gauge boson mass term $c_V \Lambda_\wp^2 \text{tr}\left[V^\mu \eta_{\mu\nu}V^\nu\right]$. To implement such a deformation, one starts with the flat spacetime null deformation

$$\begin{aligned}
\widetilde{S}[\eta] &= S[\eta] + \int d^4x \sqrt{-\eta}\, c_V \,\text{tr}\left[V^\mu([D_\mu, D_\nu] - iV_{\mu\nu})V^\nu\right] \\
&= S[\eta]
\end{aligned} \tag{3.3}$$

which is null because $[D_\mu, D_\nu]V^\nu = iV_{\mu\nu}V^\nu$, with $D_\mu = \partial_\mu + iV_\mu$ being the gauge-covariant derivative, and $V_{\mu\nu}$ being the field strength tensor of the gauge field V_μ (a vector in adjoint). The nullness ensures that the flat spacetime effective action $S[\eta]$ remains unchanged, $\widetilde{S}[\eta] \equiv S[\eta]$. But, once the general covariance map in (3.1) is applied the null deformation (3.3) gives rise to a nontrivial relation

$$\begin{aligned}
\widetilde{S}[g] &= S[g] + \int d^4x \sqrt{-g}\, c_V \,\text{tr}\left[V^\mu([\mathcal{D}_\mu, \mathcal{D}_\nu] - iV_{\mu\nu})V^\nu\right] \\
&= S[g] - \int d^4x \sqrt{-g}\, c_V \,\text{tr}\left[V^\mu R_{\mu\nu}({}^g\Gamma)V^\nu\right]
\end{aligned} \tag{3.4}$$

as the curved spacetime image of the flat spacetime effective action $S[\eta]$. In this curved spacetime effective action, curvature arises in the second line via the commutator $[\mathcal{D}_\mu, \mathcal{D}_\nu]V^\nu = (-R_{\mu\nu}({}^g\Gamma) + iV_{\mu\nu})V^\nu$ of the curved spacetime covariant derivative $\mathcal{D}_\mu = \nabla_\mu + iV_\mu$. The $R_{\mu\nu}({}^g\Gamma)$ term in (3.4) ensures via the Levi-Civita connection in (3.2) that the metric $g_{\mu\nu}$ is curved (dynamical) and invertible. But, despite all this, there is no gravity! There is no gravity because the Einstein-Hilbert term is missing in (3.4). This imperative term, proportional to the curvature scalar $g^{\mu\nu}R_{\mu\nu}(^{g}\Gamma)$, must somehow emerge from within the curved spacetime effective action (3.4).

The main objective is to determine how the Einstein-Hilbert term emerges from within the curved spacetime effective action $\widetilde{S}[g]$. In this regard, it proves useful to contrast the two mass terms: $M_I^2\text{tr}[I_\mu I^\mu]$ and $c_V\Lambda_\wp^2\text{tr}[V_\mu V^\mu]$. In the former, M_I conserves the Poincaré symmetry as it is the mass of an intermediate vector boson I_μ (like the W/Z bosons), and it can, therefore, be promoted as

$$M_I^2\text{tr}[I_\mu I^\mu] \longmapsto (I_\mu S)^\dagger I^\mu S \subset (D_\mu S)^\dagger D^\mu S \tag{3.5}$$

to a Poincaré-conserving Higgs scalar S with the gauge-covariant derivative $D_\mu = \partial_\mu + iI_\mu$ [13, 14, 15, 16]. This promotion restores the gauge symmetry at the Lagrangian level. In the latter, however, the UV cutoff $\Lambda_\wp^2$ breaks the Poincaré symmetry, and thus, it can be promoted to only a Poincaré-breaking field (not a scalar like S) for restoring the gauge symmetry [17, 22, 23, 24]. It is known that Poincaré-breaking fields are related to the spacetime curvature [30]. But, the Poincaré-breaking field promoting $\Lambda_\wp^2$ must be able to live with both the flat and curved metrics, and one particular field that has these properties is the Ricci curvature $\mathbb{R}_{\mu\nu}(\Gamma)$ of a general affine connection $\Gamma_{\mu\nu}^\lambda$. It is so because $\Gamma_{\mu\nu}^\lambda$ is independent of the metric $g_{\mu\nu}$ and yet approaches to $^{g}\Gamma_{\mu\nu}^\lambda$ in (3.2) via the affine dynamics and, as a result of this approach, $\mathbb{R}_{\mu\nu}(\Gamma)$ tends to $R_{\mu\nu}(^{g}\Gamma)$. This dynamical tendency ensures that $\mathbb{R}_{\mu\nu}(\Gamma)$ could be the sought-after Poincaré-breaking field. In this sense, the UV cutoff $\Lambda_\wp$ can be promoted as [17, 22, 23, 24]

$$\Lambda_\wp^2 g_{\mu\nu} \longmapsto \mathbb{R}_{\mu\nu}(\Gamma) \tag{3.6}$$

in parallel with the promotion in (3.5) of the intermediate vector boson masses M_I to scalars S [16]. It makes the curved spacetime effective action $\widetilde{S}[g]$ in (3.4) a metric-Palatini gravity theory [21, 31]

$$\begin{aligned}
\widetilde{S}[g,\Gamma,\psi] =\ & S_{tree}[g,\psi] + \delta S_{log}[g,\psi;\log\mu] \\
& + \int d^4x\sqrt{-g}\left\{ -\frac{c_O}{16}\mathbb{R}^2 - \frac{M_O^2}{2}\mathbb{R} - \frac{c_\phi}{4}\phi^\dagger\phi\,\mathbb{R} \right. \\
& \left. + c_V\text{tr}\left[V^\mu(\mathbb{R}_{\mu\nu} - R_{\mu\nu})V^\nu\right] \right\}
\end{aligned} \tag{3.7}$$

in which $\mathbb{R} = g^{\mu\nu}\mathbb{R}_{\mu\nu}(\Gamma)$ is the affine scalar curvature. Its coefficient M_O^2 has to be positive and much bigger than any QFT scale since Palatini dynamics will eventually reshape it as the fundamental scale of gravity [19, 20, 21]

$$M_O \equiv M_{Pl} \tag{3.8}$$

and give it this way a whole new role compared to its original definition in (2.4) and original role in (2.2) in the flat spacetime.

4 Gauge-Restoring Emergent Gravity

Being nearly classical, the stationary points of the metric-Palatini action (3.7) are expected to yield physical field configurations as the extremal points. Thus, sending the matter sector to its vacuum configuration ($\langle V_\mu \rangle = 0$, $\langle f \rangle = 0$, $\langle \phi \rangle / M_{Pl} \approx 0$), the extremal values of Γ follow from the stationarity condition

$$\frac{\delta \widetilde{S}[g,\Gamma]}{\delta \Gamma^\lambda_{\mu\nu}} = 0 \implies {}^\Gamma\nabla_\lambda \left(\mathbb{Q}^{1/3} g^{\mu\nu} \right) = 0 \tag{4.1}$$

in which ${}^\Gamma\nabla_\lambda$ is covariant derivative with respect to Γ, and the scalar

$$\mathbb{Q} = \frac{M_{Pl}^2}{2} + \frac{c_O}{8} \mathbb{R} \tag{4.2}$$

is nothing but the variation of $-\widetilde{S}[g,\Gamma]$ with $\mathbb{R}$ in the vacuum.

The first solution of the motion equation (4.1) is $\mathbb{Q} = 0$. And it leads to a homogeneous affine curvature scalar $-4M_{Pl}^2/c_O$ from (4.2). It is the extremal curvature that leaves the action (3.7) stationary. It must be equal to $4\Lambda_\wp^2$ since the affine curvature $\mathbb{R}$ is the "Higgs field" promoting $\Lambda_\wp^2$ as in (3.6), and its extremal value must therefore regenerate the UV cutoff. But, this regeneration happens if $-4M_{Pl}^2/c_O$ is positive, and this can happen only if

$$c_O < 0 \implies n_b < n_f \tag{4.3}$$

which means that nature has more fermions than bosons. This negative c_O makes the $\mathbb{R}$–potential in (3.7) maximum at the extremum $-4M_{Pl}^2/c_O > 0$. As a result, $\mathbb{Q} = 0$ solution with $c_O < 0$ leads to the extremal affine curvature scalar

$$\mathbb{R}_{\max} = -\frac{4M_{Pl}^2}{c_O} = 4\Lambda_\wp^2 \tag{4.4}$$

at which $-\widetilde{S}[g,\Gamma]$ attains its maximum, and at this maximum $\widetilde{S}[g,\Gamma]$ reduces to the curved spacetime action in (3.4) and revives therefore all the power-law UV corrections in (2.2). This maximum is where all the UV problems arise, including the explicit gauge breaking.

The second solution of the motion equation (4.1) is obtained with $\mathbb{Q} \neq 0$ and leads to the minimal configuration

$$(\Gamma_{\min})^\lambda_{\mu\nu} = {}^g\Gamma^\lambda_{\mu\nu} + \frac{1}{6\mathbb{Q}} \left(\nabla_\mu \delta^\lambda_\nu + \nabla_\nu \delta^\lambda_\mu - \nabla^\lambda g_{\mu\nu} \right) \mathbb{Q} \tag{4.5}$$

which we take to be the minimal configuration since the maximal configuration was already found in (4.4). Expanding the minimal configuration (4.5) up to $\mathcal{O}\left(M_{Pl}^{-4} \right)$ terms, the extremal affine curvature is found to be

$$(\mathbb{R}_{\min})_{\mu\nu} = R_{\mu\nu}({}^g\Gamma) - \frac{c_O}{24M_{Pl}^2} (\Box g_{\mu\nu} + 2\nabla_\mu \nabla_\nu) R \tag{4.6}$$

where $R = g^{\mu\nu} R_{\mu\nu}({}^g\Gamma)$ is the metrical curvature scalar. In this minimum, the gauge part of (3.7) reduces to

$$\int d^4x \sqrt{-g} \, c_V \text{tr} \left[V^\mu \left(\text{zero} + \mathcal{O}\left(M_{Pl}^{-2} \right) \right) V^\nu \right] \tag{4.7}$$

which is seen to vanish up to $\mathcal{O}\left(M_{Pl}^{-2}\right)$ terms. This vanishing of the gauge boson mass terms is proof that the gauge symmetries got restored in the minimum in (4.6) up to doubly Planck-suppressed terms.

For the minimal curvature value in (4.6), the metric-Palatini effective action (3.7) generates, up to $\mathcal{O}\left(M_{Pl}^{-2}\right)$ terms, a conjoint of the dimensionally-regularized QFT and $R + R^2$ gravity theory

$$
\begin{aligned}
S[g] &= S_{tree}[g, \psi] + \delta S_{log}[g, \psi; \log \mu] \\
&+ \int d^4 x \sqrt{-g} \left\{ -\frac{M_{Pl}^2}{2} R - \frac{c_O}{16} R^2 - \frac{c_\phi}{4} \phi^\dagger \phi R \right\}
\end{aligned}
\tag{4.8}
$$

up to doubly Planck-suppressed terms, including the remainder in (4.7). This action is the end result of the affine dynamics and illustrates the emergence of the curvature sector via the promotion in (3.6) of the UV cutoff to affine curvature and condensation in (4.6) of the affine curvature to the metrical curvature. It is clear that affine condensation results in both the gauge and gravitational interactions. The action (4.8) gives a gauge symmetry restoring emergent gravity framework, which we abbreviate as *symmergent gravity* [17, 22, 23]. Symmergence is the physics at the minimum. Fig. 1 contrasts the maximum and minimum item by item and makes it clear that the QFT is carried into a physically viable minimum in (4.6) via the affine dynamics in (4.1) starting from the problem-full maximum in (4.4).

Figure 1: A schematic plot of $-\widetilde{S}[g, \Gamma]$ drawn from its maximum in (4.4) and minimum in (4.6). Its minimum leads to symmergent gravity. (This figure is taken from Ref. [25].)

Symmergence possesses a number of physically important properties. The first property concerns the gravitational scale M_{Pl}, which comes out wrong in the SM given in its definition in (2.4) via (3.8). This means that new particles with heavy bosons are a necessity, but their couplings to the SM are not as because they do not need any non-gravitational couplings to SM to satisfy the mass sum rule in (2.4). In fact, one finds $n_f - n_b < 3.75 \times 10^{33}$ from (4.4) for the allowed range $\Lambda_\wp > 1$ TeV at the LHC [9]. In flat spacetime, there is no preset scale like M_{Pl}, and thus, it is possible to take $\Lambda_\wp > M_{Pl}$, which requires $n_f - n_b < 630$.

The second property is that the Higgs and other scalars can remain light, with no problem for workings of the symmergence. Indeed, having the power-law UV corrections in (2.2) converted to the curvature terms in (3.7) in symmergence, large logarithms in δS_{log} remain as the only source of destabilization for light scalars. But those logarithms

$$
\delta m_h^2 \propto \lambda_{h\Psi} M_\Psi^2 \log \frac{M_\Psi^2}{\mu^2}
\tag{4.9}
$$

do not have to be large since the Higgs coupling $\lambda_{h\Psi}$ to new fields Ψ is allowed to be small even vanish in symmergence. In fact, couplings of the size $\lambda_{h\Psi} \lesssim m_h^2/m_\Psi^2$ stabilize δm_h^2 and render the Higgs boson natural [32]. This naturalness window also allows a natural dark sector, given the fact that the dark fields do not have to couple to the SM fields. In fact, the dark sector can be all black (completely decoupled) in agreement with the current dark matter bounds [33, 34, 32]. In this sense, black stars or even galaxies become a possibility for astrophysical observations.

The third property implies that the Universe starts out flat as an extended quantum object, and gravity emerges afterward from quantum fluctuations within a time around $\Lambda_\wp^{-1}$ as expected from (3.6). Indeed, the maximum and minimum in Fig. 1 are split in energy by $\Lambda_\wp - M_{Pl} \sim \Lambda_\wp$ so that gravity emerges within a time about $\Lambda_\wp^{-1}$. Once gravity emerges, the symmergent evolution can be described by an FRW Universe with a quasi-linear scale factor [35, 36, 37]. There is a sharp contrast between this symmergent cosmology and the quantum cosmology [38] in that the Universe starts flat (no gravity) in the former and curved (quantum gravity) in the latter.

The last property is that the symmergent gravity action (4.8) gives an intertwined merge of the renormalized QFT and classical gravity. They are nearly compatible, but they can still be regarded as incompatible in the face of small remnant fluctuations of the quantum fields. However, given the Einstein field equations following from (4.8), fluctuations in quantum fields can be received as stochastic fluctuations in the spacetime metric. The irreversibility of the stochastic fluctuations can give rise to a certain compatibility between the QFT and the gravity [4, 5].

5   Conclusion

In this talk, we have discussed an emergent gravity completion of QFTs. Actually, the QFT turns out to be the SM plus new particles required by the correct value of the gravitational constant. As was proven by a detailed analysis, affine curvature condensation has led to the restoration of gauge interactions, the emergence of gravity, and the renormalization of the underlying QFT. The resulting framework, the symmergent gravity, intertwines gravity and SM with the introduction of new particles that do not have to couple to the SM directly and with the allowance to suppress couplings to heavy fields for enabling stabilization of the scalar masses. In this picture, the Universe starts out flat, where the gauge forces and gravity take shape in a time set by the UV cutoff. This symmergent gravity framework can be probed via cosmic inflation [39, 40] and reheating; black holes [41, 42, 43, 44, 45, 46, 47], wormholes and neutron stars; and collider [48] and beyond-the-collider [32] experiments.

For future prospects, one important research direction would be the symmergent evolution of the Universe in which the usual quantum gravity phase is replaced with the flat spacetime [35, 36, 37]. Another direction would be probes of symmergence in astrophysical compact objects like neutron stars and black holes. There are actually various research directions since symmergence brings in not only an emergent gravity sector but also a dark (or black) sector having its own astrophysical, cosmological, and collider signatures [17, 22, 23, 24].

Acknowledgements

I thank Orfeu Bertolami, Patrick Das Gupta, Canan Karahan, Eric Ling, Sebastian Murk, Ali Övgün, Beyhan Puliçe, Ozan Sargın, and Kai Schwenzer for fruitful discussions on different aspects of this work. This talk material parallels the recent paper [25].

References

[1] S. Weinberg, Phys. Rev. Lett. **121**, 220001 (2018).

[2] G. 't Hooft and M. J. G. Veltman, Ann. Inst. H. Poincaré Phys. Theor. A **20**, 69 (1974).

[3] S. Hollands and R. M. Wald, Phys. Rept. **574**, 1 (2015) [arXiv:1401.2026 [gr-qc]].

[4] J. Oppenheim, arXiv:1811.03116 [hep-th].

[5] T. D. Galley, F. Giacomini and J. H. Selby, Quantum **7**, 1142 (2023) [arXiv:2301.10261 [quant-ph]].

[6] S. Weinberg, Eur. Phys. J. H **46**, 6 (2021) [arXiv:2101.04241 [hep-th]].

[7] J. Polchinski, Nucl. Phys. B **231**, 269 (1984).

[8] C. P. Burgess, Ann. Rev. Nucl. Part. Sci. **57**, 329 (2007) [arXiv:hep-th/0701053 [hep-th]].

[9] E. Gibney, Nature **605**, 604 (2022).

[10] M. J. G. Veltman, Acta Phys. Polon. B **12**, 437 (1981).

[11] H. Um zawa, J. Yukawa and E. Yamada, Prog. Theor. Phys. **3** (1948) 317.

[12] W. Pauli and F. Villars, Rev. Mod. Phys. **21**, 434 (1949).

[13] F. Englert a d R. Brout, Phys. Rev. Lett. **13**, 321 (1964).

[14] P. W. Higgs, Phys. Rev. Lett. **13**, 508 (1964).

[15] G. S. Guralnik, C. R. Hagen and T. W. B. Kibble, Phys. Rev. Lett. **13**, 585 (1964).

[16] S. Weinberg, Phys. Rev. Lett. **19**, 1264 (1967).

[17] D. Demir, Phys. Rev. D **107**, 105014 (2023) [arXiv:2305.01671 [hep-th]].

[18] F. Bauer and D. A. Demir, Phys. Lett. B **665**, 222 (2008) [arXiv:0803.2664 [hep-ph]].

[19] D. A. Demir, Phys. Rev. D **90**, 064017 (2014) [arXiv:1409.2572 [gr-qc]].

[20] V. Vitagliano, T. P. Sotiriou and S. Liberati, Annals Phys. **326** (2011) 1259 [arXiv:1008.0171 [gr-qc]].

[21] D. Demir and B. Puliçe, JCAP **04** (2020) 051 [arXiv:2001.06577 [hep-ph]].

[22] D. Demir, Gen. Rel. Grav. **53** (2021) 22 [arXiv:2101.12391 [gr-qc]].

[23] D. Demir, Adv. High Energy Phys. **2019** (2019) 4652048 [arXiv:1901.07244 [hep-ph]].

[24] D. Demir, Adv. High Energy Phys. **2016** (2016) 6727805 [arXiv:1605.00377 [hep-ph]].

[25] D. Demir, [arXiv:2312.16270 [hep-th]].

[26] I. Jack and D. R. T. Jones, Nucl. Phys. B **342** (1990) 127.

[27] M. S. Al-sarhi, I. Jack and D. R. T. Jones, Z. Phys. C **55** (1992) 283.

[28] D. Demir, C. Karahan and O. Sargın, Phys. Rev. D **107**, 045003 (2023) [arXiv:2301.03323 [hep-th]].

[29] J. D. Norton, Rep. Prog. Phys. **56** (1993) 791.

[30] C. D. Froggatt and H. B. Nielsen, Annalen Phys. **517**, 115 (2005) [arXiv:hep-th/0501149 [hep-th]].

[31] D. Demir and B. Pulice, Eur. Phys. J. C **82**, 996 (2022) [arXiv:2211.00991 [gr-qc]].

[32] D. Demir, Galaxies **9**, 33 (2021) [arXiv:2105.04277 [hep-ph]].

[33] D. Perez Adan [ATLAS and CMS], arXiv:2301.10141 [hep-ex].

[34] C. Antel *et al.* arXiv:2305.01715 [hep-ph].

[35] E. Ling, Found. Phys. **50**, 385 (2020) [arXiv:1810.06789 [gr-qc]].

[36] E. Ling, Gen. Rel. Grav. **54**, 68 (2022) [arXiv:2202.04014 [gr-qc]].

[37] G. Geshnizjani, E. Ling and J. Quintin, arXiv:2305.01676 [gr-qc].

[38] M. Bojowald, Rept. Prog. Phys. **78**, 023901 (2015) [arXiv:1501.04899 [gr-qc]].

[39] İ. İ. Çimdiker, Phys. Dark Univ. **30** (2020) 100736.

[40] N. Bostan, C. Karahan and O. Sargın, arXiv:2308.04507 [astro-ph.CO].

[41] İ. Çimdiker, D. Demir and A. Övgün, Phys. Dark Univ. **34**, 100900 (2021) [arXiv:2110.11904 [gr-qc]].

[42] J. Rayimbaev, R. C. Pantig, A. Övgün, A. Abdujabbarov and D. Demir, Annals Phys. **454**, 169335 (2023) [arXiv:2206.06599 [gr-qc]].

[43] R. C. Pantig, A. Övgün and D. Demir, Eur. Phys. J. C **83**, 250 (2023) [arXiv:2208.02969 [gr-qc]].

[44] R. Ali, R. Babar, Z. Akhtar and A. Övgün, Results Phys. **46** 10630 (2023) [arXiv:2302.12875 [gr-qc]].

[45] D. J. Gogoi, A. Övgün and D. Demir, Phys. Dark Univ. **42**, 101314 (2023) [arXiv:2306.09231 [gr-qc]].

[46] İ. İ. Çimdiker, A. Övgün and D. Demir, Class. Quant. Grav. **40**, 184001 (2023) [arXiv:2308.03947 [gr-qc]].

[47] B. Puliçe, R. C. Pantig, A. Övgün and D. Demir, Class. Quant. Grav. **40**, 195003 (2023) [arXiv:2308.08415 [gr-qc]].

[48] K. Cankoçak, D. Demir, C. Karahan and S. Şen, Eur. Phys. J. C **80**, 1188 (2020) [arXiv:2002.12262 [hep-ph]].

2 On the gravitationally driven instabilities of a scalar free field quantized iniside a Schwarzschild Black Hole

Pawel Gusin, Andrzej Radosz and Romuald J. Ściborski

Abstract   We discuss the problem of the instabilities in the dynamics of a scalar free field quantized in the interior of a Schwarzschild black hole. Two kinds of possible sources of instabilities, "soft-mode-like" and "feedback-loop-like" are indicated and briefly discussed.

Keywords: Schwarzschild black hole, globally hyperbolic spacetime, anisotropic cosmological model

1 Introduction

The interior of a Schwarzschild Black Hole (BH) is a dynamically changing spacetime and it also may be regarded as a solution of the Einstein's equation [1]. This is a globally hyperbolic spacetime and due to its dynamical character may be regarded as a cosmological model. Its spatial slice is a hypercylinder of $R^1 \times S^2$ symmetry: expanding along its axis of the homogeneity R,(see also [1-3]) and contracting transversally, in the S^2 coordinates.

Variety of classical phenomena have been considered within this anisotropic cosmological model [3-7] and within its extensions [1] (hereafter "T-model"). In this report we discuss a specific aspect of a scalar field quantized in the "T-model". Namely the problem of possible instabilities of a scalar field quantized in this cosmology will be considered. The dynamics of such a system is described by means of a Hamiltonian expressed in terms of annihilation and creation operators. In this case the possible instabilities might arise only due to the presence of the background spacetime, hence they are referred to as "gravitationally driven instabilities". We are considering some sources of potential gravitationally driven instabilities.

The paper is organized as follows. In Sec. 2 we discuss the properties of the scalar field quantized within the T-model. In Sec. 3 the Heisenberg equations for annihilation/creation operators are derived and converted into ordinary differential equations. Final remarks are presented in the Sec.4.

Eric Ling and Annachiara Piubello (Eds), SPACETIME 1908-2023. Selected peer-reviewed papers presented at the *Third Hermann Minkowski Meeting on the Foundations of Spacetime Physics*, 11-14 September 2023, Albena, Bulgaria (Minkowski Institute Press, Montreal 2024). ISBN 978-1-998902-25-5 (softcover), ISBN 978-1-998902-26-2 (ebook).

2 "T-model" and quantization of a scalar field

The interior of a Schwarzschild BH may be regarded as a cosmological anisotropic model called a "T-sphere" (or simply T-model) [3]. It is described by the following line element (see also e.g. [8])

$$ds_-^2 = g_T dT^2 - g_z dz^2 - g_2(T)\left(d\theta^2 + \sin^2\theta d\varphi^2\right), \tag{2.1}$$

where, $T \in \langle r_S, 0 \rangle$, $z \in (-\infty, +\infty)$, $g_T = \frac{1}{\frac{r_S}{T}-1} = g_z^{-1}$. At the instant T_0 the spatial slice is a hypercylinder $R^1 \times S^2$, expanding longitudinally (along $z-$axis) and contracting transversally (see e.g. [2]). The system is homogeneous along cylinder's axis that is represented by the momentum z-component's conservation.

The action of the scalar free field takes in this case the following form

$$S = \frac{1}{2}\int dT \int_\Sigma dz d\Omega T^2 \left[\frac{1}{g_T}(\partial_T \Phi)^2 - \frac{1}{g_z}(\partial_z \Phi)^2 + \right.$$
$$\left. + \frac{1}{T^2}\Phi\Delta_{S^2}\Phi - \mu^2\Phi^2\right], \tag{2.2}$$

where $\Sigma = \mathbf{R}^1 \times S^2$, $d\Omega = \sin\theta d\varphi d\theta$ and we have integrated by parts in the sector S^2 which resulted in the Laplace operator Δ_{S^2} on S^2:

$$\Delta_{S^2}\Phi = \frac{1}{\sin\theta}\frac{\partial}{\partial\theta}\left(\sin\theta\frac{\partial\Phi}{\partial\theta}\right) + \frac{1}{\sin^2\theta}\frac{\partial^2\Phi}{\partial\varphi^2}. \tag{2.3}$$

The parameter μ is not the mass because this space-time is not asymptotically flat. The Klein-Gordon (KG) equation may be written here as:

$$\partial_T\left(T^2 g_z \partial_T \Phi\right) - \frac{T^2}{g_z}\partial_z^2 \Phi - \Delta_{S^2}\Phi + \mu^2 T^2 \Phi = 0, \tag{2.4}$$

One can search for the particular solutions of KG equation in a factorized form:

$$\Phi(T, z, \theta, \phi) = R(T)u(z)Y(\theta, \phi). \tag{2.5}$$

KG equation separates then into the following differential equations:

$$\Delta_{S^2}Y = -l(l+1)Y, \tag{2.6}$$

$$\frac{d^2 u_\varepsilon}{dz^2} = -\varepsilon^2 u_\varepsilon, \tag{2.7}$$

$$\frac{d}{dT}\left(T^2 g_z \frac{dR_{\varepsilon l}}{dT}\right) + T^2\left(\frac{\varepsilon^2}{g_z} + \mu^2 + \frac{l(l+1)}{T^2}\right)R_{\varepsilon l} = 0, \tag{2.8}$$

14

where ε is a (separation) constant. The solutions of Eq.(2.6) are spherical harmonics $Y_{lm}(\theta, \phi)$,

$$\int\limits_{S^2} d\Omega Y_{lm}(\theta, \varphi) Y_{l'm'}^*(\theta, \varphi) = \delta_{ll'}\delta_{mm'}, \tag{2.9}$$

$$\int\limits_{S^2} d\Omega Y_{lm}(\theta, \varphi) Y_{l'-m'}(\theta, \varphi) = \delta_{ll'}\delta_{m-m'} \tag{2.10}$$

where $m = -l, -(l-1), \ldots 0 \ldots, l$; the solution of equation (2.7) is

$$u(z) = e^{\pm i\varepsilon z}. \tag{2.11}$$

One can expand the field function Φ into the complete system of functions on $\mathbf{R}^1$ and S^2:

$$\Phi(T, z, \theta, \varphi) = \sum_{\varepsilon, l, m} [R_{\varepsilon l}(T) e^{i\varepsilon z} Y_{lm}(\theta, \varphi) A_{\varepsilon lm} + \tag{2.12}$$
$$+ R_{\varepsilon l}^*(T) e^{-i\varepsilon z} Y_{lm}^*(\theta, \varphi) A_{\varepsilon lm}^*],$$

where $R_{\varepsilon l}(T)$ are the functions of the temporal variable T satisfying Eq. (2.8) and $A_{\varepsilon lm}$ are (complex) expansion coefficients. The (Klein-Gordon) scalar product takes then the form (see [8]):

$$(\Phi, \Psi) = iT^2 g_z \int\limits_{S^2} \sin\theta d\theta d\phi \int\limits_A (\Phi^* \partial_T \Psi - \Psi \partial_T \Phi^*) dz. \tag{2.13}$$

The normalization condition

$$A_{\varepsilon lm} = \left(R_{\varepsilon l}(T) e^{i\varepsilon z} Y_{lm}(\theta, \varphi), \Phi\right) \tag{2.14}$$

is satisfied iff

$$T^2 g_z \left[R_{\varepsilon l}^* \dot{R}_{\varepsilon l} - \dot{R}_{\varepsilon l}^* R_{\varepsilon l}\right] = -i, \tag{2.15}$$

$$R_{\varepsilon l}^* \dot{R}_{-\varepsilon l}^* - R_{-\varepsilon l}^* \dot{R}_{\varepsilon l}^* = 0. \tag{2.16}$$

The momentum field is introduced as the one canonically conjugated to $\Phi(T, z, \theta, \varphi)$, i.e.

$$\pi = \frac{\partial \mathcal{L}}{\partial(\partial_T \Phi)} = \frac{T^2}{g_T} \partial_T \Phi. \tag{2.17}$$

Canonical commutation relations

$$\left[\widehat{\Phi}(t, \mathbf{x}), \widehat{\pi}(t, \mathbf{y})\right] = i\delta(\mathbf{x}, \mathbf{y}), \tag{2.18}$$
$$\left[\widehat{\Phi}(t, \mathbf{x}), \widehat{\Phi}(t, \mathbf{y})\right] = [\widehat{\pi}(t, \mathbf{x}), \widehat{\pi}(t, \mathbf{y})] = 0,$$

15

where $\mathbf{x}, \mathbf{y} \in \Sigma_t$, the slice Σ_t has the topology of the product space of the set $A \subset \mathbf{R}^1$ and the two-dimensional sphere S^2, imply the commutation relations for $\widehat{A}_{\varepsilon l m}$, $\widehat{A}^\dagger_{\varepsilon l m}$,

$$\left[\widehat{A}_{\varepsilon l m}, \widehat{A}^\dagger_{\varepsilon' l' m'}\right] = \delta_{\varepsilon \varepsilon'} \delta_{ll'} \delta_{mm'}, \tag{2.19}$$

all of the other commutators vanish.

Hamiltonian of the quantized scalar field as expressed in terms of annihilation and creation operators is:

$$H = \frac{1}{2} \sum_{\varkappa} \left[\omega_\varkappa \widehat{A}_\varkappa \widehat{A}^\dagger_\varkappa + \gamma_{\varkappa\varkappa'} \widehat{A}_\varkappa \widehat{A}_{\varkappa'} + (h.c) \right] \tag{2.20}$$

where indices $\varkappa, \varkappa'$ correspond to the appropriate three-letter $\varepsilon l m$ sets, coefficients $\omega_\varkappa, \gamma_{\varkappa\varkappa'}$ are given as

$$\gamma_{\varepsilon l m / \varepsilon' l m'} = \Big[T^2 g_z \partial_T R_{\varepsilon l}(T) \, \partial_T R_{-\varepsilon l}(T) +$$

$$+ T^2 \left\{ \frac{\varepsilon^2}{g_z} + \frac{l(l+1)}{T^2} + \mu^2 \right\} \times R_{\varepsilon l}(T) R_{-\varepsilon l}(T) \Big]_{\delta_{\varepsilon-\varepsilon'}; \delta_{m-m'}}, \tag{2.21}$$

$$\omega_{\varepsilon l m} = \Big[T^2 g_z \partial_T R_{\varepsilon l}(T) \, \partial_T R^*_{\varepsilon l}(T)$$

$$+ T^2 \left\{ \frac{\varepsilon^2}{g_z} + \frac{l(l+1)}{T^2} + \mu^2 \right\} R_{\varepsilon l}(T) R^*_{\varepsilon l}(T) \Big]_{\delta_{mm'}}. \tag{2.22}$$

3   Dynamics

The dynamics of the quantized scalar field (2.20) may be described by means of the Heisenberg equations for the operators $\widehat{A}_{\varepsilon l m}$

$$i \frac{d}{dt} \widehat{A}_{\varepsilon l m} = \left[\widehat{A}_{\varepsilon l m}, \widehat{H} \right] = \omega_{\varepsilon l m}(t) \, \widehat{A}_{\varepsilon l m}(t) + \gamma^*_{\varepsilon l m}(t) \, \widehat{A}^\dagger_{-\varepsilon l - m}(t) \tag{3.1}$$

In order to solve Eqs. (3.1) one invokes the following ansatz:

$$\widehat{A}_{\varepsilon l m}(t) = \alpha_{\varepsilon l m}(t) \, \widehat{A}_{\varepsilon l m} + \beta_{\varepsilon l m}(t) \, \widehat{A}^\dagger_{-\varepsilon l - m}, \tag{3.2}$$

where $\alpha_{\varepsilon l m}(t)$ and $\beta_{\varepsilon l m}(t)$ are some unknown complex functions and $\widehat{A}_{\varepsilon l m}$ and $\widehat{A}^\dagger_{-\varepsilon l - m}$ are time independent operators (taken at some specific time-instant). The relation (3.2) preserves the commutation relations (2.19) leading to

$$\left| \alpha_{\varepsilon l m}(t) \right|^2 - \left| \beta_{\varepsilon l m}(t) \right|^2 = 1, \tag{3.3}$$

i.e. Eq. (3.2) is the Bogolyubov transformation. The Heisenberg equations (3.1) are then converted into differential equations for the Bogolyubov coefficients,

$$i\frac{d}{dt}\alpha_{\varepsilon lm}\left(t\right) = \omega_{\varepsilon lm}\left(t\right)\alpha_{\varepsilon lm}\left(t\right) + \gamma^*_{\varepsilon lm}\left(t\right)\beta^*_{\varepsilon lm}\left(t\right), \tag{3.4}$$

$$i\frac{d}{dt}\beta_{\varepsilon lm}\left(t\right) = \omega_{\varepsilon lm}\left(t\right)\beta_{\varepsilon lm}\left(t\right) + \gamma^*_{\varepsilon lm}\left(t\right)\alpha^*_{\varepsilon lm}\left(t\right). \tag{3.5}$$

In order to solve Eqs. (3.4-5) the complex function γ is expressed in terms of its two real components γ_1 and γ_2:

$$\gamma = \gamma_1 + i\gamma_2. \tag{3.6}$$

Thus one obtains the following equations

$$i\frac{d}{dt}\left(\alpha \pm \beta^*\right) = \left(\omega \mp \gamma_1\right)\left(\alpha \mp \beta^*\right) \mp i\gamma_2\left(\alpha \pm \beta^*\right). \tag{3.7}$$

One can introduce two new functions, X and Y

$$X = \alpha + \beta^* \text{ and } Y = \alpha - \beta^*, \tag{3.8}$$

with the following initial conditions

$$X\left(t_0\right) = Y\left(t_0\right) = 1. \tag{3.9}$$

The system of the two first order equations (3.7) may be then expressed as the following second order equation for X:

$$\frac{d^2X}{dt^2} - \frac{\dot{\omega}_-}{\omega_-}\frac{dX}{dt} + \left(\Omega^2 + \dot{\gamma}_2 - \frac{\dot{\omega}_-}{\omega_-}\gamma_2\right)X = 0, \tag{3.10}$$

where

$$\omega_\pm = \omega \pm \gamma_1, \tag{3.11}$$

$$\omega_+\omega_- - \gamma_2^2 = \Omega^2. \tag{3.12}$$

The initial conditions for X are:

$$X\left(t_0\right) = 1 \text{ and } \dot{X}\left(t_0\right) = -i\left[\omega\left(t_0\right) - \gamma\left(t_0\right)\right] \tag{3.13}$$

and

$$\frac{i}{\omega_-}\left(\frac{dX}{dt} + \gamma_2 X\right) = Y. \tag{3.14}$$

The second order differential equation (3.10) reveals some interesting (in)stability properties. Namely if one of the following inequalities

$$\Omega^2 = \omega^2 - \left(\gamma_1^2 + \gamma_2^2\right) > 0 \tag{3.15}$$

$$\left(\Omega^2 + \dot{\gamma}_2 - \frac{\dot{\omega}_-}{\omega_-}\gamma_2\right) > 0 \tag{3.16}$$

17

$$\frac{\dot{\omega}_-}{\omega_-} < 0 \tag{3.17}$$

is not obeyed, the system may exhibit an instability. There are two kinds of the sources of potential instabilities there. They may be illustrated by referring to the simplified case of (a) the harmonic oscillator

$$\frac{d^2x}{dt^2} + \mu_0^2 x = 0, \tag{3.18}$$

and (b) the (under)damped, $\beta < \mu_0$ harmonic oscillator

$$\frac{d^2x}{dt^2} + 2\beta\frac{dx}{dt} + \mu_0^2 x = 0, \tag{3.19}$$

where μ_0^2 and β represent the squared frequency of harmonic oscillations and damping coefficient in the case (a) and (b), respectively. There are two possible instabilities in the cases (a) and (b). The first one, (a) "soft-mode-like" one, occurs when the squared frequency of harmonic oscillations tends to zero, $\mu_0^2 \to 0$ ("soft mode") and eventually changes its sign, $\mu_0^2 < 0$. This is manifested by the "softening", $\mu_0^2 \to 0$ of the (parabolic) potential well $V(x) = \frac{1}{2}m\mu_0^2 x^2$, and its eventual reflection, for $\mu_0^2 < 0$, into parabolic potential barrier, $V(x) = -\frac{1}{2}m|\mu_0^2|x^2$, (Reversed Harmonic Oscillator, RHO - see e.g. [9]). The second one, (b) "feedback-loop-like" one occurs when attenuation coefficient reverses its sign, $\beta < 0$ and damping converts into the feedback loop.

Therefore, there are two kinds of possible instabilities in the dynamics of the system (2.20-22) described by means of the differential equations (3.4-5):

a) "soft mode-like" instability, when one (or both) of the inequalities (3.15) and (3.16) are not obeyed

and

b) "feedback loop" instability, in the case when the inequality (39) is not obeyed.

These instabilities hereafter will be referred to as "gravitationally driven instabilities".

4 Discussion

Considering the quantization of the scalar free field in the interior of a Schwarzschild BH one can describe its dynamics within a unitary approach (see [8]). There is an interesting aspect of such a treatment: one discovers a possible instabilities of the dynamics of the system. As the scalar free field is considered any possible instability must be regarded as "gravitationally" driven. We have indicated two kinds of possible sources of gravitationally driven instabilities: "soft-mode-like" and "feedback-loop-like". The obvious questions is: what is a status of those indicators? There is no definite answer yet for that question. It is believed that when the inequalities (3.15-17) are satisfied, then the solutions of the differential equation (3.10) and consequently the equations (3.4) and (3.5) are regular and the process of spontaneous particle production (or annihilation) in the system (2.20) is smoth and non-violent. Breaking one or more of inequalities (3.15-17), would result in the system's instability, i.e. in an exponential grow of the number of spontaneously produced particles - that problem is the the subject of our current and future investigations.

One specific answer has already been found [8]. There is no "soft-mode instability" type (3.15) in this case as

$$\Omega_{\varepsilon l m}^2 = \frac{1}{g_z} \left[\frac{\varepsilon^2}{g_z} + \frac{l\,(l+1)}{T^2} + \mu^2 \right] \tag{4.1}$$

is positive.

References

[1] R. Doran, F. S. Lobo, and P. Crawford, *Interior of a Schwarzschild black hole revisited*, Foundations of Physics, vol. 38, no. 2, pp. 160 (2008)

[2] P. Gusin, A. T. Augousti, F. Formalik, and A. Radosz, *The (A)symmetry between the Exterior and Interior of a Schwarzschild Black Hole*, Symmetry (2018) 10, 366

[3] V.A. Ruban, *Spherically Symmetric T-Models in the General Theory of Relativity*, Gen. Rel. and Grav., 33, No. 2, 2001

[4] A. Radosz, P. Gusin, A.T. Augousti, F. Formalik, *Inside spherically symmetric black holes or how a uniformly accelerated particle may slow down* Eur. Phys. J. C 2019, 79, 876. https://doi.org/10.1140/epjc/s10052-019-7372-5

[5] A.V. Toporensky and O.B. Zaslavskii, *Zero-momentum trajectories inside a black hole and high energy particle collisions*, J. Cosmol. Astr. Phys. 2019(12):063-063.

[6] Augousti, A.T.; Gusin, P.; Kuśmierz, B.; Masajada, J.; Radosz, A. *On the speed of a test particle inside the Schwarzschild event horizon and other kinds of black holes*. Gen. Relativ. Gravit. 2018, 50, 131, doi:10.1007/s10714-018-2445-6.

[7] A. V. Toporensky and O. B. Zaslavskii, *Redshift of a photon emitted along the black hole horizon,*" E. Phys. J. C, vol. 77, no. 3, p. 179,

[8] P. Gusin, A. Radosz, A. T. Augousti, J. Polonyi, O. B. Zaslavskii and R.J. Ściborski, *Quantum phenomena inside a black hole: quantization of the scalar field iniside horizon in Schwarzschild spacetime*, Universe 2023, 9(7), 299

[9] B.M. Villegas-Martínez, H. M. Moya-Cessa, F. Soto-Eguibar, *Exact and approximated solutions for the harmonic and anharmonic repulsive oscillators: Matrix method*, E. Phys. J. D 2020, vol. 74 (7), p. 137

3 SCHWINGER-UNRUH-HAWKING RADIATION ON MANIFOLDS

TOMOHIRO MATSUDA

Abstract The whole picture of gauge theory is described by manifolds, while the field equation provides only a part (a section) of the manifold. Just as a three-dimensional object is reconstructed from two planar images, a monopole is constructed by combining two solutions. The Schwinger and the Unruh effects and the Hawking radiation are the production of particles out of the "vacuum". If the "vacuum" on the manifold is properly defined, these phenomena should be described as local phenomena. However, calculations using the field equations have so far resulted in unnatural extrapolations. We present a method for properly defining the "vacuum" and explain how to resolve the local particle production on manifolds. By defining the Stokes phenomena on the manifold, the Schwinger effect is naturally accompanied by the Unruh effect. Also, unlike the conventional Unruh effect, calculations on manifolds do not suffer from the entanglement between disconnected wedges.

Keywords: Schwinger-Unruh-Hawking radiation, gauge theory, vacuum

1 Introduction

If particles are created from the "vacuum" in front of us, it is natural to define that "vacuum" on the spot. If this cannot be done, the theory should be considered somehow flawed. The monopole solution is a simple illustration of the fact that the field equations in field theory are somehow flawed. They describe only a part (a section) of the theory, and the whole image can be understood naturally by analyzing the manifold. The Schwinger[1] and the Unruh (Fulling-Davies-Unruh[2, 3, 4]) effects and Hawking radiation[5], which deal with stationary particle production, have similar problems. Looking at the field equations alone, the vacuum can only be defined in the far asymptotic state. On the other hand, in manifolds, tangent spaces can be defined at any point. This tangent space is well known to be the definition of the natural local vacuum in general relativity.[1] However, when discussing Hawking radiation, such a local vacuum has never been used for local particle creation in combination with the field equations. There is something here that must be unraveled, and that is what we have found in this work.

---

[1] There are two definitions of the local "vacuum" in general relativity. One is the Lorentz frame in mathematics, which is a direct product space, and the other is Einstein's local inertial system, which is not a direct product. The starting point for the field equations is the Lorenz frame (i.e, mathematical covariant derivatives are defined in open coverings using local trivialization). Although they are giving the same metric, we see that the difference between the two has a crucial consequence in the case of Hawking radiation.

Eric Ling and Annachiara Piubello (Eds), SPACETIME 1908-2023. Selected peer-reviewed papers presented at the *Third Hermann Minkowski Meeting on the Foundations of Spacetime Physics*, 11-14 September 2023, Albena, Bulgaria (Minkowski Institute Press, Montreal 2024). ISBN 978-1-998902-25-5 (softcover), ISBN 978-1-998902-26-2 (ebook).

In this work[6, 7], stationary particle production is naturally described as "the Stokes phenomena always appearing in the vacuum". Such a picture can only be obtained by considering particle production and the vacuum definition on the manifold.

The topological properties of manifolds have been actively discussed, but the characteristic properties of manifolds concerning particle production have rarely been discussed. In particular, as far as we know, the local "vacuum" has never been defined on manifolds to explain local particle production.

To get an overview and an idea, let us first have a short look at the field equation for the Schwinger effect. The simplest case is that the electric field is spatially homogeneous and it is constant in the z-direction. We introduce a complex scalar field ϕ of mass m in the four-dimensional Minkowski spacetime. The action S_0 on the tangent space is

$$S_0 = \int d^4x \left(\partial_\mu \phi \partial^\mu \phi^* - m^2 \phi \phi^* \right). \tag{1.1}$$

Introducing covariant derivatives, the partial derivatives are replaced as

$$\partial_\mu \rightarrow \nabla_\mu \equiv \partial_\mu + iqA_\mu, \tag{1.2}$$

where A_μ is a gauge field. The "vacuum" is now defined on the tangent space attached to $A_\mu = 0$. Assuming the limit where dynamics of the gauge field itself is negligible, the external gauge field is given by

$$A^\mu = (0, 0, 0, -E(t - t_0)), \tag{1.3}$$

which explains the electric field strength $\vec{E} = (0, 0, E)$. Note that t_0 is an arbitrary parameter. Although it is not often recognized, if the electric field is defined in such a way as to restrict the degrees of freedom of the theory, then the scope of this equation is also restricted to the vicinity of $t = t_0$, since (strictly speaking) the open covering for "this equation" is only defined in the vicinity of $t = t_0$. In manifolds, such open coverings are used in bundles to cover the base space.

The field equation for the scalar field ϕ after Fourier transformation is

$$\ddot{\phi}_k + \omega_k^2(t)\phi_k = 0, \tag{1.4}$$

where $\omega_k^2(t) = m^2 + k_\perp^2 + (k_z - qE(t - t_0))^2$. We call $V(t) \equiv -\omega_k^2(t)$ a "potential". Typical Stokes lines are shown in Fig.2. Let us first think about $k_z = 0$. Since the Stokes line appears in the defined vacuum on the tangent space, it can be seen that the vacuum solutions are mixed there. The serious problem is that it is not apparent from the field equation alone why this mixing is occurring "all the time". Also, it should be hard to recognize the definition of the vacuum only by using the field equation. In order to understand the situation, we consider manifold for the field equation and the Stokes phenomena.

In the case of the Schwinger effect, the Stokes phenomenon can be seen directly from the field equation, as we have described above. Therefore, the problem can be solved when the equation is properly considered on the manifold. On the other hand, for the Unruh effect and Hawking radiation, the Stokes phenomena of particle production are not directly deducible from the field equation. The answers to these questions are the subject of this study[6, 7].

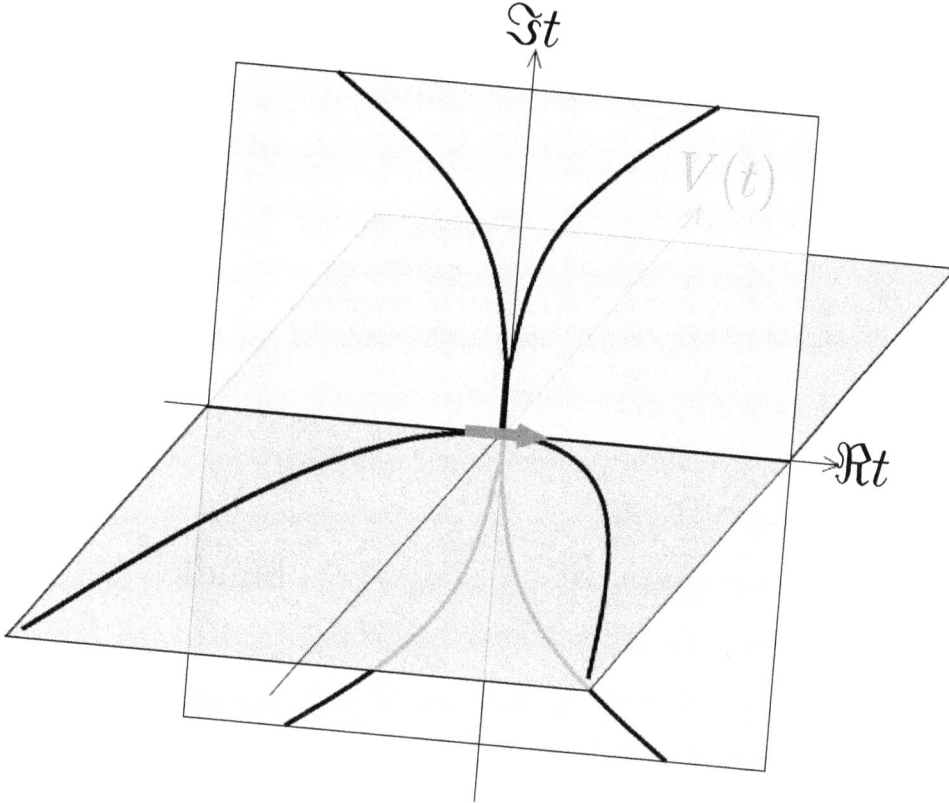

Figure 2: The potential $V(t) = -t^2$ and the Stokes lines (on the complex t plane) are shown. Particle production occurs when the real t-axis crosses the Stokes line (indicated by the arrow at the origin). For clarity, the Stokes line is written for $\omega_k^2(t) = 1 + t^2$, for which two turning points appear on the imaginary t axis. The original figure can be found in Ref.[6].

2 How to define the Schwinger effect on the manifold

Again, we consider Eq.(1.4) for the complex scalar field;

$$\ddot{\phi}_k + \omega_k^2(t)\phi_k \;=\; 0, \tag{2.1}$$

where $\omega_k^2(t) = m^2 + k_\perp^2 + (k_z - qE(t - t_0))^2$.

To describe the Bogoliubov transformation of the particle creation, we expand $\phi_k(t)$ using the solutions $\psi_k^\pm(t)$ as

$$\phi_k(t) \;=\; \alpha_k \psi_k^-(t) + \beta_k \psi_k^+(t). \tag{2.2}$$

Then the transformation matrix is

$$\begin{pmatrix} \alpha_k^R \\ \beta_k^R \end{pmatrix} \;=\; \begin{pmatrix} \sqrt{1 + e^{-2\pi\kappa}}\,e^{i\theta_1} & ie^{-\pi\kappa + i\theta_2} \\ -ie^{-\pi\kappa - i\theta_2} & \sqrt{1 + e^{-2\pi\kappa}}\,e^{-i\theta_1} \end{pmatrix} \begin{pmatrix} \alpha_k^L \\ \beta_k^L \end{pmatrix}, \tag{2.3}$$

where the indices L and R are for $t < t_0$ and $t > t_0$, respectively. If one uses the exact WKB[9, 10, 11], the constant κ can be calculated from the integral connecting the two turning points $t_*^\pm$ appearing on the imaginary axis. The exact calculation gives $\kappa = \frac{m^2+k_\perp^2}{2E}$[8, 9, 10]. Here, all the phase parameters are included in $\theta_{1,2}(k)$. It is already known[12] that Schwinger's original result (the vacuum decay rate) can be calculated from the above result by adding up all possibilities.

The obvious problem is that while the above equation describes particle production at a specific time t_0, it should be arbitrary because of the gauge degrees of freedom. Therefore, we need to consider a specific way to incorporate the degrees of freedom of the gauge into the equations of motion. Our main claim is that this can be done on the manifold. We have illustrated the situation in Fig.3. We consider the equation with k set to 0 (or $k^2 \ll !^2$), since the local "vacuum" is also defined for the rest frame of the particle.

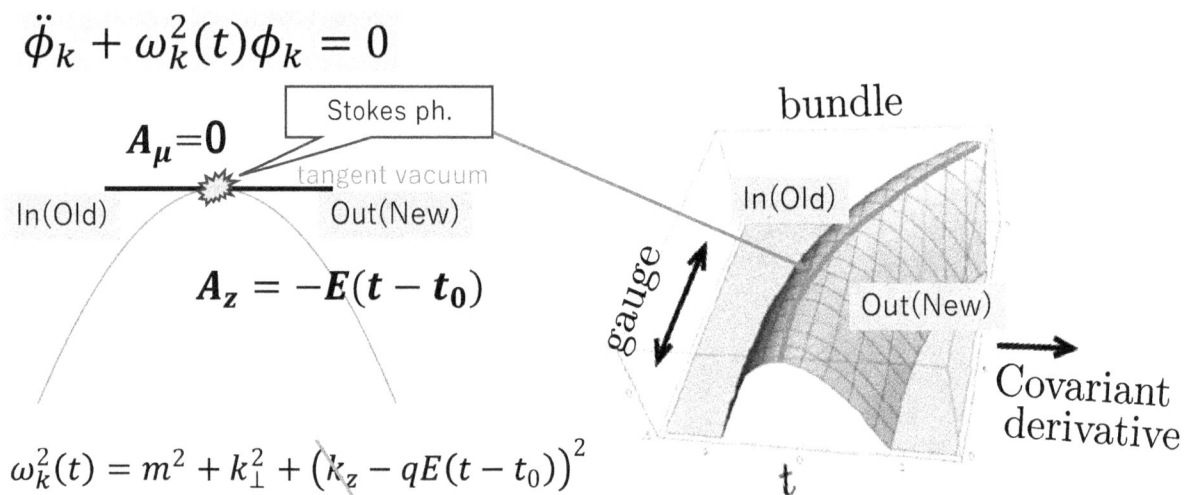

$$\ddot{\phi}_k + \omega_k^2(t)\phi_k = 0$$

$$A_\mu = 0$$

Stokes ph.

tangent vacuum

In(Old)   Out(New)

$$A_z = -E(t - t_0)$$

$$\omega_k^2(t) = m^2 + k_\perp^2 + \left(k_z - qE(t-t_0)\right)^2$$

bundle

In(Old)

Out(New)

gauge

Covariant derivative

t

Figure 3: The field equation of the Schwinger effect is illustrated on the manifold. The Stokes phenomenon occurs in the "vacuum" (the tangent space) attached to the red line.

Fig.3 can also have a different explanation. Using local trivialization, the covariant derivative is defined in the open covering (U_i) around a point $(t = t_i)$. In writing down the field equations, A^μ was fixed as in Eq.(1.3), which compromised the degrees of freedom of the theory and restricted the range of application of the equation to the vicinity of $t = t_0$. Therefore, if one wants to define the covariant derivative in the range $0 < t < 1$, one has to bundle the open coverings U_i defined around $t_i = i/N, i = 0, 1, ...N$ to have $\lim_{N\to\infty} \bigcup_{i=0,...N} U_i$. Here, for each U_i one has to define

$$A^\mu = (0, 0, 0, -E(t - t_i)). \tag{2.4}$$

This mathematical procedure of the manifold explains stationary particle production at any time. In the case of monopoles, only the two U_N and U_S are enough to describe the solution because it is about topology, but in the case of particle generation, the procedure must be followed according to the fundamentals. For comparison, Fig.4 illustrates how to construct a "clever" solution and a "stick-to-the-basics" solution for a monopole. It is by no means a self-evident issue that the left construction method coincides with the right construction method.

24

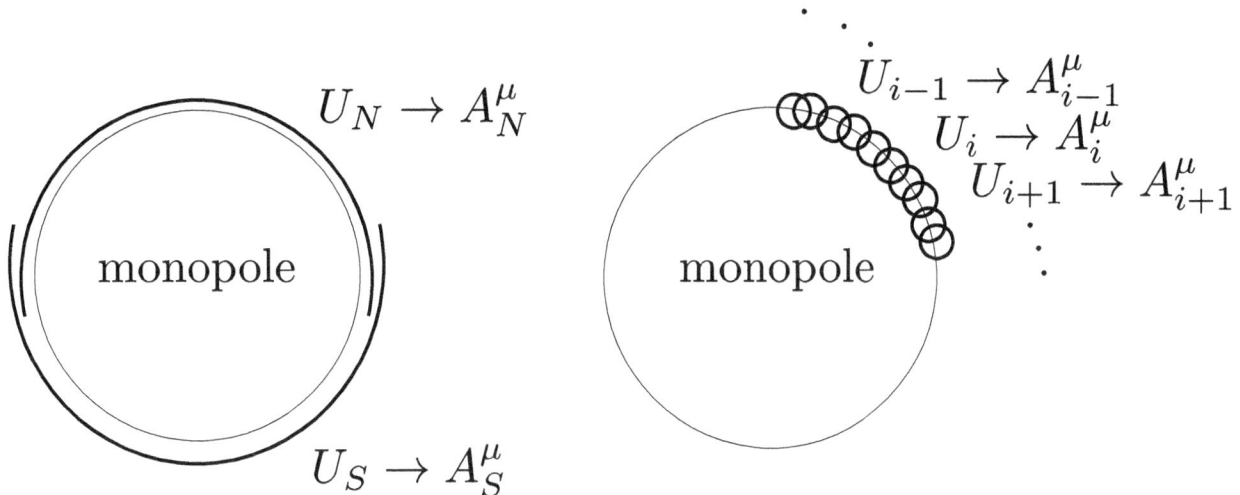

Figure 4: In the left panel, the surface of the sphere is covered by two open coverings, and two solutions are constructed in total. The two solutions are connected by a gauge transformation. In the right panel, the sphere is covered by open coverings as per the most basic definition of the covariant derivative. These are called "bundles". As local gauge transformations are used in defining the covariant derivative, gauge transformations are needed to connect the solutions obtained between neighboring open coverings. Normally, no one uses the definition on the right when constructing a monopole solution but it becomes important when discussing stationary particle production.

To understand the importance of the definition of the local vacuum on the manifold, which incorporates the freedom of the gauge and general relativity, note that $k_z \neq 0$ shifts the position of the Stokes line from $t = t_0$ (i.e, $A^\mu = 0$). This shifts the Stokes lines away from the defined "vacuum" and spoils the scenario of stationary radiation. The choice of the reference frame is crucial, even if the Schwinger effect deals with gauge theory and not obviously with general relativity. We show in Fig.5 the displacement of the Stokes line from the center-of-mass frame and how it coincides with the defined vacuum in the rest frames. We omitted explicit Lorentz factors of E and ω.

We have explained stationary particle production of the Schwinger effect using the field equation with the help of manifolds. The Stokes phenomena of the Schwinger effect are defined without relying on artificial asymptotic states. The definition of the "vacuum" on the manifold is consistent with stationary radiation. The freedom of the gauge and general relativity is thus properly incorporated into the field equation using the manifold.

We have seen that once the field equation is considered on manifolds, one can find the local vacuum of particle creation. Our definition of the vacuum is a natural extension of the vacuum in general relativity to gauge theory.

So far, our calculation is fully consistent with Schwinger's original calculation. However, after calculating the local Unruh effect, we will find a significant difference.

For the Schwinger effect, the Stokes phenomenon is obvious in the field equation. However, this is not true for Hawking radiation and the Unruh effect. The cause of such failure tells us a significant difference between the local inertial frame and the Lorentz frame.

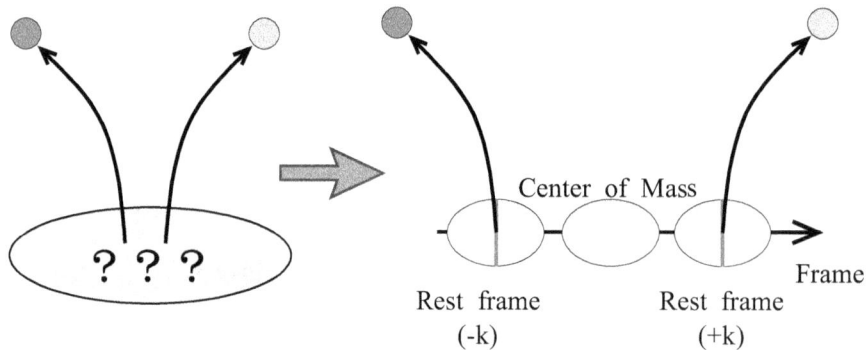

Figure 5: This figure shows the relationship between the Stokes lines and the frames when $k_z \neq 0$ in the center of the mass frame. The original figure can be found in Ref.[6].

3 How to find the Stokes phenomenon for the Unruh effect and Hawking radiation

We have seen that for the Schwinger effect, one can define a local vacuum on the manifold to find stationary radiation due to the Stokes phenomenon. The vacuum of the particle production is defined by choosing a frame for which each particle is at rest.

Let us think about the definition of the "vacuum" on the manifold. In general relativity, there are two different coordinate systems for which the "vacuum" can be defined: the Lorentz frame and the local inertial frame. Although the two frames are often considered to be physically identical (because they give the same metric), the difference is crucial for our discussion.

In order to understand the difference from the Schwinger effect, let us see what happens if the derivatives of the tangent space are replaced by covariant derivatives. This manipulation gives the Klein-Gordon equation of an accelerating observer (Unruh) or on curved spacetime (Hawking), but unlike the Schwinger effect, it cannot explain the local Stokes phenomenon[7].[2] The origin of this problem is the fact that the equations of motion of the accelerating system are constructed from the Lorenz system and they are not directly looking at the inertial system. To understand the situation, one can calculate the vierbein of the inertial system in the Rindler coordinate. Then one will find that it has off-diagonal elements. To define the covariant derivatives on manifolds, local trivialization has to be used. Since the local trivialization gives a direct product space, it cannot give the inertial system. The mathematical definition of the covariant derivatives uses the Lorenz frame.

Therefore, since the Unruh effect and Hawking radiation are described for the inertial vacuum, what we have to consider is the vierbein, not the field equations. However, there has been no analysis of the Stokes phenomenon using the vierbein until Ref.[7]. Of course, even if the field equations are used, the Bogoliubov transformation can be computed by extrapolating the solutions and analyzing them in the whole space. However, such an extrapolation has a risk that extra information may be introduced by the procedure. Our speculation is that a strong correlation between distant wedges in the conventional calculation of the Unruh effect is nothing but the "extra information". We refer

---

[2]Stokes phenomena may occur around black holes. See Ref.[7, 13] for their contribution to the glaybody factor. We discuss Stokes phenomena that are directly related to local particle production.

26

to this problem as the factor 2 problem, from the reason that will be stated below.

To understand the essence, it would be better to consider the Unruh effect before Hawking radiation. In the Unruh effect, the vacuum seen by an accelerating observer is the inertial system, not the Lorentz system. One can compute the vierbein to confirm that the vacuum of the Unruh effect is defined for the inertial system. The metric is the same whether it is a local inertial system or a Lorentz system, so we rarely distinguish between the two, but in the present case, the difference is crucial. In fact, the vierbein of the Rindler spacetime can be calculated from

$$
\begin{aligned}
t &= \frac{1 + \alpha x_r}{\alpha} \sinh(\alpha t_r) \\
x &= \frac{1 + \alpha x_r}{\alpha} \cosh(\alpha t_r),
\end{aligned}
\tag{3.1}
$$

which describes the coordinate system of an object moving at constant acceleration α through a flat space-time represented by (t, x). One will find

$$
\begin{aligned}
dt &= (1 + \alpha x_r) \cosh(\alpha t_r) dt_r + \sinh(\alpha t) dx_r \\
dx &= \cosh(\alpha t) dx_r + (1 + \alpha x_r) \sinh(\alpha t_r) dt_r.
\end{aligned}
\tag{3.2}
$$

Obviously, the vierbein has off-diagonal elements and does not support local trivialization by itself. On the other hand, the metric is calculated as

$$
g_{\mu\nu} = \eta_{mn} e_\mu^m e_\nu^n,
\tag{3.3}
$$

where η_{mn} is for the local Minkowski space. This gives the Rindler metric given by

$$
ds^2 = -(1 + \alpha x_r)^2 dt_r^2 + dx_r^2.
\tag{3.4}
$$

Note that the metric is identical for both (inertial and the Lorentz) frames. As we will see, the vierbein of the inertial system is where the Stokes phenomenon of stationary radiation on curved space-time comes from, while the field equations are defined for the Lorentz frame.

As is summarized in the textbook[14], one can calculate the Bogoliubov coefficients by considering carefully the global structure of the Rindler coordinates and the relationship between the vacuum solutions written in the two coordinate systems. The question, of course, is why it is not possible to find the Stokes phenomenon locally, as in the case of the Schwinger effect.

Let us elaborate on the global calculation first to make the reason for the distant correlation clear. We introduce the Rindler coordinates (τ, ξ) in the right wedge as

$$
\begin{aligned}
t &= \frac{e^{\alpha\xi}}{\alpha} \sinh a\tau \\
z &= \frac{e^{\alpha\xi}}{\alpha} \cosh a\tau,
\end{aligned}
\tag{3.5}
$$

where (t, z) are the coordinates of the Minkowski space. Introduce the light-cone coordinate system of the Rindler space as $u = \tau - \xi, v = \tau + \xi$. One can see that they are connected to the light-cone coordinate system of the Minkowski space (U, V) as

$$
\begin{aligned}
U &= -\frac{e^{-\alpha u}}{\alpha} \\
V &= \frac{e^{\alpha v}}{\alpha}.
\end{aligned}
\tag{3.6}
$$

27

We define the Rindler coordinates $(\tilde{\tau}, \tilde{\xi})$ in the left wedge.

We consider massless particles for simplicity and consider only the right-moving waves. In the inertial system, the field $\phi(U)$ is expanded as

$$
\begin{aligned}
\phi(U) &= \int_0^\infty dk \left[a_k f_k^{(M)}(U) + a_k^\dagger f_k^{(M)*}(U) \right], \\
f_k^{(M)}(U) &= \frac{e^{-ikU}}{\sqrt{4\pi k}},
\end{aligned}
\tag{3.7}
$$

which defines the inertial vacuum by $a_k |0_M\rangle = 0$

For the right Rindler system (τ, ξ), we have

$$
\begin{aligned}
\phi_R(u) &= \int_0^\infty dp \left[b_p^{(R)} f_p^{(R)}(u) + b_p^{(R)\dagger} f_p^{(R)*}(u) \right], \\
f_p^{(R)}(u) &= \frac{e^{-ipu}}{\sqrt{4\pi p}} = \theta(-U) \frac{(-\alpha U)^{ip/\alpha}}{\sqrt{4\pi p}},
\end{aligned}
\tag{3.8}
$$

which defines the Rindler (right) vacuum by $b_p^{(R)} |0_R\rangle = 0$, and same calculation defines the Rindler (left) vacuum. Using the above solutions one can find[15]

$$
|0_M\rangle \propto \exp\left[-\prod_p \left(e^{-\pi p/2\alpha} b_p^{L\dagger} \right) \left(e^{-\pi p/2\alpha} b_p^{R\dagger} \right) \right] |0_L\rangle \otimes |0_R\rangle,
\tag{3.9}
$$

which shows a strong correlation between distant wedges. After taking normalization and trace about the left Rindler states, one will find that the Unruh temperature is $T_U = \hbar\alpha/2\pi c k_B$. What is important is the duplication of the factor $e^{-\pi p/2\alpha}$ due to the correlation. This is the source of our factor 2 problem and of course, this factor does not appear in local calculation.

One might insist that the global calculation "revealed" a surprising correlation between two causally disconnected wedges. However, it is quite unnatural that global information is essential for the calculation when its motion can be viewed as constant acceleration only over a certain period of time.

Now let us see how the Stokes phenomenon of the Unruh effect appears. First remember that during the Unruh effect, an accelerating observer is looking at the inertial vacuum. The situation is illustrated in fig.6. In that case, the vacuum solutions of the inertial system have to be seen by an observer using the vierbein. We use $dt = \cosh(\alpha t_r) dt_r$ to write the vacuum solutions into

$$
\begin{aligned}
\phi_k^\pm(t) &= A_k e^{\pm i \int \omega dt} \\
&= A_k e^{\pm i \int \omega_k \cosh(\alpha t_r) dt_r}.
\end{aligned}
\tag{3.10}
$$

Note that we are choosing the particle's rest frame. Normally, it is quite difficult to recognize the Stokes phenomenon of these solutions. However, using the exact WKB[9, 10, 11], one can solve the problem. First, define $Q(t)_0 \equiv -\omega_k^2 \cosh^2(\alpha t_r)$ and consider the "Schrödinger" equation

$$
\left(-\frac{d^2}{dt^2} + \eta^2 Q(t, \eta) \right) \psi(t, \eta) = 0,
\tag{3.11}
$$

$$e^{\pm i \int \omega_k dt} \;\rightarrow\; e^{\pm i \int \omega_k \cosh(\alpha\, t_r)\, dt_r}$$

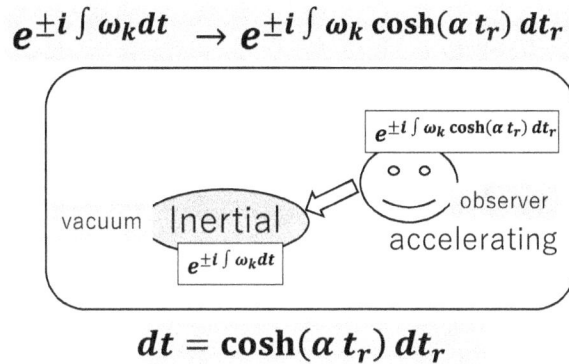

$$dt = \cosh(\alpha\, t_r)\, dt_r$$

Figure 6: This figure illustrates the inertial vacuum solutions for an accelerating observer.

where $\eta \gg 1$. $Q(t,\eta)$ is expanded as

$$Q(t,\eta) \;=\; Q_0(t) + \eta^{-1}Q_1(t) + \eta^{-2}Q(t) + \cdots . \tag{3.12}$$

Note that $\hbar \ll 1$ of quantum mechanics has been replaced by $\eta \gg 1$ according to mathematical convention. It is important to note that η is not just a large parameter, but a unique parameter that governs singular perturbations. Moreover, since η is analytically continued to the complex η-plane, the concept of the exact WKB is rather different from the conventional WKB approximation with small $\hbar$.[3]

The solution of this equation can be written as $\psi(t,\eta) \equiv e^{\int S(t,\eta)dt}$. Here, $S(t,\eta)$ can be expanded as

$$S \;=\; S_{-1}(t)\eta + S_0(t) + S_1(t)\eta^{-1} + \cdots . \tag{3.13}$$

The point of this argument is that after introducing η properly in Eq.(3.10), one can choose $Q_i(t), i \geq 1$ to find the "Schrödinger equation" of the exact WKB.

This procedure allows one to make use of a powerful analysis of the exact WKB. One can calculate the Stokes lines only by using $Q_0(t)$.[4] After drawing the Stokes lines, one can understand that the Stokes line crosses on the real axis at the origin[7]. The Stokes lines of the Unruh effect are shown in Fig.7. This allowed us to expand $Q(t)_0$ near the origin and finally, we have

$$\begin{aligned}Q(t)_0 &= -\omega_k^2 \cosh(\alpha t_r) \\ &\simeq -\omega_k^2 - \alpha^2 \omega_k^2 t_r^2,\end{aligned} \tag{3.14}$$

which gives the typical Schrödinger equation of scattering by an inverted quadratic potential[7, 9, 10]. Again, as in the case of the Schwinger effect, the Stokes lines coincide with the vacuum only when the vacuum is defined for the rest frame of the particles. This frame is not the experimenter's frame. This choice of the frame is of course consistent with experimental analyses[16].

---

[3]In the exact WKB, it is clear that the Stokes line is a discontinuous surface, since mixing of solutions occurs when the integral paths defined on the Borel panel recombine on the Stokes line.

[4]Remember that η is the unique parameter that governs the singular perturbation. The reason why $Q_0(t)$ is special is explained in Ref.[11] and the references therein in the light of singular perturbation theory.

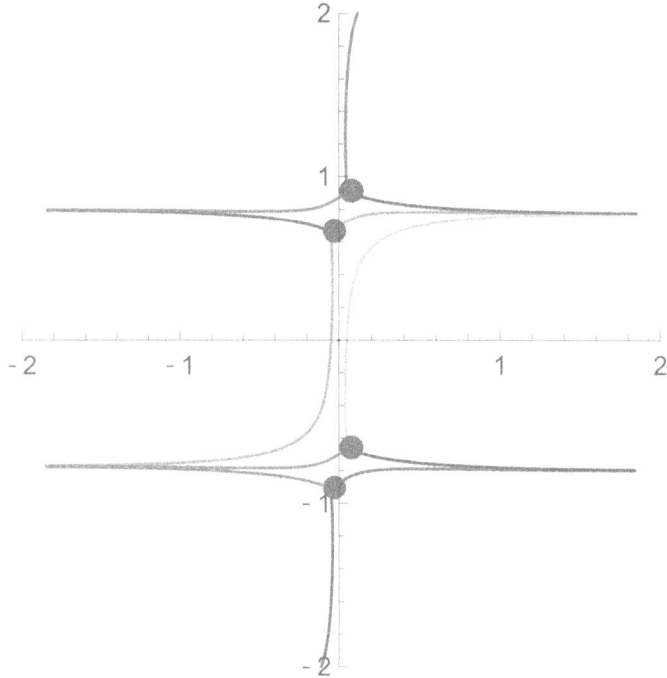

Figure 7: The Stokes lines of the Unruh effect are shown for $Q(t) = -(\cosh^2(2t) - 0.05 + 0.05i)(1 + 0.05i)$. The degenerated Stokes lines are separated introducing small parameters.

The above calculation suggests the Boltzmann factor $\sim e^{-\pi\omega_k/\alpha}$. On the other hand, $\sim e^{-2\pi\omega_k/\alpha}$ was calculated in the global(conventional) calculation. They are showing the factor 2 problem. As we have explained, duplication of the factor due to the entanglement between disconnected wedges is the origin of this problem. Whether this correlation actually exists will be verified by experimentation, since the crucial evidence appears in the Unruh temperature.

Our calculation also applies to Hawking radiation. Hawking radiation requires creation of a pair of negative and positive energy particles across the horizon. Then, only the positive energy particle outside the horizon can be observed as radiation. This situation is typical "2 for 1". Namely, two particles are produced but only one is observable. In this case, the production probability of one particle (P_1) must be discriminated from the observation probability of one particle (P_1^{obs}). To be more precise, our local analysis distinguishes "one particle production rate P_1" from "one particle observation rate P_1^{obs}(= two particle production rate $(P_1)^2$)". Considering these discrepancies, one can find that there is no factor 2 problem in Hawking radiation because the "2 for 1" particle production is essential in Hawking radiation.

4 How do the Unruh effect and the Schwinger effect coexist?

Finally, it should be mentioned that the Unruh effect[5] occur simultaneously with the Schwinger effect if they are both defined on the manifold. We will focus on the differences between them in

---

[5]This corresponds to Hawking radiation without the event horizon, because the strong electric field allows pair creation. The crucial difference is that both particles are observed.

order to avoid a superficial discussion.

The acceleration of a charged particle in a strong electromagnetic field is $\alpha = qE/m$, which, when used in "our" equation for the Unruh effect, gives the same coefficient as the Schwinger effect.

We have seen that the source of the Schwinger effect is the vector potential and it can be seen by the field equation on the manifold. On the other hand, the Unruh effect is explained by the vierbein of the inertial system, not directly by the field equation.

As noted above, the way to choose the vacuum in the Unruh effect is to "look at the local inertial system, choosing rest frame of the particle (to be produced)". Considering that the generated particles are accelerated by the electric field, the particles are generated from "the vacuum seen by an accelerating observer (particle)". This is precisely the same situation as in the case of the Unruh effect. One thing that is still not clear is whether the vacuum defined for the Schwinger effect is identical to the inertial vacuum in the Unruh effect.

Unlike the Unruh effect, pair production is possible without the event horizon, if the electric field is strong enough. In this case, since one of the two particles of a pair does not disappear into the horizon, both particles are observed. This is not a "2 for 1" particle production and the Unruh temperature observed simultaneously with the Schwinger effect should show a factor of 2 difference compared to Hawking radiation.

Perhaps the most interesting aspect of this story is whether it is possible to verify experimentally that these effects occur simultaneously. As we have discussed above, the two effects arise from separate physical phenomena. The differences between the two are listed below.

- The Stokes phenomena of the Schwinger effect appears from the field equation, while the Stokes phenomena of the Unruh effect appears from the vierbein.

- Since the Schwinger effect appears in the field equation, the vacuum of the Schwinger effect would be the Lorentz frame, while the Unruh effect looks at the local inertial frame.

Therefore, it is still not clear whether they could contribute in the same way or not. What is important for the discussion is the definition of the vacuum on the manifold. If the vacuum were defined as an asymptotic state, such an argument would never have arisen.

Let us think about possibilities. When written in the exact WKB notation, there might be $Q_0^{Schwinger}(t)$ and $Q_0^{Unruh}(t)$ to give $Q_0^{total}(t) = Q_0^{Schwinger}(t) + Q_0^{Unruh}(t)$. In this case, the coefficient of the inverted quadratic potential has to be the sum of the two contributions. On the other hand, it is not surprising that the Schwinger and Unruh effects are defined for different vacuum states and they can occur as independent phenomena. In that case, the result should be $\beta^{tot} = \beta^{Unruh} + \beta^{Schwinger}$. At this moment we cannot determine which is correct, and it has to be determined only by experimental verification.

If our considerations are correct, the experimental results must be significantly different from Schwinger's calculation. The previously mentioned problem with factor 2 of the Unruh effect can also be confirmed by experiment. If, on the other hand, the amplification noted here did not occur, then it can be concluded that the Schwinger and Unruh effects are in fact identical physical phenomena that are linked even more deeply in the theory.

5 Concluding remarks

I first read Hawking's paper on Hawking radiation when I was in a master's program. At the time, I had just moved from pharmaceutical sciences to a postgraduate degree in physics, so I didn't know what was right or wrong. Even so, I couldn't stand the unnaturalness of Hawking's calculations. His calculation was to prepare a vacuum in the infinitely distant infinite past and future, and connect the two via the collapse of a black hole. As a physical phenomenon, it deals with the production of particles on the surface of a black hole. I thought it was strange if calculations could not be done in the vicinity of what is supposed to be a black hole. For me, his calculation looked like a painstaking work of a genius with superhuman arithmetic skills. As a newcomer from another field, I thought there was something deeper that I had not yet been able to recognize. So I investigated how similar physical phenomena were handled in physics. I first looked at the Unruh effect. What I understood was that the Unruh effect also requires the same extrapolation as the Hawking radiation, and that there seems to be entanglement at a distance in the Rindler space. Having wondered about these situations, after many years, I looked into the Schwinger effect. While studying the Schwinger effect, I came across the calculation that I had hoped for. It was the Stokes phenomenon from the field equation. But it still seemed incomplete to me. In the discussion of the field equations, the vacuum is still defined as a distant asymptotic state, and the unnaturalness I felt with Hawking radiation was present here too. Schwinger's calculation, which claims to have calculated the decay rate of the vacuum, did not involve a "real" vacuum transitioning into another vacuum, in the sense that no domain walls are formulated around the new vacuum. It looked to me like there was some kind of deception going on.

Then I came across a way of constructing a monopole solution. When the monopole solution is constructed from the equation of motion, the two solutions are pasted together and connected by a gauge transformation between them. I realized that the reason for this cumbersome procedure is that the original mathematical manifold is not viewed as it is in the field equation. I also thought that the vacuum must be properly defined on the manifold that represents the whole theory.

The reason why similar calculations (the Stokes phenomena) could not be done with the accelerating system was also a mystery. After noticing a clue, further years were needed before the Stokes phenomenon was discovered in the accelerating system. It is unfortunate for students that no textbook mentions why in the Unruh effect (Hawking radiation) naive use of field equations cannot explain particle production, since many students have to learn unnatural extrapolations and complicated calculations without knowing why such calculations are necessary.

We believe we have solved these problems with our study. This research began with a simple question from my student days and hopefully, this research will help students struggling with the same questions.

6 Acknowledgments

The author would like to thank all the participants and the organizers of the conference, and he is deeply grateful to Professor Vesselin Petkov. The author was particularly encouraged by the comments given by Professor Orfeu Bertolami.

References

[1] J. S. Schwinger, "On gauge invariance and vacuum polarization," Phys. Rev. **82** (1951), 664-679

[2] S. A. Fulling, "Nonuniqueness of canonical field quantization in Riemannian space-time," Phys. Rev. D **7** (1973), 2850-2862

[3] P. C. W. Davies, "Scalar particle production in Schwarzschild and Rindler metrics," J. Phys. A **8** (1975), 609-616

[4] W. G. Unruh, "Notes on black hole evaporation," Phys. Rev. D **14** (1976), 870

[5] S. W. Hawking, "Particle Creation by Black Holes," Commun. Math. Phys. **43** (1975), 199-220 [erratum: Commun. Math. Phys. **46** (1976), 206]

[6] T. Matsuda, "Nonperturbative particle production and differential geometry," Int. J. Mod. Phys. A **38** (2023) no.28, 2350158 [arXiv:2303.11521 [hep-th]].

[7] S. Enomoto and T. Matsuda, "The Exact WKB analysis and the Stokes phenomena of the Unruh effect and Hawking radiation," JHEP **12** (2022), 037 [arXiv:2203.04501 [hep-th]].

[8] L. Kofman, A. D. Linde and A. A. Starobinsky, "Towards the theory of reheating after inflation," Phys. Rev. D **56** (1997) 3258 [hep-ph/9704452].

[9] S. Enomoto and T. Matsuda, "The exact WKB for cosmological particle production," JHEP **03** (2021), 090 [arXiv:2010.14835 [hep-ph]].

[10] S. Enomoto and T. Matsuda, "The exact WKB and the Landau-Zener transition for asymmetry in cosmological particle production," JHEP **02** (2022), 131 [arXiv:2104.02312 [hep-th]].

[11] N. Honda, T. Kawai and Y. Takei, "Virtual Turning Points", Springer (2015), ISBN 978-4-431-55702-9.

[12] J. Haro, "Topics in Quantum Field Theory in Curved Space," [arXiv:1011.4772 [gr-qc]].

[13] C. K. Dumlu, "Stokes phenomenon and Hawking radiation," Phys. Rev. D **102** (2020) no.12, 125006 [arXiv:2009.09851 [hep-th]].

[14] N. D. Birrell and P. C. W. Davies, "Quantum Fields in Curved Space,"

[15] S. Iso, "Quantum Field Theory", ISBN 978-4-320-03487-7

[16] A. Di Piazza, C. Muller, K. Z. Hatsagortsyan and C. H. Keitel, "Extremely high-intensity laser interactions with fundamental quantum systems," Rev. Mod. Phys. **84** (2012), 1177 [arXiv:1111.3886 [hep-ph]].

Part II

Cosmological Models

Eric Ling and Annachiara Piubello (Eds), SPACETIME 1908-2023. Selected peer-reviewed papers presented at the *Third Hermann Minkowski Meeting on the Foundations of Spacetime Physics*, 11-14 September 2023, Albena, Bulgaria (Minkowski Institute Press, Montreal 2024). ISBN 978-1-998902-25-5 (softcover), ISBN 978-1-998902-26-2 (ebook).

4 ANISOTROPIC EXAMPLES OF INFLATION-GENERATING INITIAL CONDITIONS FOR THE BIG BANG

ERIC LING AND ANNACHIARA PIUBELLO[1]

Abstract The inflationary scenario, which states that the early universe underwent a brief but dramatic period of accelerated spatial expansion, has become the current paradigm of early universe cosmology. Although inflationary cosmology has its many successes, it does not (as of yet) have the status of an established physical theory. In this paper, we provide mathematical support for the inflationary scenario in a class of anisotropic spacetimes by generalizing the work in [21]. These anisotropic spacetimes satisfy certain initial conditions so that they are perfectly isotropic at the big bang but become less isotropic as time progresses. The resulting inflationary eras are a consequence of the initial conditions which force the energy-momentum tensor to be dominated by a cosmological constant at the big bang.

Keywords: cosmology, inflation, initial conditions, anisotropy, cosmological constant

1 Introduction

1.1 Cosmic inflation

The inflationary scenario has become the current paradigm of early universe cosmology. Roughly, it states the following.

The inflationary scenario: In the early universe, before the radiation-dominated era, there was a brief but dramatic period of accelerated spatial expansion.

The inflationary scenario was proposed in the late 1970s and early 80s [13, 29, 18] as a solution to some problems in the standard big bang model, e.g., the flatness and horizon problems. It was soon realized that inflation can provide a framework for generating the seeds of the large-scale structures in our universe [24]. Observations of the anisotropies in the CMB radiation performed by COBE, WMAP, and most recently by Planck[8] support these claims.

Given the many successes of the inflationary scenario, it is perhaps not too surprising that most papers on early universe cosmology give the impression that inflation has been firmly established and observationally proven. However there are many inflationary models that can be in agreement

[1]Research supported by the DFG Project ME 3816/3-1, part of the SPP2026.

Eric Ling and Annachiara Piubello (Eds), SPACETIME 1908-2023. Selected peer-reviewed papers presented at the *Third Hermann Minkowski Meeting on the Foundations of Spacetime Physics*, 11-14 September 2023, Albena, Bulgaria (Minkowski Institute Press, Montreal 2024). ISBN 978-1-998902-25-5 (softcover), ISBN 978-1-998902-26-2 (ebook).

with observation [23]. In fact, any theory which predicts an almost flat universe with a nearly scale-invariant curvature power spectrum, small tensor-to-scalar ratio, and small Gaussian fluctuations would be in agreement with current data, e.g., [5].

Moreover, although phenomenologically successful, current realizations of inflationary models suffer from conceptual problems, perhaps none more so than the problem of initial conditions [4, 17]. In fact there are conflicting opinions on the naturalness of initial conditions for inflation [16, 14].

Most papers on initial conditions for inflation begin in an inhomogeneous universe with an energy-momentum tensor dominated by an inflaton scalar field in a slow-roll potential and see if the resulting dynamics can produce an inflationary era followed by a homogeneous universe. This is not our approach. Our approach is purely geometrical. Quantities of interest are described solely in terms of a special unit timelike vector field u (whose integral curves represent the comoving observers in the universe) and the spacetime metric g.

In this paper we provide mathematical support for the inflationary scenario. In section 3, we show that inflation arises for a class of anisotropic spacetimes from special geometrical initial conditions. Our initial conditions are stated informally in section 1.3 and formally in section 2. These anisotropic spacetimes are examples of our main result, Theorem 2.4, which concludes from the special initial conditions that the Ricci tensor (and hence also the energy-momentum tensor) is dominated by a cosmological constant at the big bang. Theorem 2.4 is a generalization of the main result in [21]. In fact, a major inspiration for this paper was to find anisotropic examples of the main result in [21].

The benefit of our geometrical approach is its conceptual clarity: we will describe precisely which comoving observers experience inflation and how fast they are accelerating solely in terms of the unit timelike vector field u and spacetime metric g.

Our geometrical initial conditions can be thought of as a certain type of fine-tuning condition for the big bang. As briefly reviewed in the next section, a Boltzmannian viewpoint on the arrow of time suggests that some type of fine-tuning initial condition for the big bang should exist.

1.2 Inflation and the arrow of time

An obvious feature of our universe is the existence of an arrow of time. We observe certain processes in our everyday experience, but we hardly ever observe those same processes time-reversed. A vase shatters into a multitude of pieces, but we never observe these pieces spontaneously arranging themselves perfectly together into a vase. The second law of thermodynamics is postulated to explain the arrow of time, and a modern Boltzmannian mindset of the second law leads to the conclusion that the universe began with special, non-generic, fine-tuned initial conditions.

It was Penrose who originally argued [27] that the overall arrow of time we observe is linked to special initial conditions for the universe that are drastically far away from the dynamical trend towards gravitational collapse. He calculated that the entropy of the radiation-dominated early universe is around 30 orders of magnitude smaller than the Beckenstein-Hawking entropy of its corresponding black hole state. See also [1, 6].

With this understanding, the homogeneous and isotropic assumptions of the standard FLRW models of cosmology are a reasonable choice of initial conditions as they match exceedingly well with current observations. However, some inflationary cosmologists instinctively take a different perspective. They seek an explanation for the large-scale isotropy of the universe from dynamical processes during inflation. But special initial conditions – by their nature – go against dynamical trends. That is, the creation of special initial conditions from dynamics beginning with generic initial conditions seems contradictory. Summarizing, some inflationary cosmologists seek generic initial conditions for the universe, but those who adopt a Boltzmannian point of view of the second law of thermodynamics (as we do) argue that initial conditions should instead be special in order to explain the arrow of time.

While inflationary theory alone may not suffice to explain the large-scale isotropy of the universe, it still has many successes and remains the prevailing paradigm in early universe cosmology. The simplest way to generate inflation is to introduce an inflaton scalar field in a slow-roll potential – a methodology that is somewhat adhoc since it simply postulates the existence of a scalar field for which we have no direct evidence for. So a natural inquiry is to ask if there is other evidence to support the inflationary scenario. A primary motivation of this paper is to demonstrate that there is mathematical evidence in support of the inflationary scenario. We will see that inflation is inevitable provided certain geometrical initial conditions are assumed at the big bang, and, as discussed in this section, some degree of fine-tuning in the initial conditions is anticipated.

1.3 Geometrical initial conditions for the big bang

In this section we describe, informally, the primary geometrical initial conditions we will be considering in our main result, Theorem 2.4. These initial conditions are supposed to mimic – without assuming isotropy – the geometrical properties at the big bang.

Let's first clarify what we mean by the "big bang" as there are conflicting view points in the literature. For us the big bang refers to a time when the scale factor limits to zero. For example, if the scale factor is $a(\tau) = \tau$ (as in the Milne model), then the big bang corresponds to $\tau = 0$. If the scale factor is $a(\tau) = e^\tau$ (as in the flat de Sitter model), then the big bang corresponds to $\tau = -\infty$.

To motivate the type of geometrical initial conditions we will be considering, we focus on scale factor perturbations of the Milne model, which have been dubbed "Milne-like spacetimes" in [10]. These models were extensively studied in [20], detailing possible applications to fundamental problems in cosmology. See also [22, 25, 7]. They are $k = -1$ FLRW spacetimes whose scale factor satisfies $a(\tau) \approx \tau$ for τ near $\tau = 0$. (An inflating example would be $a(\tau) = \sinh(\tau)$.) Interestingly, for Milne-like spacetimes, the big bang appears as a coordinate singularity, and so they extend into a larger spacetime.

Recall that the comoving observers are the integral curves of the vector field u given by $u = \partial_\tau$ in comoving coordinates. As illustrated in Figure 8, the comoving observers for a Milne-like spacetime all emanate from a single point $\mathcal{O}$ in the extended spacetime, which is just the origin $(0, 0, 0, 0)$ in the conformal Minkowskian coordinates (t, x, y, z). We refer to this property as "$\mathcal{O}$ being an origin point for u," see Definition 2.1. The existence of an origin point $\mathcal{O}$ for u is a highly fine-tuned and non-generic assumption. Recall that some fine-tuning is to be expected from the discussion on the arrow of time in section 1.2.

An origin point $\mathcal{O}$ for u is the first main assumption in Theorem 2.4. The other main assumption is that the energy-momentum tensor T approaches that of a perfect fluid at $\mathcal{O}$. See Definition

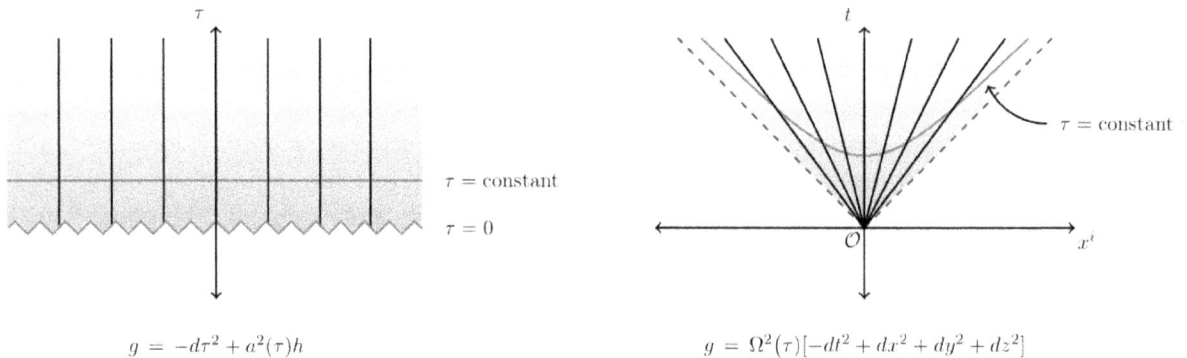

$$g = -d\tau^2 + a^2(\tau)h \qquad\qquad g = \Omega^2(\tau)[-dt^2 + dx^2 + dy^2 + dz^2]$$

Figure 8: A Milne-like spacetime represented in two different coordinate systems. On the left, standard comoving coordinates are used; the metric is degenerate at $\tau = 0$. On the right, conformal Minkowskian coordinates are used; the metric is nondegenerate at $\tau = 0$ which corresponds to the lightcone at $\mathcal{O}$. The black lines depict the comoving observers.

2.2. This assumption is more physically convincing than assuming that T is exactly a perfect fluid (as in the FLRW models) since we expect small deviations from perfect isotropy in our universe. Therefore the perspective taken here is that the universe began in a state of perfect isotropy at the big bang. This is the crux of Definition 2.2. Moreover, this perspective is reinforced in our examples since the shear vanishes towards the big bang, see eq. (3.23).

An "origin point $\mathcal{O}$ for u" and "T approaching a perfect fluid at $\mathcal{O}$" are the two primary assumptions in Theorem 2.4. There are three other assumptions that are purely technical. The conclusion of Theorem 2.4 is that the energy-momentum tensor is precisely given by a cosmological constant at $\mathcal{O}$. This fact will be used in section 3.2 to prove the existence of inflationary eras in our anisotropic examples.

2 The main theorem

The initial conditions stated informally in section 1.3 will be stated formally in this section. Our main result, Theorem 2.4, is a generalization of the main result (Theorem 2.2) in [21]. Anisotropic examples of our main theorem are provided in section 3.1, and we prove the existence of inflationary eras for these examples in section 3.2.

We set our conventions. Our definition of a spacetime (M, g) will follow [19]. (Except that, for simplicity, we will assume that all spacetimes are four-dimensional.) The manifold M is always assumed to be smooth. A C^k spacetime is one where the metric g is C^k, that is, its components $g_{\mu\nu} = g(\partial_\mu, \partial_\nu)$ are C^k functions with respect to any coordinates $(x^0, \dots, x^3)$. A *continuous* spacetime is one where the metric is continuous, that is, its components are continuous functions with respect to any coordinates. Our definitions of timelike curves and the timelike future I^+ will also follow [19].

Let (M, g) be a C^k spacetime. A C^0 spacetime $(M_{\text{ext}}, g_{\text{ext}})$ is said to be a *continuous spacetime extension* of (M, g) provided there is an isometric embedding

$$(M, g) \hookrightarrow (M_{\text{ext}}, g_{\text{ext}})$$

40

preserving time orientations such that $M \subset M_{\mathrm{ext}}$ is a proper subset. (M is in fact an open submanifold of M_{ext} since they are both four-dimensional.) Note that we are identifying M with its image under the embedding. We remark that g_{ext} is C^2 in the examples constructed in the next section.

Definition 2.1 (Origin point). *Let $(M_{\mathrm{ext}}, g_{\mathrm{ext}})$ be a continuous spacetime extension of a C^k spacetime (M, g). Let u be a unit future directed timelike vector field on M. We say that a point $\mathcal{O}$ **is an origin point for u** if $\mathcal{O} \in M_{\mathrm{ext}} \setminus M$ and $\mathcal{O}$ is a past endpoint for each integral curve of u, and each extended integral curve is C^1 at $\mathcal{O}$. (Clearly this implies $\mathcal{O}$ lies in the closure $\overline{M}$ within M_{ext}.) In other words, $\mathcal{O}$ is an origin point for u if each integral curve of u, parameterized as $\gamma\colon (0, b) \to M$, satisfies*

(i) $\lim_{\tau \to 0} \gamma(\tau) = \mathcal{O}$,

(ii) $\widetilde{\gamma}'(0)$ *exists and* $\widetilde{\gamma}'(0) = \lim_{\tau \to 0} \gamma'(\tau)$,

where $\widetilde{\gamma}\colon [0, b) \to M_{\mathrm{ext}}$ is the extended curve defined by $\widetilde{\gamma}(0) = \mathcal{O}$ and $\widetilde{\gamma}(\tau) = \gamma(\tau)$ for $\tau > 0$. Continuity of the metric implies $\widetilde{\gamma}'(0)$ is a unit future directed timelike vector.

Remarks. Definition 2.1 is supposed to model the behavior of the comoving observers in Figure 8 (right). It is essentially the same as assumption (b) in [21, Thm. 2.2]. Actually, Definition 2.1 is slightly stronger; we assume this stronger assumption since it's easier to state and all the examples in section 3 will satisfy it.

We recall some terminology from section 2 of [21]. Let $\mathcal{O} \in M_{\mathrm{ext}} \setminus M$ be an origin point for u. A C^k function $f\colon M \to \mathbb{R}$ *extends continuously* to $M \cup \{\mathcal{O}\}$ if there is a continuous function $\widetilde{f}\colon M \cup \{\mathcal{O}\} \to \mathbb{R}$ such that $\widetilde{f}|_M = f$. In this case, we call $\widetilde{f}$ the *continuous extension* of f. A C^k tensor T defined on M *extends continuously* to $M \cup \{\mathcal{O}\}$ if there is a coordinate neighborhood U of $\mathcal{O}$ with coordinates $(x^0, \dots, x^3)$ such that each of the components of T extends continuously to $(U \cap M) \cup \{\mathcal{O}\}$. (This definition does not depend on the choice of coordinate system by the usual transformation law for tensor components.) This defines a continuous tensor $\widetilde{T}$ on $M \cup \{\mathcal{O}\}$, called the *continuous extension* of T, which satisfies $\widetilde{T}|_M = T$. For example, the metric tensor g extends continuously to $M \cup \{\mathcal{O}\}$ (by definition of a continuous extension). Trivially, if T is a smooth tensor defined on all of M_{ext}, then clearly $T|_M$ extends continuously to $M \cup \{\mathcal{O}\}$.

Definition 2.2 (Limiting to a perfect fluid near $\mathcal{O}$). *Let (M, g) be a C^2 spacetime, and let $\mathcal{O} \in M_{\mathrm{ext}} \setminus M$ be an origin point for u. Let T be the energy-momentum tensor on M (i.e., $T = \frac{1}{8\pi} G$ in suitable units where $G = Ric - \frac{1}{2} Rg$ is the Einstein tensor). Let $\rho_0, p_0 \in \mathbb{R}$. We say that T **limits to a perfect fluid (u, ρ_0, p_0)** at $\mathcal{O}$ if*

(i) $\rho := T(u, u)$ *extends continuously to $M \cup \{\mathcal{O}\}$ and $\widetilde{\rho}(\mathcal{O}) = \rho_0$,*

(ii) for any unit spacelike vector field e on M, which is orthogonal to u, the function $p_e := T(e, e)$ extends continuously to $M \cup \{\mathcal{O}\}$ and $\widetilde{p}_e(\mathcal{O}) = p_0$,

(iii) $T - T_{\mathrm{perfect}}$ extends continuously to $M \cup \{\mathcal{O}\}$ and its continuous extension is zero at $\mathcal{O}$, where T_{perfect} is the tensor on M given by

$$T_{\mathrm{perfect}} = (\rho_0 + p_0) u_* \otimes u_* + p_0 g,$$

where $u_* = g(u, \cdot)$ is the one-form metrically equivalent to u.

41

Remark. Definition 2.2 relaxes the requirement that T is identically a perfect fluid in assumption (a) of [21, Thm. 2.2]. Moreover, it's more physically convincing: FLRW models have perfect fluid energy-momentum tensors, and we expect that an FLRW model approximates our universe better as we go back in time towards the big bang.

Lastly, we require a mild, technical timelike convexity assumption:

Definition 2.3 (Locally timelike convex near $\mathcal{O}$). *Let $\mathcal{O} \in M_{\text{ext}} \setminus M$ be an origin point for u. Let $\gamma \colon (0, b) \to M$ be an integral curve of u. We say $\boldsymbol{M}$ **is locally timelike convex about** $\boldsymbol{\gamma}$ **near** $\boldsymbol{\mathcal{O}}$ if there is an $\varepsilon > 0$ and a coordinate neighborhood $U \subset M_{\text{ext}}$ centered at $\mathcal{O}$ with coordinates $(x^0, \ldots, x^3)$ satisfying*

(i) $g_{\mu\nu}(\mathcal{O}) = \eta_{\mu\nu}$ and $|g_{\mu\nu}(p) - \eta_{\mu\nu}| < \varepsilon$ for all $p \in U$,

(ii) $\partial_0|_{\mathcal{O}} = \tilde{\gamma}'(0)$,

(iii) $I_{\eta_\varepsilon}^+(\mathcal{O}, U) \subset M$,

where η_ε is the narrow Minkowskian metric given by $\eta_\varepsilon = -\frac{\varepsilon}{2-\varepsilon}(dx^0)^2 + \delta_{ij}dx^i dx^j$.

Remarks. In [21, Thm. 2.2], it was assumed that the manifold M satisfies $M = I^+(\mathcal{O}, M_{\text{ext}})$. Definition 2.3 relaxes this requirement and is a much weaker assumption. It will hold for the examples constructed in section 3. Also, conditions (i) and (ii) in Definition 2.3 will always be satisfied by continuity of the metric and applying the Gram-Schmidt orthogonalization process appropriately. The heart of Definition 2.3 is condition (iii) and is the motivation for the terminology "timelike convex near $\mathcal{O}$."

We are now ready to state our main theorem which generalizes [21, Thm. 2.2].

Theorem 2.4. *Let $(M_{\text{ext}}, g_{\text{ext}})$ be a continuous spacetime extension of a C^2 spacetime (M, g). Let u be a unit future directed timelike vector field on M. Assume the following.*

(a) $\mathcal{O} \in M_{\text{ext}} \setminus M$ is an origin point for u.

(b) The energy-momentum tensor T on M limits to a perfect fluid (u, ρ_0, p_0) at $\mathcal{O}$.

(c) M is locally timelike convex about γ near $\mathcal{O}$ for some integral curve γ of u.

(d) The Ricci tensor Ric on M extends continuously to $M \cup \{\mathcal{O}\}$.

(e) $(M_{\text{ext}}, g_{\text{ext}})$ is strongly causal at $\mathcal{O}$.

Then

$$\rho_0 = -p_0.$$

Moreover, the continuous extension of Ric at $\mathcal{O}$ is given by

$$\widetilde{Ric}|_{\mathcal{O}} = 8\pi\rho_0\, g_{\text{ext}}|_{\mathcal{O}}.$$

Remark. Assumptions (a), (b), and (c) are Definitions 2.1, 2.2, and 2.3, respectively. Assumption (d) will be satisfied whenever $(M_{\text{ext}}, g_{\text{ext}})$ is a C^2 extension of (M, g), which is the case for the examples constructed in the next section. Assumption (e) is a technical assumption needed for the proof; it's satisfied, for example, whenever M_{ext} is a subset of a globally hyperbolic spacetime.

Proof. Seeking a contradiction, assume $\rho_0 \neq -p_0$. Then

$$u_* \otimes u_* = \frac{1}{\rho_0 + p_0}(T_{\text{perfect}} - p_0 g)$$

$$= \frac{1}{\rho_0 + p_0}((T_{\text{perfect}} - T) + T - p_0 g).$$

By assumption (b), $T_{\text{perfect}} - T$ extends continuously to $M \cup \{\mathcal{O}\}$, and its continuous extension is zero at $\mathcal{O}$. Also T extends continuously to $M \cup \{\mathcal{O}\}$ by assumption (d). Therefore $u_* \otimes u_*$ extends continuously to $M \cup \{\mathcal{O}\}$. As in the proof of [21, Thm. 2.2], this implies that the vector field u extends continuously to $M \cup \{\mathcal{O}\}$. However, assumptions (c) and (e) prove that u does *not* extend continuously. Heuristically, this can be seen in Figure 8 (right). Rigorously, this follows from an analogous contradiction argument used in the proof of [21, Thm. 2.2]. Thus we have $\rho_0 = -p_0$.

Next we prove that $\widetilde{\text{Ric}}|_{\mathcal{O}} = 8\pi\rho_0 \, g_{\text{ext}}|_{\mathcal{O}}$. The Einstein equations imply

$$\text{Ric} = 8\pi T + \frac{1}{2}Rg$$

$$= 8\pi(T - T_{\text{perfect}}) + 8\pi T_{\text{perfect}} + \frac{1}{2}Rg$$

$$= 8\pi(T - T_{\text{perfect}}) + 8\pi T_{\text{perfect}} - 4\pi(\text{tr}\,T)g.$$

Since $\rho_0 = -p_0$, we have $T_{\text{perfect}} = -\rho_0 g$, and so T_{perfect} extends continuously to $M \cup \{\mathcal{O}\}$. Also $\text{tr}\,T$ extends continuously to $M \cup \{\mathcal{O}\}$, and its continuous extension is $-\rho_0 + 3p_0 = -4\rho_0$ at $\mathcal{O}$. Therefore evaluating the above expression at $\mathcal{O}$ gives

$$\widetilde{\text{Ric}}|_{\mathcal{O}} = 0 - 8\pi\rho_0 \, g_{\text{ext}}|_{\mathcal{O}} + 16\pi\rho_0 \, g_{\text{ext}}|_{\mathcal{O}} = 8\pi\rho_0 \, g_{\text{ext}}|_{\mathcal{O}}. \qquad \square$$

3 Anisotropic examples of the main theorem

In section 3.1 we construct explicit examples of spacetimes satisfying the hypotheses of Theorem 2.4. Clearly any Milne-like spacetime with a C^2 spacetime extension will satisfy the hypotheses of the theorem. But the goal of this section is to construct anisotropic examples as well, i.e., examples that are not FLRW spacetimes. (Recall Milne-like spacetimes are $k = -1$ FLRW spacetimes and hence are isotropic.) Briefly, to achieve this, we generalize Milne-like spacetimes in the following way: In spherical coordinates (t, r, θ, φ), the comoving observers in a Milne-like spacetime are parameterized by the curves $t = \mu r$ for $1 < \mu \leq \infty$, see Figure 8. ($\mu = \infty$ corresponds to the comoving observer traveling along $r = 0$.) In our anisotropic examples, we stipulate that the comoving observers follow the trajectories $t = \mu f(r)$, where $f(r) \approx r$ for r small. Like Milne-like spacetimes, the metric is still conformally flat and the conformal factor is a function of the foliation of the spacelike hypersurfaces orthogonal to the comoving observers, i.e., the conformal factor is a function of the rest spaces of u.

In section 3.2, we use the conclusion of Theorem 2.4 (that the Ricci tensor, and hence also the energy-momentum tensor, is dominated by a cosmological constant) to show that those comoving observers with μ-value greater than some critical number μ_{crit} will experience inflationary eras, lending support to the inflationary scenario. Our analysis depends on investigating the terms in the Raychaudhuri equation as they approach the origin point $\mathcal{O}$.

3.1 The examples

In this section we construct explicit examples of spacetimes satisfying the hypotheses of Theorem 2.4. Our examples will depend on only two functions $f(r)$ and $\Phi(\zeta)$.

Let $f(r)$ be a smooth positive function on $[0, \infty)$ satisfying $f(r) = r + O(r^3)$ as $r \to 0$ and $f'(r) \geq 1$ for all $r \geq 0$.[2] A simple example of such a function is $f(r) = \sinh(r)$. Our manifold of interest is

$$M := \{(t, x, y, z) \mid t > f(r), \text{ where } r = \sqrt{x^2 + y^2 + z^2}\}, \tag{3.1}$$

equipped with the metric

$$g = e^{2\Phi(\zeta)}[-dt^2 + dx^2 + dy^2 + dz^2] \tag{3.2}$$

for some arbitrary smooth[3] function $\Phi(\zeta)$ on $\mathbb{R}$. Here $\zeta = \zeta(t, r)$ is given by

$$\zeta(t, r) = \frac{t^2}{2} - \int_0^r \frac{f(s)}{f'(s)} \, ds. \tag{3.3}$$

The spacetime extension $(M_{\text{ext}}, g_{\text{ext}})$ of (M, g) is simply defined by extending g to all $\mathbb{R}^4 \approx M_{\text{ext}}$. In fact the metric is C^2 on M_{ext}, which follows from the assumptions on $f(r)$.

Remark. The simple case $f(r) = r$ corresponds to (a subclass of) Milne-like spacetimes [20]. This follows since the conformal factor is a function of $t^2 - r^2$.

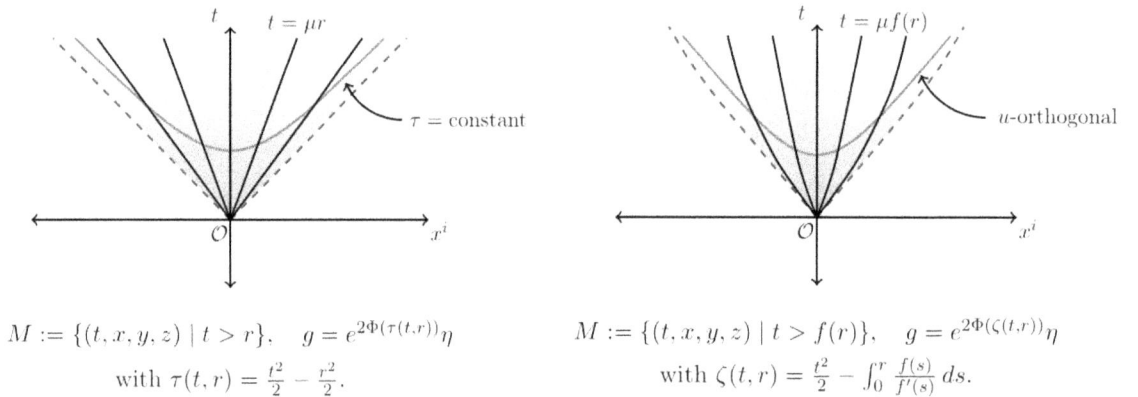

$$M := \{(t, x, y, z) \mid t > r\}, \quad g = e^{2\Phi(\tau(t,r))}\eta$$
$$\text{with } \tau(t,r) = \frac{t^2}{2} - \frac{r^2}{2}.$$

$$M := \{(t, x, y, z) \mid t > f(r)\}, \quad g = e^{2\Phi(\zeta(t,r))}\eta$$
$$\text{with } \zeta(t,r) = \frac{t^2}{2} - \int_0^r \frac{f(s)}{f'(s)} \, ds.$$

Figure 9: On the left is a Milne-like spacetime represented in conformal Minkowskian coordinates. On the right, the anisotropic examples constructed in this section. They are constructed to look like a Milne-like spacetime around the origin $\mathcal{O}$. The comoving observers (i.e., the integral curves of u) still emanate from the origin, and the manifold is still foliated by slices orthogonal to the comoving observers.

---

[2]We have $f(0) = 0$, $f'(0) = 1$, and $f''(0) = 0$. In fact $f(r) = r + O_2(r^3)$.

[3]More generally $\Phi(\zeta)$ only needs to be a C^2 function on a neighborhood of $\zeta = 0$, but, for simplicity, we assume $\Phi(\zeta)$ is smooth on all of $\mathbb{R}$.

The unit future directed timelike vector field u (whose integral curves are the comoving observers) will be given by normalized gradient of ζ:

$$u := -\frac{\nabla\zeta}{|\nabla\zeta|_g} = \frac{e^{-\Phi(\zeta)}}{\sqrt{t^2 f'(r)^2 - f(r)^2}} \left(t\, f'(r)\, \partial_t + f(r)\partial_r \right). \tag{3.4}$$

By construction the integral curves of u emanate from the origin $\mathcal{O} = (0,0,0,0)$ in (t,x,y,z)-coordinates. Each integral curve of u follows the trajectory of the curve $t = \mu f(r)$ for $1 < \mu \leq \infty$ (with $\mu = \infty$ corresponding to $r = 0$). To see this, recognize that the curve

$$r \mapsto (\mu f(r), r, \theta_0, \varphi_0) \tag{3.5}$$

in (t,r,θ,φ)-coordinates has tangent vector parallel to u. By rewriting these curves in (t,x,y,z)-coordinates, it's clear that they extend as C^1 curves through the origin $\mathcal{O}$. Thus $\mathcal{O}$ is an origin point for u. Hence part (a) of Theorem 2.4 is verified.

Now we verify properties (b) through (e) of Theorem 2.4. Property (c) is evidently satisfied; simply consider the integral curve along the t-axis given by $r = 0$. Property (e) holds since M_{ext} is conformal to Minkowski spacetime. Property (d) holds since the metric is C^2 on all of M_{ext}.

The remainder of this section will be dedicated to proving property (b), namely, that the energy-momentum tensor converges to that of a perfect fluid. However, to gain control over the terms appearing in the energy-momentum tensor, we found it easier to work with the following subset of our original manifold:

$$M_\varepsilon := \{(t,x,y,z) \mid t > (1+\varepsilon)f(r), \text{ where } r = \sqrt{x^2 + y^2 + z^2}\}, \tag{3.6}$$

where $\varepsilon > 0$ is arbitrary. Note that M_ε approaches M as $\varepsilon \to 0$. Moreover, for any $\varepsilon > 0$, we see that M_ε also satisfies properties (a) and (c) - (e) of Theorem 2.4.

Now we prove property (b) of Theorem 2.4 with M_ε playing the role of M in statement of Theorem 2.4. And this will hold for any $\varepsilon > 0$. The following fact will be used.

<u>Fact:</u> We have the following bound on M_ε:

$$\left| \frac{tf(r)}{t^2 f'(r)^2 - f(r)^2} \right| < \frac{1+\varepsilon}{2\varepsilon + \varepsilon^2}. \tag{3.7}$$

Proof of fact. Since M_ε is only defined for $t > (1+\varepsilon)f(r)$, we have

$$\begin{aligned}
\left| \frac{tf(r)}{t^2 f'(r)^2 - f(r)^2} \right| &= \frac{f(r)}{tf'(r)^2}\left(1 + \frac{f(r)^2}{t^2 f'(r)^2 - f(r)^2} \right) \\
&< \frac{1}{(1+\varepsilon)f'(r)^2}\left(1 + \frac{1}{(1+\varepsilon)^2 f'(r)^2 - 1} \right) \\
&\leq \frac{1}{1+\varepsilon}\left(1 + \frac{1}{(1+\varepsilon)^2 - 1} \right),
\end{aligned}$$

where we used the positivity of $f(r)$ and the fact that $f'(r) \geq 1$. $\qquad \square$

We start by showing property (i) in Definition 2.2. From conformal geometry, the Ricci tensor is given by

$$R_{\alpha\beta} = -2(\text{Hess}\Phi)_{\alpha\beta} + 2\nabla_\alpha\Phi\nabla_\beta\Phi - (\Box\Phi + 2|d\Phi|^2_\eta)\eta_{\alpha\beta}. \tag{3.8}$$

Here, all operators on the right-hand side are taken with respect to the Minkowski metric η. Using (3.8), we have

$$8\pi T_{\alpha\beta} = R_{\alpha\beta} - \frac{1}{2}Rg_{\alpha\beta}$$
$$= -2(\text{Hess}\Phi)_{\alpha\beta} + 2\nabla_\alpha\Phi\nabla_\beta\Phi + \Box\Phi\,\eta_{\alpha\beta} + 2|d\Phi|^2_\eta\,\eta_{\alpha\beta},$$

Straightforward computation shows

$$\rho = T(u,u)$$
$$= \frac{e^{-2\Phi(\zeta)}}{8\pi}\left[\left(2 + \frac{4f(r)}{rf'(r)} - \frac{2t^2 f(r)f''(r)}{t^2 f'(r)^2 - f(r)^2}\right)\Phi'(\zeta) + \frac{3(t^2 f'(r)^2 - f(r)^2)}{f'(r)^2}\Phi'(\zeta)^2\right]. \tag{3.9}$$

We are interested in showing that ρ extends continuously to $M_\varepsilon \cup \{\mathcal{O}\}$ and finding its limit at the origin $\mathcal{O}$. Both the first and second terms will contribute to $\tilde{\rho}(\mathcal{O})$ since, for small r, we have:

$$2 + \frac{4f(r)}{rf'(r)} = 6 + O(r^2).$$

Additionally, by utilizing (3.7), we see that the third term vanishes at the origin $\mathcal{O}$. Finally, the fourth term in (3.9) also vanishes since $f'(r) = 1 + O(r^2)$.

Hence, ρ extends continuously to the origin and

$$\tilde{\rho}(\mathcal{O}) = \frac{3}{4\pi}e^{-2\Phi(0)}\Phi'(0). \tag{3.10}$$

This shows property (i) in Definition 2.2. To show (ii), consider the vector field

$$v = \frac{e^{-\Phi(\zeta)}}{\sqrt{t^2 f'(r)^2 - f(r)^2}}\left(f(r)\,\partial_t + t\,f'(r)\partial_r\right).$$

By construction v is unit spacelike and orthogonal to u. Let e_θ, e_φ be the standard orthonormal vectors on the sphere so that $\{u, v, e_\theta, e_\varphi\}$ forms an orthonormal basis on M_ε (modulo some spherical coordinate singularities). Straightforward computations show

$$p_v = T(v,v)$$
$$= \frac{e^{-2\Phi(\zeta)}}{8\pi}\left[\left(-2 - \frac{4f(r)}{rf'(r)} - \frac{2f(r)^3 f''(r)}{f'(r)^2(t^2 f'(r)^2 - f(r)^2)}\right)\Phi'(\zeta)\right.$$
$$\left. - \frac{t^2 f'(r)^2 - f(r)^2}{f'(r)^2}(2\Phi''(\zeta) + \Phi'(\zeta)^2)\right],$$

$$p_{e_\theta} = T(e_\theta, e_\theta)$$
$$= \frac{e^{-2\Phi(\zeta)}}{8\pi}\left[\left(-4 - \frac{2f(r)}{rf'(r)} + \frac{2f(r)f''(r)}{f'(r)^2}\right)\Phi'(\zeta) - \frac{t^2 f'(r)^2 - f(r)^2}{f'(r)^2}(2\Phi''(\zeta) + \Phi'(\zeta)^2)\right],$$

$$p_{e_\varphi} = p_{e_\theta}.$$

Using (3.7) again, we see that these functions extend continuously to $M_\varepsilon \cup \{\mathcal{O}\}$ and

$$\tilde{p}_v(\mathcal{O}) = \tilde{p}_{e_\theta}(\mathcal{O}) = \tilde{p}_{e_\varphi}(\mathcal{O}) = -\frac{3}{4\pi}e^{-2\Phi(0)}\Phi'(0).$$

To finish our analysis, we need to compute all the cross terms of T. These cross terms are

$$T(u,v) = -\frac{2e^{-2\Phi(\varsigma)}}{8\pi}\frac{tf(r)^2 f''(r)}{f'(r)(t^2 f'(r)^2 - f(r)^2)}\Phi'(\varsigma),$$

$$T(u,e_\theta) = T(u,e_\varphi) = T(v,e_\theta) = T(v,e_\varphi) = T(e_\theta,e_\varphi) = 0.$$

Moreover, using (3.7), we have

$$T(u,v) \to 0. \tag{3.11}$$

as we approach $\mathcal{O}$ within M_ε.

Let e_0, e_1, e_2, e_3 denote $u, v, e_\theta, e_\varphi$ respectively. If e is any unit spacelike vector field orthogonal to u, then it can be written as $e = \sum_{i=1}^{3} a_i e_i$ with $\sum_{i=1}^{3} a_i^2 = 1$. Then vanishing of the cross terms implies

$$p_e = T(e,e) = \sum_{i=1}^{3} a_i^2\, T(e_i,e_i),$$

and so in the limit

$$\tilde{p}_e(\mathcal{O}) = -\frac{3}{4\pi}e^{-2\Phi(0)}\Phi'(0). \tag{3.12}$$

It is only left to show property (iii) in Definition 2.2, i.e., that $T - T_{\text{perfect}}$ extends continuously to $M_\varepsilon \cup \{\mathcal{O}\}$ and is zero at $\mathcal{O}$. Recall that

$$T_{\text{perfect}} = (\rho_0 + p_0)u_* \otimes u_* + p_0 g,$$

where ρ_0 and p_0 are given by (3.10) and (3.12).

We work in (t,x,y,z)-coordinates as they clearly cover the origin $\mathcal{O}$. We have

$$\partial_t = \frac{e^{\Phi(\varsigma)}}{\sqrt{t^2 f'(r)^2 - f(r)^2}}(tf'(r)u + f(r)v)$$

$$\partial_x = e^{\Phi(\varsigma)}\left[\frac{\sin(\theta)\cos(\varphi)}{\sqrt{t^2 f'(r)^2 - f(r)^2}}(tf'(r)u + f(r)v) + \cos(\theta)\cos(\varphi)e_\theta - \sin(\varphi)e_\varphi\right]$$

$$\partial_y = e^{\Phi(\varsigma)}\left[\frac{\sin(\theta)\sin(\varphi)}{\sqrt{t^2 f'(r)^2 - f(r)^2}}(tf'(r)u + f(r)v) + \cos(\theta)\sin(\varphi)e_\theta + \cos(\varphi)e_\varphi\right]$$

$$\partial_z = e^{\Phi(\varsigma)}\left[\frac{\cos(\theta)}{\sqrt{t^2 f'(r)^2 - f(r)^2}}(tf'(r)u + f(r)v) - \sin(\theta)e_\theta\right].$$

Then

$$T(\partial_t, \partial_t) = \frac{e^{2\Phi(\varsigma)}}{t^2 f'(r)^2 - f(r)^2}[t^2 f'(r)^2 \rho + 2tf(r)f'(r)T(u,v) + f(r)^2 p_v].$$

47

On the other hand,

$$T_{\text{perfect}}(\partial_t, \partial_t) = \frac{e^{2\Phi(\zeta)}}{t^2 f'(r)^2 - f(r)^2} \left[(\rho_0 + p_0)t^2 f'(r)^2 + p_0(-t^2 f'(r)^2 + f(r)^2)\right]$$

$$= \frac{e^{2\Phi(\zeta)}}{t^2 f'(r)^2 - f(r)^2} \left[\rho_0 t^2 f'(r)^2 + p_0 f(r)^2\right].$$

Therefore

$$T(\partial_t, \partial_t) - T_{\text{perfect}}(\partial_t, \partial_t)$$

$$= \frac{e^{2\Phi(\zeta)}}{t^2 f'(r)^2 - f(r)^2} \left[t^2 f'(r)^2 (\rho - \bar{\rho}) + 2tf(r)f'(r)T(u,v) + f(r)^2(p_v - \bar{p})\right].$$

Combining (3.7), (3.10), (3.11), and (3.12), we obtain

$$T(\partial_t, \partial_t) - T_{\text{perfect}}(\partial_t, \partial_t) \to 0,$$

as we approach the origin $\mathcal{O}$ within M_ε. In a similar manner, all other components of $T - T_{\text{perfect}}$ in (t, x, y, z)-coordinates also converge to 0. Consequently, assumption (b) in Theorem 2.4 is satisfied as well. Hence the conclusions of the Theorem 2.4 hold:

$$\rho_0 = -p_0 = \frac{3}{4\pi} e^{-2\Phi(0)} \Phi'(0) \qquad \text{and} \qquad \text{Ric}|_{\mathcal{O}} = 8\pi \rho_0 \, g|_{\mathcal{O}}. \tag{3.13}$$

Remark. We emphasize that we have applied Theorem 2.4 to the spacetime M_ε and not M; this is sufficient for the analysis in the next section.

3.2   Existence of inflationary eras in the examples

In this section we show how the conclusion of our main result, Theorem 2.4, proves the existence of inflationary eras for the examples constructed in section 3.1. A majority of the analysis in this section was outlined in section 3 of [21]. (At the time of writing [21], we had not yet found anisotropic examples of our main theorem which is a main inspiration for writing this paper.)

To gain some familiarity with the problem at hand, let's consider the FLRW setting. Friedmann's second equations is

$$3\frac{a''(\tau)}{a(\tau)} = -4\pi(\rho(\tau) + 3p(\tau)). \tag{3.14}$$

Therefore

$$\rho(0) = -p(0) > 0 \implies a''(\tau) > 0 \text{ for } \tau \text{ near } \tau = 0. \tag{3.15}$$

The assumption in (3.15) is what we mean by "the cosmological constant appears as an initial condition." It holds for a class of Milne-like spacetimes, see [21, eq. (1.11)]. In fact, our main result, Theorem 2.4, is essentially an anisotropoic generalization of this.

In this section, we generalize (3.15) to our anisotropic examples. Specifically what we demonstrate is the following. Let (M, g) be the spacetime defined by equations (3.1), (3.2), and (3.3). Let $\gamma(\tau)$ denote a comoving observer in M (i.e., γ is an integral curve of u). Here τ is the proper time of the comoving observer, and we fix it so that $\gamma(0) = \mathcal{O}$. We will define a "generalized scale factor"

$\mathfrak{a}(\tau)$ associated with $\gamma(\tau)$ and show that this generalized scale factor is accelerating, $\mathfrak{a}''(\tau) > 0$, for proper times τ near $\tau = 0$ (i.e., near the big bang). However, we only prove that some comoving observers experience inflation. Recall that the comoving observers follow the trajectories $t = \mu f(r)$, see (3.5). We find that only those comoving observers with a μ-value above a certain threshold μ_{crit} experience an inflationary era. This threshold is given by (3.27); it's completely determined by the functions $f(r)$ and $\Phi(\zeta)$ appearing in the previous section, and hence depends solely on the spacetime metric.

Remark. Throughout this section, we have in mind a fixed comoving observer. Since the comoving observers travel along the trajectories $t = \mu f(r)$, we can assume any fixed comoving observer is contained in some M_ε (see (3.6)) by choosing $\varepsilon > 0$ small enough. Therefore the bound (3.7) can be utilized.

Figure 10: We prove the existence of inflationary eras along a fixed comoving observer by computing the terms on the right-hand side of the Raychaudhuri equation (3.19) in a small neighborhood in M about the origin.

Recall u is given by (3.4). By construction u is orthogonal to the spacelike hypersurfaces of constant ζ. In Figure 8 (right), one should image that the spacelike hypersurfaces $\tau = $ constant are replaced with $\zeta = $ constant. In the terminology of [26, p. 359], u is "synchronizable," but it is not necessarily "proper time synchronizable." The latter occurs if and only if u is geodesic which occurs if and only if $f(r) = r$, see eq. (3.24). (Recall $f(r) = r$ corresponds to a Milne-like spacetime, and we know u is geodesic in this case.)

Set $H = \frac{1}{3}\text{div}u$ so that H coincides with the mean curvature of the spacelike hypersurfaces orthogonal to u, i.e., H is one-third the trace of the second fundamental form K.[4] Let τ denote the proper time of the flow lines of u (i.e., the proper time of the comoving observers). If $c(r)$ denotes the curve $r \mapsto (\mu f(r), r, \theta_0, \varphi_0)$ along the trajectory $t = \mu f(r)$, then the proper time τ is simply

$$\tau(r) = \int_0^r \sqrt{-g(c'(s), c'(s))}\, ds = \int_0^r e^{\Phi(\zeta)}\sqrt{\mu^2 f'(s)^2 - 1}\, ds. \tag{3.16}$$

When $c(r)$ is reparameterized by τ, it yields a comoving observer $\gamma(\tau)$.

Along each comoving observer $\gamma(\tau)$, we define a *generalized scale factor* $\mathfrak{a}(\tau)$ by[5]

$$\frac{\mathfrak{a}'}{\mathfrak{a}} = H. \tag{3.17}$$

We have $H(\tau) \approx \frac{1}{\tau}$ for τ small along each comoving observer $\gamma(\tau)$, see eq. (3.21) below. Since $\mathfrak{a}(\tau) = \exp(\int_{\tau_0}^\tau H)$ for some arbitrary time τ_0, it follows that

$$\mathfrak{a}(\tau) \to 0 \quad \text{as} \quad \tau \to 0 \tag{3.18}$$

---

[4]In the physics literature, $\text{div}u$ is often called the *expansion* θ of the congruence formed by the integral curves of u.

[5]The generalized scale factor $\mathfrak{a}(\tau)$ is also known as an "average length scale", see [9, eq. (2.14)]. In the isotropic FLRW setting, $\mathfrak{a}(\tau)$ is simply the scale factor $a(\tau)$, and the mean curvature H corresponds to the Hubble parameter. It's a happy coincidence that the notation for the mean curvature and the Hubble parameter happen to coincide.

along each comoving observer. Recall that, for us, the big bang corresponds to the time when the scale factor limits to 0. Therefore (3.18) suggests that the origin point $\mathcal{O}$ represents the big bang in these models.

For FLRW spacetimes, Friedmann's second equation (3.14) is used to analyze the acceleration of the scale factor. In the anisotropic setting, the generalization of Friedmann's second equation is the Raychaudhuri equation [15, eq. (4.26)],

$$3\frac{a''}{a} = -\text{Ric}(u, u) - 2\sigma^2 + \text{div}(\nabla_u u). \tag{3.19}$$

(The vorticity term vanishes since u is hypersurface orthogonal.)

Our goal is to compute all the terms on the right-hand side of (3.19) for points along a comoving observer near $\mathcal{O}$ (see Figure 10). First, using the conclusions of Theorem 2.4 and eq. (3.13), sufficiently close to the origin $\mathcal{O}$, we have

$$-\text{Ric}(u, u) \approx 8\pi\rho_0 = 6e^{-2\Phi(0)}\Phi'(0). \tag{3.20}$$

The $\approx$ in the above expression is understood in the following way: $-\text{Ric}(u, u)$ can be made arbitrarily close to $8\pi\rho_0$ by choosing points in M arbitrarily close to $\mathcal{O}$.

The shear term is defined by $2\sigma^2 = \sum_{i,j=1}^{3} \sigma(e_i, e_j)\sigma(e_i, e_j)$ where $\{e_1, e_2, e_3\}$ is an orthonormal basis spanning $u^\perp$ and

$$\sigma(e_i, e_j) = K(e_i, e_j) - H\delta_{ij},$$

where $K(X, Y) = g(\nabla_X u, Y)$ is the second fundamental form of the hypersurfaces orthogonal to u. (Recall $H = \frac{1}{3}\text{tr}K$.) Choosing the orthonormal basis $\{v, e_\theta, e_\varphi\}$ from the previous section, the only nonvanishing terms for K are

$$K(v, v) = \frac{e^{-\Phi(\zeta)}}{\sqrt{t^2 f'(r)^2 - f(r)^2}}\left[f'(r) - \frac{t^2 f(r)f'(r)f''(r)}{t^2 f'(r)^2 - f(r)^2} + \Phi'(\zeta)\frac{t^2 f'(r)^2 - f(r)^2}{f'(r)}\right]$$

$$K(e_\theta, e_\theta) = K(e_\varphi e_\varphi) = \frac{e^{-\Phi(\zeta)}}{\sqrt{t^2 f'(r)^2 - f(r)^2}}\left[\frac{f(r)}{r} + \Phi'(\zeta)\frac{t^2 f'(r)^2 - f(r)^2}{f'(r)}\right].$$

Therefore the mean curvature H is

$$3H = \frac{e^{-\Phi(\zeta)}}{\sqrt{t^2 f'(r)^2 - f(r)^2}}\left[f'(r) + \frac{2f(r)}{r} - \frac{t^2 f(r)f'(r)f''(r)}{t^2 f'(r)^2 - f(r)^2} + 3\Phi'(\zeta)\frac{t^2 f'(r)^2 - f(r)^2}{f'(r)}\right].$$

Along $t = \mu f(r)$, we have $H\big|_{t=\mu f(r)} = \frac{e^{-\Phi(0)}}{r\sqrt{\mu^2-1}} + O(r)$. Using (3.16), we reparameterize in terms of τ giving

$$H\big|_{\gamma(\tau)} = \frac{1}{\tau} + o(1). \tag{3.21}$$

Direct computation shows that

$$2\sigma^2 = \frac{2e^{-2\Phi(\zeta)}}{3[t^2 f'(r)^2 - f(r)^2]}\left[f'(r) - \frac{f(r)}{r} - \frac{t^2 f(r)f'(r)f''(r)}{t^2 f'(r)^2 - f(r)^2}\right]^2. \tag{3.22}$$

For small r, we have $f'(r) - \frac{f(r)}{r} = O(r^2)$ which combined with (3.7) yields

$$2\sigma^2 \to 0. \tag{3.23}$$

In other words $2\sigma^2$ extends continuously to $M_\varepsilon \cup \{\mathcal{O}\}$ and takes on the value 0 at $\mathcal{O}$. Geometrically, this "isotropization" effect is a consequence of the u-orthogonal hypersurfaces becoming more hyperbolic as we approach the origin $\mathcal{O}$.

The last term in (3.19) to compute is $\text{div}(\nabla_u u)$. We have

$$\nabla_u u = -\frac{t f(r)^2 f''(r) e^{-2\Phi(\zeta)}}{(t^2 f'(r)^2 - f(r)^2)^2}\Big(f(r)\partial_t + t f'(r)\partial_r \Big). \tag{3.24}$$

Hence

$$\text{div}(\nabla_u u) = -\frac{f(r) e^{-2\Phi(\zeta)}}{(t^2 f'(r)^2 - f(r)^2)^2}\bigg[2t^2 f'(r)^2 f''(r) + t^2 f(r) f''(r)^2 + t^2 f(r) f'(r) f'''(r)$$
$$+ f(r)^2 f''(r) + \frac{2t^2 f(r) f'(r) f''(r)}{r} - \frac{4t^4 f(r) f'(r)^2 f''(r)^2}{t^2 f'(r)^2 - f(r)^2} \bigg].$$

Evaluating along $t = \mu f(r)$ and taking the limit $r \to 0$, we find

$$\text{div}(\nabla_u u)\big|_{t=\mu f(r)} = -e^{-2\Phi(0)}\bigg[\frac{f'''(0)(5\mu^2 + 1)}{(\mu^2 - 1)^2} + O(r) \bigg]. \tag{3.25}$$

Using (3.20), (3.23), and (3.25), the Raychaudhuri equation (3.19), for points along the comoving observer sufficiently close to the origin $\mathcal{O}$, becomes

$$3\frac{\mathfrak{a}''}{\mathfrak{a}}\bigg|_{t=\mu f(r)} \approx 6 e^{-2\Phi(0)}\bigg[\Phi'(0) - \frac{\frac{1}{6} f'''(0)(5\mu^2 + 1)}{(\mu^2 - 1)^2} \bigg]. \tag{3.26}$$

Similar to (3.20), the $\approx$ symbol in the above expression is understood in the following way: $3(\mathfrak{a}''/\mathfrak{a})$ can be made arbitrarily close to the right-hand side of (3.26) by choosing points along $t = \mu f(r)$ that are sufficiently close to the origin $\mathcal{O}$.

From (3.26) we can determine which comoving observers experience an inflationary era, i.e., which comoving observers experience $\mathfrak{a}''(\tau) > 0$ arbitrarily close to $\tau = 0$. Assuming $\Phi'(0) > 0$ (which is equivalent to $\tilde{\rho}(\mathcal{O}) > 0$), it's precisely those comoving observers with μ-values satisfying

$$\mu > \mu_{\text{crit}} := \sqrt{\frac{12\Phi'(0) + 5 f'''(0) + \sqrt{144\Phi'(0) f'''(0) + 25 f'''(0)^2}}{12\Phi'(0)}}. \tag{3.27}$$

Moreover, we see that if $f'''(0) = 0$ and $\Phi'(0) > 0$, then all the comoving observers experience an inflationary era. This reproduces the results for Milne-like spacetimes, see (3.15).

51

3.3 Remarks on proving anisotropy

In this section we show that the examples constructed in section 3.1 are generally anisotropic. Although this is heuristically evident, a formal mathematical proof is not immediately clear.

First, the definition of an "isotropic spacetime" is not consistent throughout the literature. See [2] and [28] and references therein. We will adopt the definition in [26, Ch. 12] since, as discussed in [2], this definition is the optimal one as it implies that the spacetime is isometric to a subset of an FLRW spacetime, see [26, Prop. 12.6] and [2, Thm. 2.1].

Therefore any spacetime that is not isometric to a subset of an FLRW model is anisotropic according to [26]. For the examples constructed in section 3.1, if $f(r) = r$ then they are isometric to a subclass of Milne-like spacetimes which are a subclass of $k = -1$ FLRW models, and hence they are isotropic. Moreover, regardless of the form of $f(r)$, if the conformal factor is identically 1, then the spacetime is isometric to a subset of Minkowski spacetime which is clearly isotropic. (This shows that it is not sufficient to simply recognize that the shear term (3.22) is nonzero. However, in this case, the vector field defining the comoving observers changes.) This suggests that if $f(r)$ is not identically r and the conformal factor is not constant, then the resulting spacetime is not a subset of an FLRW spacetime and hence is anisotropic. We believe such a statement can be proven rigorously. However, in this section, we will content ourselves with the following algorithm: Pick functions $f(r)$ and $\Phi(\zeta)$. The steps below show how to verify that the corresponding spacetime (M, g) from section 3.1 is anisotropic.

Seeking a contradiction, suppose (M, g) is in fact isometric to a subset of an FLRW spacetime. Since FLRW spacetimes satisfy the Einstein equations with a perfect fluid [26, Thm. 12.11], there is a unit future directed timelike vector field $\tilde{u}$ on M such that T is a perfect fluid with respect to $\tilde{u}$. There exist functions a, b, c, d such that

$$\tilde{u} = au + bv + ce_\theta + de_\varphi,$$

where $\{u, v, e_\theta, e_\varphi\}$ is the orthonormal frame constructed in section 3.1. Consider the unit spacelike vectors orthogonal to $\tilde{u}$

$$\tilde{v} = \frac{bu + av}{a^2 - b^2}, \qquad \tilde{e}_\theta = \frac{cu + ae_\theta}{a^2 - c^2}, \qquad \tilde{e}_\varphi = \frac{du + ae_\varphi}{a^2 - d^2}.$$

From section 3.1, we know how T acts on the orthonormal frame $\{u, v, e_\theta, e_\varphi\}$, and so we know how T acts on $\{\tilde{u}, \tilde{v}, \tilde{e}_\theta, \tilde{e}_\varphi\}$.

Fix a point $p_0 \in M$ given by $(t_0, r_0, \theta_0, \varphi_0)$. At p_0, the following equations set up an overdetermined system for (a, b, c, d) at p_0.

$$-a^2 + b^2 + c^2 + d^2 = -1$$
$$T(\tilde{u}, \tilde{v}) = 0$$
$$T(\tilde{u}, \tilde{e}_\theta) = 0$$
$$T(\tilde{u}, \tilde{e}_\varphi) = 0$$
$$T(\tilde{v}, \tilde{v}) = T(\tilde{e}_\theta, \tilde{e}_\theta) = T(\tilde{e}_\varphi, \tilde{e}_\varphi).$$

For most choices of $f(r)$ and $\Phi(\zeta)$, this system does not have any solutions, giving a contradiction. However, even if there are solutions, one can still obtain a contradiction by other means, e.g., showing that the orthogonal subspace to $\tilde{u}$ does not have constant sectional curvature. Lastly, we remark that the point p_0 must lie away from $r = 0$. Indeed, points along $r = 0$ will past the above tests. This is due to the spacetime being spherically symmetric and hence spatially isotropic precisely at points along $r = 0$.

We remark that if Φ is constant or $f(r) = r$, then (M, g) passes the above tests. In the first case, the metric is homothetic to the Minkowski metric (and hence isometric to a subset of a $k = 0$ FLRW spacetime), and in the second case, the spacetime is given by a Milne-like spacetime and hence is isometric to a $k = -1$ FLRW spacetime.

4 Summary and outlook

The inflationary scenario has become the current paradigm of early universe cosmology. Roughly, it states that scale factor underwent a brief but dramatic period of acceleration after the big bang but before the radiation dominated era. Although inflationary theory has many successes (e.g., solutions to the horizon and flatness problems along with providing a framework for generating the seeds of large-scale structures in our universe), it does not carry the status of an established physical theory. In this work, we provide mathematical support for the inflationary scenario by showing that a class of anisotropic spacetimes experience inflationary eras after the big bang.

Our main result, Theorem 2.4, says that if the universe began with special initial conditions at the big bang, then the energy-momentum tensor was dominated by a cosmological constant at the big bang. These special initial conditions are (1) the existence of an origin point $\mathcal{O}$ for a unit timelike vector field u (whose integral curves represent the comoving observers in the universe) and (2) the energy-momentum tensor approaches a perfect fluid at $\mathcal{O}$. An informal discussion of these special initial conditions is given in section 1.3.

In section 3.1, we construct anisotropic spacetimes which satisfy the hypotheses of Theorem 2.4. These examples can be thought of as "quasi Milne-like spacetimes." In section 3.2, we define a generalized scale factor $\mathfrak{a}(\tau)$ along each comoving observer (τ denotes the proper time of the comoving observer), and we show that $\mathfrak{a}(\tau) \to 0$ as $\tau \to 0$, see (3.18). Consequently, we associate $\tau = 0$ (and hence also the origin point $\mathcal{O}$) with the big bang. Lastly, we describe which comoving observers experience inflation, $\mathfrak{a}''(\tau) > 0$, immediately after the big bang $\tau = 0$. See equations (3.26) and (3.27).

Our examples exhibit isotropization towards the past and, in fact, are perfectly isotropic at the big bang $\mathcal{O}$, see (3.23). Our isotropization-towards-the-past result is consistent with a universe starting from special initial conditions. This is unlike results related to the cosmic no-hair conjecture (see Wald's original paper [30] or some more recent work, e.g., [3]), where isotropization occurs towards the future.

A limitation of our approach is that we only show accelerated expansion immediately after the big bang. For example reheating does not appear in our analysis. For an analysis of the physics after the accelerated expansion, our geometrical initial conditions should be supplemented with, for example, appropriate scalar field matter models.

We believe that differential geometry (and geometric analysis in particular) has a role to play in the investigation of initial conditions for the big bang. The work presented in this paper should be thought of as a "proof of concept" of this proposal. Our work can be generalized in many ways. In particular, although our examples are not necessarily isotropic, they are still spherically symmetric. So a natural generalization is to reproduce the analysis in sections 3.1 and 3.2 with non-spherically symmetric spacetimes. Also, our examples are anisotropic versions of $k = -1$ FLRW spacetimes. What about $k = 0$ FLRW spacetimes? In this case one would want to apply [12, Thm. 5.2] or a suitable generalization thereof. Lastly, it remains to be seen if the results in [11] can be used to generate comoving observers with an origin point $\mathcal{O}$.

Acknowledgments

Eric Ling was supported by Carlsberg Foundation CF21-0680 and Danmarks Grundforskningsfond CPH-GEOTOP-DNRF151. Annachiara Piubello was supported by the DFG Project ME 3816/3-1, part of the SPP2026. We thank Jerome Quintin for helpful comments on an earlier draft and are grateful to the Minkowski institute where this project began to take shape.

References

[1] Andreas Albrecht, *Cosmic inflation and the arrow of time*, 2003, arXiv:astro-ph/0210527v3.

[2] Rodrigo Avalos, *On the rigidity of cosmological space-times*, Letters of Math Phys. **113** (2023), no. 98.

[3] Ferez Azhar and David I. Kaiser, *Flows into de Sitter space from anisotropic initial conditions: An effective field theory approach*, Phys. Rev. D **107** (2023), no. 4, 043506.

[4] Robert Brandenberger, *Initial conditions for inflation — A short review*, Int. J. Mod. Phys. D **26** (2016), no. 01, 1740002.

[5] Robert Brandenberger and Patrick Peter, *Bouncing Cosmologies: Progress and Problems*, Found. Phys. **47** (2017), no. 6, 797-850.

[6] Sean M. Carroll and Jennifer Chen, *Spontaneous inflation and the origin of the arrow of time*, 2004, arXiv:hep-th/0410270.

[7] Sidney Coleman and Frank De Luccia, *Gravitational effects on and of vacuum decay*, Phys. Rev. D **21** (1980), 3305–3315.

[8] Planck Collaboration, *Planck 2018 results. X. Constraints on inflation*, Astron. Astrophys. **641** (2020), A10.

[9] George F.R. Ellis, *Relativistic cosmology*, Proc. Int. Sch. Phys. Fermi **47** (1971), 104–182.

[10] Gregory Galloway and Eric Ling, *Some Remarks on the C^0-(in)extendibility of Spacetimes*, Annales Henri Poincare **18** (2017), no. 10, 3427–3447.

[11] Ya Gao and Jing Mao, *Inverse mean curvature flow for spacelike graphic hypersurfaces with boundary in Lorentz-Minkowski space $\mathbb{R}_1^{n+1}$*, 2021, arXiv:2104.10600v5.

[12] Ghazal Geshnizjani, Eric Ling, and Jerome Quintin, *On the initial singularity and extendibility of flat quasi-de Sitter spacetimes*, Journal of High Energy Physics **2023** (2023), no. 10.

[13] Alan H. Guth, *The Inflationary Universe: A Possible Solution to the Horizon and Flatness Problems*, Phys. Rev. D **23** (1981), 347–356.

[14] Alan H. Guth, David I. Kaiser, and Yasunori Nomura, *Inflationary paradigm after planck 2013*, Physics Letters B **733** (2014), 112–119.

[15] Stephen W. Hawking and George F. R. Ellis, *The Large Scale Structure of Space-Time*, Cambridge Monographs on Mathematical Physics, Cambridge University Press, 2023.

[16] Anna Ijjas, Paul J. Steinhardt, and Abraham Loeb, *Inflationary paradigm in trouble after Planck2013*, Phys. Lett. B **723** (2013), 261–266.

[17] Andrei Linde, *On the problem of initial conditions for inflation*, Foundations of Physics **48** (2018), no. 10, 1246–1260.

[18] Andrei D. Linde, *A New Inflationary Universe Scenario: A Possible Solution of the Horizon, Flatness, Homogeneity, Isotropy and Primordial Monopole Problems*, Phys. Lett. B **108** (1982), 389–393.

[19] Eric Ling, *Aspects of C^0 causal theory*, Gen. Rel. Grav. **52** (2020), no. 6, 57.

[20] Eric Ling, *The Big Bang is a Coordinate Singularity for $k = -1$ Inflationary FLRW Spacetimes*, Found. Phys. **50** (2020), no. 5, 385–428.

[21] Eric Ling, *Remarks on the cosmological constant appearing as an initial condition for Milne-like spacetimes*, Gen. Rel. Grav. **54** (2022), no. 7, 68.

[22] Eric Ling and Annachiara Piubello, *On the asymptotic assumptions for Milne-like spacetimes*, Gen. Rel. Grav. **55** (2023), no. 4, 53.

[23] Jerome Martin, Christophe Ringeval, and Vincent Vennin, *Encyclopædia Inflationaris*, Phys. Dark Univ. **5-6** (2014), 75–235.

[24] Viatcheslav F. Mukhanov and G. V. Chibisov, *Quantum Fluctuations and a Nonsingular Universe*, JETP Lett. **33** (1981), 532–535.

[25] Kimihiro Nomura and Daisuke Yoshida, *Past extendibility and initial singularity in Friedmann-Lemaître-Robertson-Walker and Bianchi I spacetimes*, JCAP **07** (2021), 047.

[26] Barrett O'Neill, *Semi-Riemannian geometry*, Pure and Applied Mathematics, vol. 103, Academic Press Inc. [Harcourt Brace Jovanovich Publishers], New York, 1983.

[27] Roger Penrose, *Singularities and Time Asymmetry*, pp. 581–638, 1980, General Relativity: An Einstein Centenary Survey.

[28] Miguel Sánchez, *A class of cosmological models with spatially constant sign-changing curvature*, Portugaliae Mathematica **80** (2023), no. 3/4.

[29] Alexei A. Starobinsky, *A New Type of Isotropic Cosmological Models Without Singularity*, Phys. Lett. B **91** (1980), 99–102.

[30] Robert M. Wald, *Asymptotic behavior of homogeneous cosmological models in the presence of a positive cosmological constant*, Phys. Rev. D **28** (1983), 2118–2120.

5 An alternative solution for the Milne Model

Seokcheon Lee

Abstract   The Milne model, a special-relativistic cosmological model, is mathematically equivalent to a particular case of the Robertson-Walker (RW) model with negative spatial curvature in the limit of vanishing energy density. Solving the Friedmann equations allows for the linear time-dependent scale factor. As a specific instance of the RW model, the Milne model adheres to the cosmological principle, ensuring that the three-dimensional space remains isotropic and homogeneous throughout. From this condition, one can derive the cosmological redshift relation. Interestingly, our findings demonstrate that this principle holds even under the assumption that the speed of light varies at cosmological scales. This result is in agreement with recent findings reported in [16, 17, 18].

Keywords:keywords Milne model, varying speed of light model, and cosmological redshiftkeywords Milne model, varying speed of light model, and cosmological redshift Milne model, varying speed of light model, cosmological redshift

1   Introduction

Milne proposed a specific relativistic cosmological model to explain the phenomenon of the recession of the spiral nebulae on purely kinematic grounds [1, 2, 3]. Since it corresponds to the empty universe with a spatial curvature, it is just the inside of the light cone of Minkowski space in an incomplete coordinate system. It is mathematically identical to a specific case of the negative spatial curvature Robertson-Walker (RW) model in which the energy density, pressure, and cosmological constant are all zero. The RW metric uses the *Cosmological Principle* (CP) that states that the Universe looks the same from all positions in space *at a particular time*, and that all directions in space at any point are equivalent. Thus, the 3-space in the Milne model is comoving space. It is also possible to adopt the standard form of the Robertson-Walker (RW) metric in cosmology for the cosmic time t [4, 5, 6, 7].

In Sec. 2, we review how to derive the Milne metric from the Minkowski space. We show that the solution of the Milne model for Einstein's field equation is a linearly increasing scale factor for all time in section 3. As a specific solution of RW metric, the 3-spatial line element of the Milne model should be a function of the comoving coordinates only. From this condition, one can derive the cosmological redshift relation. We show that the solution is still valid when we allow the time variation of the speed of light (VSL) at cosmological scales in section 4. We conclude in Sec. 5.

Eric Ling and Annachiara Piubello (Eds), Spacetime 1908-2023. Selected peer-reviewed papers presented at the *Third Hermann Minkowski Meeting on the Foundations of Spacetime Physics*, 11-14 September 2023, Albena, Bulgaria (Minkowski Institute Press, Montreal 2024). ISBN 978-1-998902-25-5 (softcover), ISBN 978-1-998902-26-2 (ebook).

2 From Minkowski space to Milne spacetime

The Milne spacetime is a patch of Minkowski space in an incomplete expanding coordinate system. Mathematically it is the interior of the future light cone of some fixed point in Minkowski, foliated by negatively curved hyperboloids [2, 3]. One can prove this by replacing the Minkowski spherical coordinates (τ, r, θ, ϕ) with the new coordinates (t, χ, θ, ϕ) defined as

$$\tau = t \cosh \chi \quad , \quad r = ct \sinh \chi \, . \tag{2.1}$$

Then, the square of the line element of the Minkowski space becomes

$$ds^2 = -c^2 d\tau^2 + dr^2 + r^2 d\Omega^2 = -c^2 dt^2 + c^2 t^2 \left(d\chi^2 + \sinh^2 \chi d\Omega^2 \right) \, , \tag{2.2}$$

where $d\Omega^2 = d\theta^2 + \sin^2 \theta d\phi^2$ is the element of the solid angle. The metric in the second equality is the so-called Milne spacetime. One can reexpress the 3-space section in Eq. (2.2) as the hyperbolic one by substituting $\sinh \chi$ as a dimensionless comoving radial coordinate σ

$$ds^2 = -c^2 dt^2 + S(t)^2 \left(\frac{d\sigma^2}{1 - k\sigma^2} + \sigma^2 d\Omega^2 \right) \, , \tag{2.3}$$

where the curvature constant $k = -1$ is a dimensionless number and $S(t) = ct$ is the scale factor with dimensions of $[L]$. This is the Robertson-Walker (RW) metric for a homogeneous and isotropic universe with an open 3-space.

3 Milne Universe

The metric in Eq. (2.3) should satisfy Einstein's field equation. The first Friedmann equation is

$$\frac{\dot{S}^2}{S^2} + \frac{kc^2}{S^2} = \frac{8\pi G}{3} \rho + \frac{\Lambda c^2}{3} \, , \tag{3.1}$$

where the dot denotes the derivative for the cosmic time t, ρ is the total mass density, and Λ is the cosmological constant. The Milne model corresponds to $\rho = \Lambda = 0$ and $k = -1$. Thus, Eq. (3.1) becomes

$$\frac{\dot{S}^2}{S^2} = \frac{c^2}{S^2} \, . \tag{3.2}$$

The general solution for this equation (3.2) is

$$S(t) = ct \, , \tag{3.3}$$

which is consistent with Eq. (2.3). Therefore, we can conclude that the Milne spacetime is mathematically equivalent to the specific case of the RW metric in the limit that the energy density, pressure, and cosmological constant all equal zero, and the spatial curvature is negative. This metric describes the Hubble expansion correctly. Suppose that our galaxy, comoving in the space-time continuum, is located at radial distance $\sigma = 0$ and another one is at an arbitrary distance σ. From Eq. (2.3), its proper distance L from us at cosmic time t is

$$L = S(t) \int_0^\sigma \frac{d\sigma'}{\sqrt{1 + \sigma'^2}} = S(t) \sinh^{-1}(\sigma) \equiv S(t) f(\sigma) \, . \tag{3.4}$$

The velocity of recession is

$$v_{\text{rec}} \equiv \frac{dL}{dt} = \frac{\dot{S}}{S}L \equiv HL = c\sinh^{-1}(\sigma)\,, \tag{3.5}$$

satisfying Hubble's law.

4 The Milne spacetime with varying speed of light

Even though the Milne model is incompatible with cosmological observations, it provided the intuition for general solutions for the expanding homogeneous and isotropic universe [6, 7]. Therefore, it is interesting to probe the possibility of another solution for this model. For this purpose, we slightly change our previous expression for (t, χ, θ, ϕ) in (4.1) as

$$c_{\text{M}}\tau = X^0 \cosh\chi \quad , \quad r = X^0 \sinh\chi \quad , \text{ where } \quad X^0 = ct\,. \tag{4.1}$$

Now, we can rewrite the metric in (2.3) as

$$\begin{aligned} ds^2 &= -\left(dX^0\right)^2 + S(t)^2 \left(\frac{d\sigma^2}{1 - k\sigma^2} + \sigma^2\Omega^2\right) \\ &\equiv -\left(dX^0\right)^2 + S(t)^2 dl(\sigma, \theta, \phi)^2\,, \end{aligned} \tag{4.2}$$

where the spatial part of the metric dl^2 measures the comoving distance between two neighboring points at the same time. The time coordinate is called cosmic time. It is easily seen that the spacelike hypersurfaces in Weyl's postulate are the surfaces of simultaneity with respect to cosmic time [8, 9, 10, 11, 12, 13, 14]. In this metric, the light signal propagates along the geodesic $ds^2 = 0$, and from the metric (4), we obtain for outgoing light signals

$$ds^2 = 0 \quad \Rightarrow \quad dl(\sigma, \theta, \phi) = \frac{dX^0}{S(t)}\,. \tag{4.3}$$

The infinitesimal line element dl is a function of the comoving coordinates (σ, θ, ϕ) only. Thus, one can obtain the traditional cosmological redshift relation from this condition (4.3)

$$\begin{aligned} \frac{dX^0}{S(t)}\Big|_{t=t_1} &= \frac{dX^0}{S(t)}\Big|_{t=t_2} \quad \Rightarrow \quad \frac{dt_1}{a(t_1)} = \frac{dt_2}{a(t_2)} \\ &\Rightarrow \quad \lambda_1 = cdt_1 = \frac{a(t_1)}{a(t_2)}\lambda_2\,, \end{aligned} \tag{4.4}$$

where we define $a(t) \equiv S(t)/c$ and denote the wavelength of a light λ_i at t_i with a frequency $(dt_i)^{-1}$. This solution is obtained under the assumption that the speed of light is constant at cosmological scales. However, the local Lorentz invariance has to be replaced by General Relativity at cosmological scales. Therefore, the quibble about whether special relativity is generally adaptable at cosmological distances and time scales should be determined by observations[15].

If we allow the possibility of a variable speed of light (VSL), then the above condition (4.3) provides

$$\begin{aligned} \frac{dX^0}{S(t)}\Big|_{t=t_1} &= \frac{dX^0}{S(t)}\Big|_{t=t_2} \quad \Rightarrow \quad \frac{\tilde{c}_1 dt_1}{c_1 a_1} = \frac{\tilde{c}_2 dt_2}{c_2 a_2} \\ &\Rightarrow \quad \lambda_1 = \tilde{c}_1 dt_1 = \frac{S(t_1)}{S(t_2)}\lambda_2\,, \end{aligned} \tag{4.5}$$

where $c_i \equiv c(t_i)$, $a_i \equiv a(t_i)$, and we define

$$dX^0 = d\left(c[t]t\right) = \left(\frac{d\ln c}{d\ln t} + 1\right)cdt \equiv \tilde{c}[t]dt \quad \text{and}$$
$$\delta c \equiv \frac{\tilde{c}}{c} = \left(\frac{d\ln c}{d\ln t} + 1\right). \tag{4.6}$$

We show the detailed derivation of Eq. (4.5) in the appendix 5. The scale factor $S(t)$ includes the time variation of the speed of light $c(t)$. Thus, the cosmological redshift still holds even if we allow the speed of light can vary as a function of cosmic time.

Even though we show that equation (4.3) is still valid when we include the possibility of the VSL in the scale factor $S(t)$ of the Milne model, we need to show that this solution satisfies the Einstein field equations [16, 17, 18]. Thus, if we include the VSL, we need to rewrite the Friedamm equation (3.1) as

$$\frac{\dot{S}^2}{S^2} + k\frac{\tilde{c}^2}{S^2} = \frac{8\pi\tilde{G}}{3}\sum\rho_i + \frac{\Lambda\tilde{c}^2}{3} \quad \Rightarrow \dot{S} = \tilde{c}. \tag{4.7}$$

The solution for this Milne mode allowing VSL is $S = X^0 = ct$ and $\dot{S} = (d\ln c/d\ln t + 1)c \equiv \tilde{c}$ being consistent with Eq. (4.6).

5 Conclusion

The Milne model is known as a specific case of the RW metric. Thus, its spatial coordinates satisfy the cosmological principle. One can derive the cosmological redshift relation from this condition. The traditional derivation is based on the assumption that the speed of light is constant. However, we do not know the speed of light is constant at cosmological distances and time scales. Thus, we investigate the possibility of the Milne model when we adopt the variation of the speed of light as a function of cosmic time. We show that the Milne model is still valid with the VSL assumption.

Appendix A: Proof of cosmological redshift

We derive the cosmological redshift in Eq. (4.5)

$$\lambda_1 = \tilde{c}_1 dt_1 = c_1\frac{\tilde{c}_2}{c_2}\frac{a_1}{a_2}dt_2 = \tilde{c}_2\frac{c_1 a_1}{c_2 a_2}dt_2 = \frac{S_1}{S_2}\lambda_2\,, \tag{A1}$$

where we use

$$\frac{\tilde{c}_1 dt_1}{c_1 a_1} = \frac{\tilde{c}_2 dt_2}{c_2 a_2} \quad \text{from} \quad \left.\frac{d(ct)}{S(t)}\right|_{t=t_1} = \left.\frac{d(ct)}{S(t)}\right|_{t=t_2}. \tag{A2}$$

Acknowledgments

SL is supported by the Basic Science Research Program through the National Research Foundation of Korea (NRF) funded by the Ministry of Science, ICT, and Future Planning (Grant No. NRF-2019R1A6A1A10073079, NRF-RS202300243411). This manuscript is based on the presentation delivered at the third Minkowski meeting in Albena, held on September 11-14, 2023.

References

[1] E. A. Milne, nature **1932**, 9-10 (1932)
https://doi.org/10.1038/130009a0.

[2] V. Mukhanov, *Physical Foundations of Cosmology* (Cambridge University Press, 2005).

[3] S. Carroll, *Spacetime and Geometry : An Introduction to General Relativity* (Pearson Education Limited, 2014).

[4] Robertson, H. P. (1929), Proceedings of the National Academy of Sciences. **15**, (11): 822–829 (1929) doi:10.1073/pnas.15.11.822. PMC 522564. PMID 16577245.

[5] H. P. Robertson, Rev. Mod. Phys. **5**, 62 (1933)
https://doi.org/10.1103/RevModPhys.5.62.

[6] A. G. Walker, E. A. Milne, Mon. Not. R. Astron. Soc. **95**, 263 (1935)
https://doi.org/10.1093/mnras/95.3.263.

[7] Walker, A. G. (1937), Proceedings of the London Mathematical Society, Series **42**, (1): 90–127, (1937) doi:10.1112/plms/s2-42.1.90.

[8] H. Weyl, Phys. Z. **24**, 230 (1923) (English translation: H. Weyl, "Republication of: On the general relativity theory" Gen. Rel. Grav. **41**, 1661 (2009)).

[9] J. V. Narlikar,*An Introduction to Cosmology* (Cambridge University Press, 3rd Ed 2002).

[10] M. P. Hobson, G. P. Efstathiou, and A. N. Lasenby, *General Relativity: An Introduction for Physicists* (Cambridge University Press, 2006).

[11] L. Ryder, *Introduction to General Relativity* (Cambridge University Press, 2009).

[12] J. N. Islam, *An Introduction to Mathematical Cosmology* (Cambridge University Press, 2001).

[13] M. Guidry, *Modern General Relativity: Black Holes, Gravitational Waves, and Cosmology* (Cambridge University Press, 2019).

[14] V. Ferrari, L. Gualtieri, and P. Pani, *General Relativity and its Applications: Black Holes, Compact Stars and Gravitational Waves* (CRC Press, 2021).

[15] M. Roos, *Introduction to Cosmology* (Wiley, 4th Ed 2015).

[16] S. Lee, JCAP **08**, 054 (2021) doi:10.1088/1475-7516/2021/08/054 [arXiv:2011.09274 [astro-ph.CO]].

[17] S. Lee, Found. Phys. **53**, 40 (2023) doi:10.1007/s10701-023-00682-1 [arXiv:2303.13772 [gen-ph]].

[18] S. Lee, Mon. Not. Roy. Astron. Soc. **522**, 3248-3255 (2023) doi:10.1093/mnras/stad1190 [arXiv:2301.06947 [astro-ph.CO]].

6   Concealed Mass and Gravitation within Whitehead's conception of observability in space and time suggest a Big Bounce Universe

Guido J.M. Verstraeten & Willem W. Verstraeten

Abstract   According to Whitehead, nature is disclosed to mind by an ensemble of events characterized by both unobservable intrinsic factors (e.g. mass and gravitation) as well as observable extrinsic factors (e.g. motion and density)[1]. Mass is not the substratum endowed of energy and linear momentum[1], in particular it implies – spatial – extension and – temporal – duration in order to be disclosed by corresponding extrinsic factors. Extension and duration are both necessary conditions of observable natural phenomena. Specifically, space is the relation between bits of matter while Time is the relation between a bit of matter with itself. As such, an instant, deprived of duration, is immeasurable[1]. Whitehead's claims on mass, space and time corroborate Verlinde's alternative conception of quantum gravitation[2]. Within the de Sitter spacetime this conception starts from the competition of the short distance degrees of freedom of the RynTakanayagi tensor of emergent spacetime[2] with long distance thermalized excitations. This enables the creation of baryonic mass and the decrease of the de Sitter entropy[2]. It is the memory effect of the original baryon creation that leads to gravitation and the production of thermodynamic extensive entropy. Despite the high degeneracy of the metastable de Sitter quantum states and the ultra-slow dynamics that prevent relaxation to the ground state, at long time scale the microscopic de Sitter space satisfies the eigenstate thermalization hypothesis and implies an extensive entropy contribution[2,3,4]. Consequently, the baryon production shrinks and the dissipating space-time transforms into a space-timeless cold sink with an accompanying strong contracting memory effect. This is equivalent to the alternately motion of a giant Carnot engine[5] where heat source and sink are sharing their function for producing the eternal periodic dynamics of the Universe. This suggests that the Universe did not start from the Big Bang[6], but oscillates eternally as a Big Bounce Universe[7].

Keywords: Mass, Space, Time, Gravitation, Big Bounce

Experimental support for Verlinde's conception of quantum gravitation is given by the TullyFisher relation[8]. This relation connects the luminosity of galaxies to their mass and diameter. Absolute magnitude, global profile and diameter are function of the velocity of relative motion between galaxy and observer. In the scope of Whitehead's concepts the observable galaxies are natural events. The corresponding relative velocity and luminosity are extrinsic observable characters, but mass, profile, diameter and eventually gravitation are intrinsic[1]. Instrumental observations in a laboratory setting are clearly the result of intelligible discursive processes, thereby not detecting events but entities represented by observable and experimental data, resulting from (mental) thought[1]. Observable events make room for the set-up of a countable database of experimental facts[1]. All observed

Eric Ling and Annachiara Piubello (Eds), Spacetime 1908-2023. Selected peer-reviewed papers presented at the *Third Hermann Minkowski Meeting on the Foundations of Spacetime Physics*, 11-14 September 2023, Albena, Bulgaria (Minkowski Institute Press, Montreal 2024). ISBN 978-1-998902-25-5 (softcover), ISBN 978-1-998902-26-2 (ebook).

data of events e under consideration make part of the ensemble $\mathscr{E}$ containing series E_i of observed events. The limit of the above mentioned series with the intrinsic property of e' s is the abstract of the converging series $E_i{}^1$ and is projected on abstract dimensionless instants of time of a four dimensional (4D) compact space-time manifold $\mathscr{F}$. Instants of time are dimensionless. Instead of a countable ensemble, instants of time make part of space-timelike space, however they are the ideal of the non-entity and therefore neither a database nor entities[1] as further explained in the Supplementary Information 1.

What is disclosed to mind by nature, and what is disclosed of nature by mind[1], that is the question. According to Whitehead, the ensemble $\mathscr{E}$ discloses nature to mind, contrary to the EinsteinFriedmann cosmological theory where the formal compact space-time $\mathscr{F}$ discloses nature. Einstein started from the Newtonian conception of space and time as an a priori 4D Minkovski manifold containing matter as substratum of elementary particles endowed by dynamical characteristics. According to the Noether's first theorem[9], homogeneity of space and time implies conservation of linear momentum and energy, respectively, while isotropy of space involves conservation of angular momentum. The symmetry properties are put a priori on the Minkovski container and the three physical parameters (linear and angular momentum, and energy) are part of the dynamics. However, the dynamic equations are derived from d'Alembert variation principle of minimum action so that any dynamic event is reduced to a static substratum[10]. In the scope of relativity, those patterns governed by dynamical gravitation laws are formally equivalent within any inertial reference system linked by a Lorentztransformation, the so called equivalence principle[11]. Constant gravitation is equivalent with a noninertial reference system. Matter is considered as the intrinsic factor of any substratum while the extensive factor of any substratum is reflected in the curvature of its occupied Space-time. In consequence, only geometric deviations of the 4D Minkovski space-time give experimental support for the action of gravitation. Formally, the Einstein Equation connects the substratum of the stressmomentum-energy-pseudo-tensor T_{ij} and G, which represents the new universal gravitation constant, to the curvature R_{ij} of Space-time container in the following way:

$$R_{ij} + \frac{1}{2}Rg_{ij} = \frac{8\pi G}{c^4}T_{ij} \tag{Eq.1}$$

Here, R is the determinant of the R_{ij} curvature tensor, and g_{ij} is the metric tensor. The indices i and j vary between 1 and 4, corresponding to the time and space coordinates, respectively.

According to Whitehead's conceptions the action T and reaction R are completely turned back. Instead of the observability of the extrinsic dynamical factor of any event to give evidence for its intrinsic gravitation and mass factor, experimental evidence – in the scope of relativity – evolves from the not observable intrinsic factors within a priori static Space-time: the universal gravitation constant G and mass are causes of the varying geometry of its environment. According to Whitehead, however, elements of geometry, i.e. space and time, are respectively the a priori conditions of the observable effects of gravitation. In consequence, Whitehead's boundary conditions of observation become the final relativistic results of experimental observation. Stated otherwise, this is putting the cart before the horse. The cosmological consequences of the Einstein-Friedmann model follow directly from the above mentioned a priori symmetry boundary conditions of Space-time and from the a priori dual conception of mass. On the one hand mass is the substratum of any dynamic event as intrinsic factor, on the other hand it makes part of dynamic observable extrinsic factors of events such as conservation of linear momentum and energy. Furthermore, two additional corollaries about the nature of the Universe evolve from the Friedman model. Firstly, the calculated average density of matter of the Universe ($10 \times e^{-29}$ g/cm^3) is 100% of the estimated matter density ($10 \times e^{-31}$ g/cm^3) in the environment of a galaxy. Secondly, the negative open radius curvature gives support for an expanding Universe from time zero to infinity with just one singularity, while the positive closed radius curvature implies an oscillating Universe with periodic singularities. The first corollary endowed the Universe with unobservable Black Matter, the second makes room for the Big Bang, provided the Universe has the curvature of the Antide Sitter Space[10]. Both substantial elements of the Universe are evolving from mathematical products of mind without disclosure of nature to mind.

What are the consequences for basic experiments of relativity such as the experiment of Pound-Rebka[12] and the redshift deviation of electromagnetic rays in the environment of Mercury[13] given the mentioned controversy? Pound-Rebka's set-up was based on the Mossbauer effect measurement of travelling γ-rays produced by a 57Fe source along a tower of distance of 22.56 m in order to give experimental evidence to the above mentioned equivalence principle of gravitation and inertial reference systems. The gravitational energy shift of massless photons during their downwards and upwards motion with increasing wavelength demonstrates the gravitational action and in consequence the length contraction and time dilatation as predicted by the Lorenz-transformation between two inertial reference systems, provided the Sun spectrum is supposed to be standard. At first glance it concerned a measurement of the extrinsic factor of the gamma-decay shift while the intrinsic gravitation factor and the equivalence principle was taken for granted. The same remark is valid on classical tests of deflection of light by the sun and the gravitational redshift in the environment of Mercury. The shift to infrared, however, is measuring acceleration and motion. These are extrinsic characters of events, and moreover, the equivalence principle is also taken for granted. At face value these experiments corroborate the formal equivalence of gravitation and inertia, on the other hand both experiments give evidence for the varying curvature of the 4D Minkovski space-time. But, this varying curvature of the 4D Minkovski space-time is precisely the a priori condition of observability of any event. Moreover, the ensemble of particle events disclosing the extrinsic energy shift of gamma particles and redshift of light in the Mercury environment is the result of interactions of the involved series of events, but it is not the result of the hidden intrinsic characters of the events. Thus, the data of these experiments only corroborate the geometric a priori conditions of their observability. They cannot disclose the deep essence of their intrinsic characters such as mass and gravitation Hence, is the interpretation of these experiments from the principle "what is disclosed of nature by mind?" shedding more light on the deep nature of gravitation or are they just obscuring some cosmological phenomena such as the accelerated expansion of our Universe by presupposing the non-observable dark matter? Evidence for the dark matter claim in order to explain excess gravity is supported by recent references in[14]. However, in this reference, measurements of hydrogen profiles beyond and within the optical discs of galaxies, dynamics of galaxies in clusters and other new methods such as weak gravitational lensing, are just demonstrating baryon acoustic oscillations and cosmic microwave oscillations. What is more, Brouwer et al.[14] give more evidence for Milgrom's non-general relativistic Modified Newtonian Dynamics[15,16], although the experimental support is disclosing the a priori conditions of observability rather than shedding light on the essential nature of mass and gravitation. These attempts, based on the equivalence principle of gravitation and inertia, discover pretty good the extrinsic characters of the whole Universal Event but they do not disclose the deep essence of the intrinsic characters such as mass and gravitation. As such they just corroborate the varying conditions of observability. We provide answers to the above stated question, analysing Verlinde's claims about the nature of gravitation and apparent dark matter in the scope of Whitehead's concepts of matter, space and time.

In order to explain the accelerated expansion of our Universe, Verlinde suggests the existence of apparent positive dark matter. Contrary to the widely accepted theoretical approach involving dark matter, Verlinde's conception of matter does not reduce this factor to the substrate of energy and linear momentum. Matter is deeply connected to pressure-less fluid revealing the emergent nature of gravitation as the intrinsic elastic response of space-time to excitations of the meta-stable microgroundstate of the de Sitter Space. Verlinde's propositions about the deep origin and the essential nature of gravitation start from the de Sitter space in which a part of the microscopic degrees of freedom are being thermalized. A non-locally stored thermodynamic entropy $St(V)$ emerges besides the BekensteinHawking area law entropy, proportional to the cosmological horizon[2]. According to the holographic principle the stored k-logic qubits at the surface spread out over n physical qubits in the bulk. The area law due to short distance entanglement of neighbouring degrees of freedom of the Ryu-Takanayagi tensor formulation of emergent spacetime generates the entanglement entropy $S(V)$[17]. In d dimensions of $S(V)$, the entanglement entropy in quantum field theory takes the form:

$$S(V) = g_{d-1}[\vartheta V]\varepsilon^{-(d-1)} + \ldots + g_1[\vartheta V]\varepsilon^{-1} + g_0[\vartheta V]\log(\varepsilon) + S_0(V) \qquad \text{(Eq.2)}$$

Here the g_k 's are local and extensive functions on boundary $[\vartheta V]_{k\text{-bits}}$, homogeneous to degree k, ϑV is the result of short range cut-off ε of surface area $A(r)$). The leading term $g_{d-1}[\vartheta V]$ is proportional to the (d-1) power of the size of V, well-known as the area law for entanglement entropy[18]. $S_0(V)$ is a finite part, g_i are local and extensive functions on boundary ϑV which are homogeneous of degree i. The shortdistance cut-off is ε.

The contributing microstates are an ensemble of metastable quantum states, highly degenerated and governed by ultraslow dynamics preventing the relaxation to the ground-state. This long-time scale implies eigenstate thermalization and as such a thermal volume law contribution of entropy $S(V)$ that competes with the Bekenstein-Hawking entanglement entropy $S_{\text{BH.}}$. Gravitation connected to the Hubble constant H_o is then a spacetime reaction of this competition after the production of baryons. Given the surface acceleration $\kappa = cH_0 = a_0$, the created baryon mass $\delta Mb = T\delta S$ is extensive and connected to the thermalized entropy $S(V)$ and the energy of vector by $\delta Mb = a_0 \delta Ht$ (Hamiltonian at instant t, a_0 the surface gravity on the cosmological horizon))[2] while the surface mass density is $\Sigma b(r) = Mb/A(r)$. It must be noticed that, according to the Ryu-Takanayagi conjecture about contributing short-range and long-range areas, entanglement entropy is the entropy of the non-observable[18]. In consequence, this entropy is out of the range of Whitehead's observable ensemble of events. The extensive thermal entropy $S_t(V)$, on the contrary, implies extension and duration and consequently $S_t(V)$ implies observability of the involved events. Furthermore, given de volume V_0 of the escaped baryons and V_r the original volume of the prebaryonic universe Verlinde defines a strain tensor: $\varepsilon(r) = NV_0/V(r_r)^2$. N is the number of unit volumes included in spacetime under consideration, V_0 (volume per unit of entropy) and $V(r)$ (volume deformation) are defined according to the metric of the pre-baryonic Ryu-Takanayagi tensor formulation of emergent spacetime. From the first law of horizon thermodynamics where the Hamiltonian is expressed in terms of strain tensor $\varepsilon(r)$ and in the regime of Newtonian potential we express the apparent positive dark mass surface density $\sum_D(r)$ in function of the stress tensor:

$$\sum_D(r) = \frac{a_0}{8\pi G} \cdot \varepsilon(r) \text{and} \sum_B = \frac{M_B}{A(r)} \qquad \text{(Eq.3a \& 3b)}$$

$\sum_B$ is the baryonic mass surface density[2].

This means that the so called Coulomb branch mass and gravitation (see also Supplementary Information 2) is evolving from Higg's branch stress-strain response of the pre-baryonic universe. The corresponding entropy decline by mass M gives $S_M = 2\pi M_B/\hbar a_0$. So Verlinde recovers the equivalent of the scaling Tully Fisher relation between baryonic and apparent positive dark mass[2]:

$$\sum_D r = \frac{a_0}{8\pi G} \cdot \frac{\sum_b(r)}{(d-1)} \tag{Eq.4}$$

G is the Newtonian gravity constant, measured by Cavendish in 1798. Furthermore, the thermal entropy is decreasing and the reason is that thermal entropy contributes to the de Sitter entanglement entropy. In consequence, thermal volume entropy turns the space-time into a reactive elastic medium. Moreover, the gravitation – fed by the created positive dark energy – emerges from the elastic reaction of the de Sitter space within the cosmological horizon in competition of both long and short range entanglements while the horizon stabilizes.

Verlinden's claims about a gravitation as the memory effect of entanglement entropy producing and competing short and long range dynamic tensor energy involve a Universe which behaves as a condense elastic medium.

$$\sigma_{ij} = \lambda \cdot \varepsilon_{kk} \cdot \delta_{ij} + 2.\mu.\varepsilon_{ij} \text{ where } \mu \geq 0, \lambda + 2.\mu \geq 0 \tag{Eq.5}$$

Here σ is the stress and ε is the strain, δ is the Kronecker delta, the indices i and j vary between 1 and 4, corresponding to the time and space coordinates, and k indicates the diagonal terms of the matrix. μ and λ are the Lamé parameters. Stress is followed by entropic elasticity due to the adiabatic expansion and, by elastic reaction, adiabatic compression. The reactive action of gravitation to the baryon mass escape minimises the memory effects to external perturbations in condensed matter. Verlinde identifies the latter as the emergence of apparent positive dark energy. The hysteresis curve describing the stress and strain is a measure for the entropy production in glassy materials identified as positive black matter in the scope of Verlinde's concept of gravitation. The relation between strain and stress is pretty well illustrated in glassy materials[19].

Since Verlinde does not put a substratum status on mass, and since he does not make curvature of space equivalent with gravitation, he rather reformulates cosmology in a Whitehead's concept of nature. Whitehead's conceptions tacitly imply the creation of an appropriate gravitational environment by mass excitation escape to settle down energy and momentum in the web of space- and time-like events. Consequently, space and time just emerge after the escape of baryonic mass and the reactive force of gravitation. Verlinde's attempt requires the experimental evidence of additional positive energy to explain border flattering of galaxies. Therefore, he links his results about visible baryonic matter density with positive apparent matter by means of the Tully-Fisher scaling relation. Given the acceleration g_D due to the apparent positive dark mass with $g_D = G.M_D(r)/r^2$, the baryonic Tully-Fisher relation becomes[2] :

$$v_f^4 = a_M \cdot G \cdot M_B \tag{Eq.6a}$$

$$a_M = \frac{a_0}{6} \tag{Eq.6b}$$

$$g_D = \frac{v_g^2}{r} \tag{Eq.6c}$$

Here v_f is the asymptotic velocity of the flattered galaxy (see[20] for recent experimental support).

We emphasize, however, the deep difference between the classical Einstein-Friedmann conclusions from the Tully-Fisher scaling relation and Verlinde's experimental gauge claims for his elastic response of the nature of the essence of gravitation. Within the Einstein-Friedmann conception, luminosity is linked to the intrinsic factor of gravitation, the conclusion of the k-shift is considered as giving evidence for mass as substratum, notwithstanding the fact that mass as substratum is already tacitly assumed, because the observability conditions of extension and duration is presupposed. According to Verlinde, gravitation and mass correspond to the expanding or shrinking space-time and indirectly it is this space-time variation that is measured, not the motion of the hypothetical substratum mass. Gravitation as the memory effect of the mass escape due to the competition of long distance and short distance quantum-states can be considered as an intrinsic character of the escape event.

Observability and experimental databases (see Supplementary Information 1) start when space and time is created by the mass escape due to competition of short range and long range entanglement of degrees of freedom of stress tensors. By baryonic mass creation the de Sitter entropy is decreasing and the measurement of this event implies that the observable creation of baryonic mass makes part of a closed set of event-particles while the observed event is mapped on a compact set of experimental data, provided that on this compact set of data the Kelvin-Planck thermodynamic theory is valid (see Supplementary Information 2). Whitehead's conception of matter, space and time does not imply a unique "first energy-momentum entity" nor an absolute beginning of the universe. It has, however, a very remarkable logic collorary: it does not exclude speculations about partial short range as well as long range entanglements within the pre-space-time energy-momentum entity that evolves to the above mentioned excitations of the meta-stable micro-groundstate of the de Sitter Space. Consequently, the origin of baryonic matter escape can be monistic by a unique entanglement or plural by plural partial entanglements. Obviously, the probability of plural baryonic creation evolving from a unique cause is rather negligible unless a fundamental uncertainty which is a basic property of the space-timeless entity and governs the different baryonic creation.

The "space-timeless entity" is not an infinite source of baryonic mass creation. Thermodynamic entropy appears in the dissipating and expanding universe. The baryon part of the universe produces space-time from energy taken from the space-timeless entity and its free energy is dissipating into the energy sink of the produced space-time. This mechanism is sustained by the "Heat" source of the spacetimeless entity. This space-time, however, is shrinking when mass and gravity production is accelerated beyond the cosmological horizon. Indeed, the gravitation response on baryonic creation as memory effect of the pre-baryonic universe can exceed the stress-factor. In that case the hysteresis-energy is negative notwithstanding the positive dark matter that provokes the expansion of the universe. Instead of expansion, contraction is the right response to the created stress by baryonic production. Dissipation increase by the overwhelming mass production while the memory effects of the original baryon creation is disappearing and space-time is shrinking into a space-timeless cold sink. In addition, also entropy as extensive physical parameter decreases so that temperature is raising in the sink. Short and long distance entanglements arise with thermalized excitations and the "cold" sink becomes the "heat" source and vice versa, the residue of the former space-timeless heat source becomes the cold sink. It looks like the periodic motion of a giant Carnot engine where heat source and sink are alternately sharing their function in order to produce the eternal periodic dynamics of the Universe. The universe did not start from the Big

Bang, but oscillates eternally (see Supplementary Information 3). Thus, the alternative conception of gravitation formulated by Verlinde is supported by Whitehead's view on matter, space and time, and this results in the decline of the Big Bang in favour of the Big Bounce Universe. Within this Giant Carnot Universe different stages of the expansion of the Universe exist. This implies that the reported values of the Hubble constant, derived from different experiments[21], could all be exact acceleration values, though only representative for different stages of the oscillating Universe.

Supplementary information

1. Database of observable facts

Celestial bodies luminosity appears by discerness of place through a period of time (duration), related to a complex of events such as its velocity and the solar global neutral hydrogen line width, both by two relations of extension and congredience[8]. How to produce a database of observable facts and what are the consequences for scientific claims?

Within an ensemble of events $\mathscr{E}$ a series of congredient events $E_1, E_2 \dots E_i \dots E_n, E_{n+1}$ is given, all making part of the same duration with index i indicating countable extensions[1]. The event-particle e is the result of converging extrinsic properties connected to surrounding events of the ensemble $\mathscr{E}$. Within the ensemble $\mathscr{E}$, particle-events require a "Moment" as duration of minimum extension representing all nature at an instant "now" as well as a spatial "Station" at a location "here" 1. Given event-particle e as the intersection of a particular Moment and Station three possibilities exists:

i. Other event-particles are all disjunctive to e.

ii. All events involved with the particular event-particle are inside the Moment containing e.

iii. All involved events overlap e by a duration containing those event-particles which are covered by corresponding bounding moments. Boundary consists of two 3D spaces. Such subset $\mathscr{M}$ of events occupies the aggregate of the event-particles within. Consequently, all subsets $\mathscr{M}$ of $\mathscr{E} (\mathscr{M} \subseteq \mathscr{E})$ containing the series of a particular event e are closed. The set $\mathscr{E}$ of experimental countable outcomes form a topological space.

This implies the definition of a Borel algebraic set on Whitehead's set of particle-events. If a collection $\mathscr{M}$ of duration subsets of the set $\mathscr{E}$ of event-particles including $\mathscr{E}$ is containing (i) all countable unions of subsets of $\mathscr{E}$, and (ii) all countable subsets of intersections of subsets of $\mathscr{E}$, and (iii) all subsets of event-particles included in the countable union of subsets, then $(\mathscr{M}, \mathscr{E})$ is a Borel Space.

In order to be distinguishable the ensemble of event-particles must be endowed by a Hausdorff topology so that any extrinsic character (e.g. velocity, position, time, luminosity) can be mapped by a homeomorphism on $\mathscr{F}$. Furthermore, a measurement protocol produces a countable ensemble of experimental data $\mathscr{E}$ as separable final results of a cyclic process starting from and ending within the experimental setting. In addition, there is always transfer of energy Q and in consequence a Kelvin-Planck thermodynamic theory is involved. The limiting points of the converging series e_i of $\mathscr{E}$ are mapped on a data base within $\mathscr{F}$. As containing all limiting points this database of observable events is a compact set. Within any time-like cone containing the ensemble of events $\mathscr{E}$, the experimental setting detects a physical phenomenon $\mathcal{M}$ (i.e. luminosity) and collects irreversibly the data in an ensemble $\mathcal{E}$. In consequence, a finite regular Borel measure B is defined on the set $\mathcal{E}$ of event-particles in order to connect the event-particles irreversibly to the compact data $\mathcal{F}$: presuppose the intervention of the existence of a Kelvin-Planck thermodynamic theory which is a necessary and sufficient condition for the Clausius-Duhem inequality representing the second law of thermodynamics[22]. The proof is based on the separation theorem of Hahn-Banach that admit a hyperplane within a Hausdorff ensemble, including one closed ensemble $\mathcal{E}$ and one compact $\mathcal{F}$ so that there exist canonical adjoint functionals T and S, called thermodynamic temperature and entropy so that $dQ/T < dS$ is applicable on the closed set of eventparticles $\mathcal{E}$. This implies that at the separation hyperplane the entropy is zero, and in consequence the system has a condition characterized with a low entropy state $dQ/T > dS$ on the compact set $\mathcal{F}$.

2. Branch off

This observed baryonic mass creation matches the branch off within the fact-like thermodynamic reduction of the Arrow of Time by Reichenbach[23] and Grünbaum[24]. The branch offs were firstly mentioned by Reichenbach[23] and afterwards by Grünbaum[24]. There branch states are nothing more than emergence of baryonic matter of ensemble E. Consequently, the entropy of a separated galaxy or a solar system increases after gravity and separation started. The cut-off by hyperplanes explains the apparent paradox of asymmetric evolution of galaxies and our solar system while the interactions on microscale are completely symmetric.

While Reichenbach and Grünbaum define their branch off as a factlike reduction of the Arrow of Time, Verlinde's claims imply nevertheless a lawlike reduction of the Arrow of Time based on the emergence of baryonic matter. There is however an important restriction to add. The Arrow of Time is not a universal unique time direction but directly connected to the respective Higg's branch entity from which the baryonic escape creates the respective low entropy branch off. Consequently, time asymmetry does not reduce to thermodynamic entropy unless it is a relative time within the separated sub-ensemble of the Universe and there is no evidence for an absolute Arrow of Time. Yet, this branch cut-off or the creation of the Coulomb branch out of the Higg's branch is presented as an instant of time. But according to Whitehead an instant is the ideal real of all nature without any temporal extension, so the ideal of immeasurable non-entity. And yet, physical theories admit concepts such as dimensionless instants of "now" and "here", notwithstanding that nature does not allow absolute instants of time "zero" and dimensionless particle-events of spatial 'here' because they are a non-event, and hence immeasurable[1].

3. Big Bang?

The concept that the Universe did not start from the Big Bang but oscillates eternally just like the Big Bounce is corroborated by Popper's claims about the nonexistence of a central source causing spontaneously the singularity of the outgoing and contacting waves[25−27]. According to Popper's claims even the singularity of the Big Bang evolving from a central source sounds rather odd.

References

1. Whitehead, A.N. (2007). The Concept of Nature. Cosimo Classics, New York.

2. Verlinde, E.P. (2017). Emergent Gravity and the dark Universe. SciPost Phys. 2, 016 (2017), http://arxiv.org/abs/1611.02269v2.

3. Deutsch, J.M. (1991). Quantum Statistical Dynamics in closed systems. Physical Review A 43(4), 10.1103/PhysRevA.43.2046.

4. Srednicki, M. (1994). Chaos and Quantum Thermalization. Physical Review E 50(2), https://doi.org/10.1103/PhysRevE.50.888.

5. Verstraeten,G.J.M. and Verstraeten, W.W. (2022). Accessing a Big Bounce Universe with Concealed Mass and Gravitation. Philosophy and Cosmology, 28,32-41. https://doi.org/10.29202/phil-cosm/28/3.

6. Lemaitre, G. (1931). The Beginning of the World from the Point of View of Quantum Theory. Nature, 127 (3210), 706, 10.1038/127706b0.

7. Penrose, R. (2010). Cycles of Time: An Extraordinary New View of the Universe. Bodley Head, London.

8. Tully, R.B and Fisher, J.R., (1977). A New Method of Determining Distances to Galaxies. Astronomy and Astrophysics, 54, 661-673.

9. Noether, E. (1971). Invariant variation problems. Transport Theory and Statistical Physics, 1(3), 186-207, 10.1080/00411457108231446. M. A. Tavel's English translation of "Invariante Variationsprobleme," Nachr. d. Köonig. Gesellsch. d. Wiss. zu Göttingen, Math-phys. Klasse, 235-257 (1918).

10. Misner, C.W, Thorne, K.S. & Wheeler, J.A. (1973). Gravitation. Freeman and Company, San Francisco.

11. Udwadia, F.E., Kalaba, R.E. (2002). On the Foundations of Analytical Dynamics. Intl. Journal Nonlinear Mechanics, 37(6), 1079- 090.

12. Pound, R.V. & Rebka Jr, G.A. (1959). Gravitational Red Shift in Nuclear Resonance. Physical Review Letters, 3(9), 439, https://doi.org/10.1103/PhysRevLett.3.439.

13. Treschman, K.J. (2014). Early Astronomical Tests of General relativity: the anomalous advance in the perihelion of Mercury and the gravitational redshift. Asian Journal of Physics, 23 (1 & 2), 171-188. ISSN 0971-3093.

14. Brouwer, M.M., Oman, K.A., Valentijn, E.A. et al. (2021). The weak lensing radial acceleration relation: Constraining modified gravity and cold dark matter theories with KiDS1000. Astronomy and Astrophysics 650, A113, 10.1051/0004-6361/202040108

15. Milgrom, M. (1983). A modification of the Newtonian dynamics as a possible alternative to the hidden mass hypothesis. Astrophysical Journal, 270, 365-370, 10.1086/161130.

16. Milgrom, M. (2013). Testing the MOND Paradigm of Modified Dynamics with GalaxyGalaxy Gravitational Lensing. Phys Rev Lett, 111(4), 10.1103/PhysRevLett.111.041105.

17. Vidal, G. (2007). Entanglement renormalization. Phys. Rev. Lett. 99, 220405.

18. Casini, H. & Huerta, M. (2009). Entanglement Entropy in free Quantum Field Theory. J. Phys. A42504007, 10.1088/1751-8113/42/50/504007.

19. Radhakrishnan, R., Divoux, T., Manneville, S., Fielding, S.M. (2017). Understanding rheological hysteresis in soft glassy materials. Soft Matter, 1834-1852, 10.1039/C6SM02581A.

20. Papastergis, E., Adams, E.A.K. & van der Hulst, J.M. (2016). An accurate measurement of the baryonic Tully-Fisher relation with heavily gas-dominated ALFALFA galaxies. Astronomy and Astrophysics, 593, 10.1051/0004-6361/201628410.

21. Di Valentino et al. (2021). In the realm of the Hubble tension-a review of solutions. Class. Quantum Grav. 38, 153001, 10.1088/1361-6382/ac086d.

22. Truesdell, C. (1983). Rational thermodynamics. Springer, New York.

23. Reichenbach, H. (1956). The direction of time. University of California Press, Berkeley California.

24. Grünbaum, A. (1973). Philosophical Problems of Space and Time. Boston Studies in the Philosophy of Science, 2nd edition. Reidel, Dordrecht, Boston.

25. Popper, K.R. (1956a). The arrow of time. Nature, 177, 538.

26. Popper, K.R. (1956b). Irreversibility and Mechanics. Nature, 178, 382.

27. Popper, K.R. (1957). Irreversible Processes in Physical Theory. Nature, 179, 1297.

7

AN INVESTIGATION OF THE LOGOTROPIC EQUATION OF STATE FOR THE CHAPLYGIN GAS, A GENERALIZED HOLOGRAPHIC MODEL, AND THERMODYNAMIC ANALYSIS USING SCALAR FIELDS

SANJEEDA SULTANA AND SURAJIT CHATTOPADHYAY

Abstract   Based on a logotropic equation of state, this work suggests a technique for reconstructing a generalized form of holographic dark energy and analyzing its cosmology. One can compute a scale factor for the total energy density of the universe, which is supplied by baryons, dark matter, and dark energy density, by assuming that the dark energy density is created by holographic dark energy with Granda–Oliveros cutoff. In the case of logarithmic internal energy and logotropic equation of state, modified Chaplygin gas is also investigated, and corresponding reconstruction is done. After taking into account the phantom and quintessence models using a logotropic approach, the generalized second law of thermodynamics is subsequently proven for event, particle, and Hubble horizons

Keywords: Holographic dark energy with Granda–Oliveros cutoff; modified Chaplygin gas; logotropic equation of state; logarithmic internal energy; generalized second law of thermodynamics

1   Introduction

Cosmologists today generally agree that the presence of the so-called dark energy (DE), which is responsible for the universe's late time acceleration of expansion, exists. Numerous cosmological measurements have confirmed this occurrence, however it was first noted in 1998 by two separate research teams, the High-z Supernova Search Team and the Supernova Cosmology Project [1, 2]. However, the DE problem still has no satisfactory answer, and it is still unclear how basic physics theories can account for it.

One theory that has been proposed as a solution to the DE problem is the holographic principle [3, 4], which has significant implications for quantum gravity. According to the holographic principle, a system's entropy scales with its surface area rather than its volume [5]. Motivated by this proposition, Cohen et al. [6] proposed that, within a quantum field theory, a short-distance cutoff is associated with a long-distance cutoff because of the boundary established by the black hole's formation; specifically, if ρ represents the quantum zero-point energy density resulting from a short-distance cutoff, then the overall energy within a region of size L must not surpass the mass of an

Eric Ling and Annachiara Piubello (Eds), SPACETIME 1908-2023. Selected peer-reviewed papers presented at the *Third Hermann Minkowski Meeting on the Foundations of Spacetime Physics*, 11-14 September 2023, Albena, Bulgaria (Minkowski Institute Press, Montreal 2024). ISBN 978-1-998902-25-5 (softcover), ISBN 978-1-998902-26-2 (ebook).

identical black hole; as a result, $L^3\rho \leq LM_{pl}^2$. To saturate this inequality, the largest infrared cutoff L_{IR} allowed is

$$\rho = 3M_p^2 c^2 L_{IR}^{-2}, \tag{1.1}$$

where M_p^2 is the reduced Planck mass and c is an arbitrary parameter. Holographic dark energy (HDE) is the term for the holographic consideration that has been frequently used in cosmology, particularly for the explanation of the late time dark energy epoch (see [7] for a thorough overview). From this perspective, the infrared cutoff L_{IR} has a cosmological origin. The authors of [8, 9, 10, 11] introduced the most general form for the infrared cutoff, known as the generalized HDE, which includes a combination of the FRW parameters, such as the cosmological constant, the particle and future horizons, the universe lifetime—if finite, and the Hubble constant.

The observed current acceleration of the universe is well predicted by a model with R_h as the length scale; however, this type of model also exhibits the causality problem, i.e., DE currently appears to depend on the future evolution of the scale factor, which in turn violates causality [12]. Therefore, we use the Granda–Oliveros (G–O) cutoff to analyze the cosmological history at late times within the framework of the HDE model. This cutoff incorporates the time derivative of the Hubble parameter in addition to the square of the Hubble parameter, i.e.,

$$L_{IR} = \left(\alpha H^2 + \beta \dot{H}\right)^{-\frac{1}{2}}, \tag{1.2}$$

where α and β are arbitrary dimensionless parameters [13]. By taking into account only dimensional considerations, the authors of [14, 15] proposed this cutoff and were able to overcome the causality issue. We also conduct an investigation utilizing updated measurements from the dynamics of the universe's expansion with the goal of fitting the parameters in the HDE model with the G-O cutoff.

The nature of the dark energy and dark matter (DM) i.e., dark sector of the universe is still unknown to us. An eco-nomical and satisfying idea to unify the dark sector of the universe, is to consider it as the only one component that acts as both DE and DM. To acquire the unification [16] of DE and DM, the so-called Chaplygin gas (CG) is used. Benaoum [17] proposed a model called modified Chaplygin gas (MCG) [18] within the framework of Friedmann–Robertson–Walker (FRW) cosmology. Based on the following EoS:

$$P = A_0\rho - \frac{B}{\rho^\alpha}, \tag{1.3}$$

where A, B and α are constant parameters, an early stage of radiation is incorporated in this model. The pure CG or generalized CG, which is a perfect fluid, acts like a pressureless fluid in the early stage of the universe and like a cosmological constant in the later stage. The Pure CG with an exotic EoS is characterized by a negative pressure [19] $p = -\frac{B}{\rho}$, where ρ is the energy density, p is the pressure, and B is a positive parameter. The pure CG has been expanded into the so-called generalized CG with the following EoS [20] $p = -\frac{B}{\rho^\alpha}$, where $0 \leq \alpha \leq 1$. For the situation $\alpha = 1$, it is evident that the pure CG is recovered.

The validity of the generalized second law (GSL) of thermodynamics for HDE with G-O cutoff and MCG models has been studied and presented in this paper. The GSL states that as time passes, the total entropy of the matter contained within the horizon and the entropy of the horizon never decreases [21, 22]. In 1973, Bekenstein [23] proposed a relationship between the black hole's thermodynamics and event horizon and inferred that the black hole's event horizon is a measure of its entropy. The connection between each horizon and entropy is the outcome of generalizing this idea to the horizons of cosmological models. The validity of the GSL for cosmological models that somewhat depart from de Sitter space is investigated in [24]. Nonetheless, since entropy measures what is occurring beyond the horizon, it is conventional to associate it with the region.

The structure of the paper is as follows: The cosmological effects of HDE with G-O cutoff in a logotropic equation of state and in the case where the internal energy is in the logarithmic form have been illustrated in Section II. The cosmology of MCG in the interacting scenario, with a logotropic equation of state and logarithmic internal energy, has been presented in Section III. In Section IV, discussed about dark energy in quintom universe. The GSL of thermodynamics for the models HDE with G-O cutoff and MCG with various universe enveloping horizons in quintessence and phantom scenarios has been validated in section V and in Section VI we have concluded.

2 Implementation of HDE with Granda-Oliveros cutoff in logotropic equation of state

As mentioned in Section I, the HDE density in Eq.(1.1) has been implemented with the G-O cutoff in Eq.(1.2) and in such a model the HDE density becomes

$$\epsilon_\Lambda = 3\left(\alpha H^2 + \beta \dot{H}\right),$$ (2.1)

where $H = \frac{\dot{a}}{a}$ is the Hubble parameter and a is the scale factor. The reduced Planck mass is defined as $M_p = \sqrt{\frac{c\hbar}{8\pi G}} = 2.4 \times 10^{18} GeV$, where $8\pi G = 1$ and G is the Gravitational constant. We use the reduced Planck units in this paper, with $c = \hbar = M_p = 1$. All quantities lose their dimension in these units. For a flat universe, the Friedmann equations without cosmological constant are given by

$$H^2 = \frac{8\pi G}{3\kappa^2}\epsilon,$$ (2.2)

and

$$\dot{H} + H^2 = \frac{\ddot{a}}{a} = -\frac{4\pi G}{3\kappa^2}(3P + \epsilon),$$ (2.3)

where ϵ is the total energy density of the universe which includes baryons, DM and DE, and let $\kappa^2 = 1$.

The energy conservation equation is

$$\dot{\epsilon} + 3H(\epsilon + P) = 0.$$ (2.4)

We assume that there is only one dark fluid in the universe, which is characterized by a logotropic equation of state [25, 26]

$$P = A\ln\left(\frac{\epsilon}{\rho_P \kappa^2}\right),$$ (2.5)

where A is a new fundamental constraint of Physics superseding Einstein's cosmological constant Λ and $\rho_P = 5.16 \times 10^{99} gm^{-3}$ is the Planck density. The universe's total energy density [25, 26] is

$$\epsilon = \frac{\epsilon_{m0}}{A^3} + \epsilon_\Lambda,$$ (2.6)

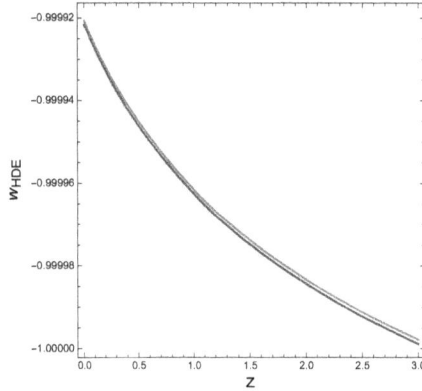

Figure 11: Evolution of EoS parameter w_{HDE} Eq.(2.10) with respect to redshift z for HDE with G–O cutoff in logotropic equation of state in an interacting scenario. We have chosen $\epsilon_{m0} = 0.004$, $\alpha = 0.9$, $C_1 = 0.003$ and $\beta = 1.1$. The red, blue and green lines correspond to $A = 0.11530707$, 0.11530710 and 0.11530713 respectively.

where ϵ_{m0} is the present density of baryons and DM.

Using Eq.(2.1) in Eq.(2.6) and then putting the same in Eq.(2.2), we get the expression for Hubble parameter as

$$H = \sqrt{a^{\frac{2-2\alpha}{\beta}} C_1 - \frac{\epsilon_{m0}}{3A^3(-1+\alpha)}}. \tag{2.7}$$

So, we have the total energy density of the universe as

$$\epsilon = 3a^{\frac{2-2\alpha}{\beta}} C_1 - \frac{\epsilon_{m0}}{A^3(-1+\alpha)}, \tag{2.8}$$

and by substituting Eq.(2.8) in Eq.(2.5), we have the expression for DE pressure as

$$P = A Log\left[5.81395 \times 10^{-100} a^{\frac{2-2\alpha}{\beta}} C_1 - \frac{1.93798 \times 10^{-100}\epsilon_{m0}}{A^3(-1+\alpha)}\right]. \tag{2.9}$$

Hence, we can express the equation of state (EoS) parameter $w = \frac{P}{\epsilon}$ for HDE with G–O cutoff in logotropic equation of state as

$$w_{HDE} = \frac{A Log\left[5.81395 \times 10^{-100} a^{\frac{2-2\alpha}{\beta}} C_1 - \frac{1.93798 \times 10^{-100}\epsilon_{m0}}{A^3(-1+\alpha)}\right]}{3a^{\frac{2-2\alpha}{\beta}} C_1 - \frac{\epsilon_{m0}}{A^3(-1+\alpha)}}. \tag{2.10}$$

The EoS parameter w_{HDE} Eq.(2.10) for HDE with G–O cutoff in logotropic equation of state in an interacting scenario is plotted in Fig.11 with respect to redshift z. From the Fig.11, we can conclude that it is showing a quintessence like behaviour and also increasing with the evolution of the universe which is consistent with the observations of the models depicting late time accelerating expansion of the universe.

In the next subsection, we have demonstrated HDE with Granda–Oliveros cutoff in logotropic equation of state where the internal energy is in the logarithmic form.

76

2.1 HDE with Granda–Oliveros cutoff in Logotropic equation of state with logarithmic internal energy

The EoS for Logotropic equation of state with logarithmic internal energy [27] is

$$P = A \ln \left(\frac{\rho_{DM}}{\rho_P} \right),$$ (2.11)

where ρ_{DM} is the density of DM. By substituting Eqs. (2.7), (2.8) and (2.11) in Eq.(2.4), we have ρ_{DM} as

$$\rho_{DM} = 5.16 \times 10^{99} e^{\frac{\epsilon_{m0}}{A^4(-1+\alpha)} - \frac{a^{\frac{2-2\alpha}{\beta}} C_1(2-2\alpha+3\beta)}{A\beta}},$$ (2.12)

and from Eqs. (2.12) and (2.11), we have the DE pressure in the scenario of logarithmic internal energy as

$$P_{HDE,log} = \frac{ALog \left[e^{\frac{\epsilon_{m0}}{A^4(-1+\alpha)} - \frac{a^{\frac{2-2\alpha}{\beta}} C_1(2-2\alpha+3\beta)}{A\beta}} \right]}{Log[2]}.$$ (2.13)

Hence, we can express the EoS parameter $w_{HDE,log}$ for HDE with G–O cutoff in logotropic equation of state where the internal energy is in the logarithmic form as

$$w_{HDE,log} = \frac{ALog \left[e^{\frac{\epsilon_{m0}}{A^4(-1+\alpha)} - \frac{a^{\frac{2-2\alpha}{\beta}} C_1(2-2\alpha+3\beta)}{A\beta}} \right]}{\left(3a^{\frac{2-2\alpha}{\beta}} C_1 - \frac{\epsilon_{m0}}{A^3(-1+\alpha)} \right) Log[2]}.$$ (2.14)

The EoS parameter $w_{HDE,log}$ Eq.(2.14) for HDE with G–O cutoff in logotropic equation of state with logarithmic internal energy in an interacting scenario is plotted against redshift z in Fig.12. From the Fig.12, we can infer that it is showing a phantom like behaviour and also increasing with the evolution of the universe but it is not crossing the phantom boundary -1.

In the next section, we have illustrated cosmology of MCG with logotropic equation of state in the interacting scenario.

3 MCG with logotropic equation of state in interacting scenario

The EoS for MCG [28] is given by Eq.(1.3). We have the DE density for MCG with logotropic equation of state from the energy conservation equation Eq.(2.4) as

$$\epsilon_{\Lambda,MCG} = \left(\frac{B + a^{-3(1+A_0)(1+\alpha)} e^{(1+A_0)C_1(1+\alpha)}}{1 + A_0} \right)^{\frac{1}{1+\alpha}},$$ (3.1)

We have the total energy density of the universe by substituting Eq.(3.1) in Eq.(2.6) as

$$\epsilon_{MCG} = \left(\frac{B + a^{-3(1+A_0)(1+\alpha)} e^{(1+A_0)C_1(1+\alpha)}}{1 + A_0} \right)^{\frac{1}{1+\alpha}} + \frac{\epsilon_{m0}}{A^3}.$$ (3.2)

77

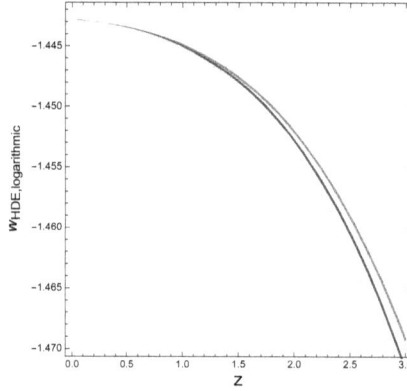

Figure 12: Evolution of EoS parameter $w_{HDE,log}$ Eq.(2.14) with respect to redshift z for HDE with G–O cutoff in Logotropic equation of state with logarithmic internal energy in an interacting scenario. We have chosen $\epsilon_{m0} = 0.2$, $C_1 = 0.003$, $\beta = 0.9$, $A = 0.11530713$. The red, blue and green lines correspond to $\alpha = 2.6$, 2.62 and 2.64 respectively.

We have the expression for Hubble parameter in this case as

$$H_{MCG} = \frac{\sqrt{\left(\frac{B+a^{-3(1+A_0)(1+\alpha)}e^{(1+A_0)C_1(1+\alpha)}}{1+A_0}\right)^{\frac{1}{1+\alpha}} + \frac{\epsilon_{m0}}{A^3}}}{\sqrt{3}}. \tag{3.3}$$

The pressure for MCG with logotropic equation of state from Eqs. (3.2) and (1.3) is

$$P_{MCG} = A Log\left[1.93798 \times 10^{-100}\left(\left(\frac{B+a^{-3(1+A_0)(1+\alpha)}e^{(1+A_0)C_1(1+\alpha)}}{1+A_0}\right)^{\frac{1}{1+\alpha}} + \frac{\epsilon_{m0}}{A^3}\right)\right]. \tag{3.4}$$

Hence, from Eqs. (3.2) and (3.4), we can express the EoS parameter for MCG with logotropic equation of state as

$$w_{MCG} = \frac{A Log\left[1.93798 \times 10^{-100}\left(\left(\frac{B+a^{-3(1+A_0)(1+\alpha)}e^{(1+A_0)C_1(1+\alpha)}}{1+A_0}\right)^{\frac{1}{1+\alpha}} + \frac{\epsilon_{m0}}{A^3}\right)\right]}{\left(\frac{B+a^{-3(1+A_0)(1+\alpha)}e^{(1+A_0)C_1(1+\alpha)}}{1+A_0}\right)^{\frac{1}{1+\alpha}} + \frac{\epsilon_{m0}}{A^3}}. \tag{3.5}$$

We have plotted the EoS parameter w_{MCG} Eq.(3.5) against redshift z for MCG with logotropic equation of state in an interacting scenario in Fig.13. Fig.13 shows that it is showing a phantom like behaviour and it is decreasing with the evolution of the universe.

In the next subsection, we have studied the cosmological consequences of MCG with logotropic equation of state in the scenario where internal energy is in the logarithmic form.

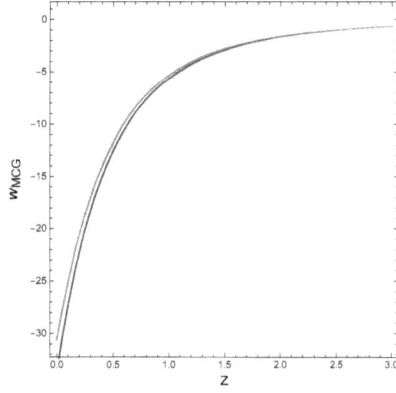

Figure 13: Evolution of EoS parameter w_{MCG} Eq.(3.5) against redshift z for MCG with logotropic equation of state in an interacting scenario. We have chosen $\epsilon_{m0} = 0.004$, $B = 0.001$, $\alpha = 0.002$, $C_1 = 0.003$, and $A_0 = 0.005$. The red, blue and green lines correspond to $A = 0.2$, 0.21 and 0.22 respectively.

3.1 MCG in Logotropic equation of state with logarithmic internal energy

The density of DM for MCG in this interacting scenario is given by

$$\epsilon_{MCG,DM} = 5.16 \times 10^{99} e^{-\frac{\left(\frac{B+a^{-3(1+A_0)(1+\alpha)}e^{(1+A_0)C_1(1+\alpha)}}{1+A_0}\right)^{\frac{1}{1+\alpha}}\left(a^{3(1+A_0)(1+\alpha)}B-A_0e^{(1+A_0)C_1(1+\alpha)}\right)}{a^{3(1+A_0)(1+\alpha)}B+e^{(1+A_0)C_1(1+\alpha)}}+\frac{\epsilon_{m0}}{A^3}}{A_0}}. \quad (3.6)$$

and the corresponding pressure is

$$P_{MCG,logarithmic} = \frac{A_0 Log\left[e^{-\frac{\left(\frac{B+a^{-3(1+A_0)(1+\alpha)}e^{(1+A_0)C_1(1+\alpha)}}{1+A_0}\right)^{\frac{1}{1+\alpha}}\left(a^{3(1+A_0)(1+\alpha)}B-A_0e^{(1+A_0)C_1(1+\alpha)}\right)}{a^{3(1+A_0)(1+\alpha)}B+e^{(1+A_0)C_1(1+\alpha)}}+\frac{\epsilon_{m0}}{A^3}}{A_0}}\right]}{Log[2]}. \quad (3.7)$$

Hence we get the EoS parameter for MCG in Logotropic equation of state with logarithmic internal energy as

$$w_{MCG,logarithmic} = \frac{A_0 Log\left[e^{-\frac{\left(\frac{B+a^{-3(1+A_0)(1+\alpha)}e^{(1+A_0)C_1(1+\alpha)}}{1+A_0}\right)^{\frac{1}{1+\alpha}}\left(a^{3(1+A_0)(1+\alpha)}B-A_0e^{(1+A_0)C_1(1+\alpha)}\right)}{a^{3(1+A_0)(1+\alpha)}B+e^{(1+A_0)C_1(1+\alpha)}}+\frac{\epsilon_{m0}}{A^3}}{A_0}}\right]}{\left(\left(\frac{B+a^{-3(1+A_0)(1+\alpha)}e^{(1+A_0)C_1(1+\alpha)}}{1+A_0}\right)^{\frac{1}{1+\alpha}}+\frac{\epsilon_{m0}}{A^3}\right)Log[2]}. \quad (3.8)$$

In Fig.14, we have plotted the EoS parameter $w_{MCG,logarithmic}$ Eq.(3.8) against redshift z for MCG in Logotropic equation of state with logarithmic internal energy in an interacting scenario. Fig.14

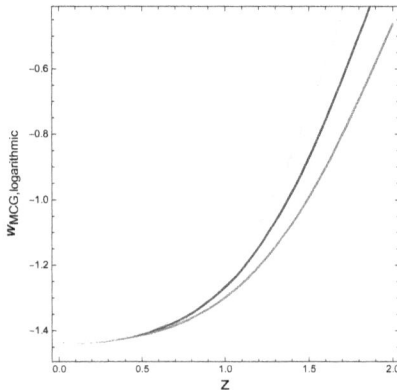

Figure 14: Evolution of EoS parameter $w_{MCG,logarithmic}$ Eq.(3.8) against redshift z for MCG in Logotropic equation of state with logarithmic internal energy in an interacting scenario. We have chosen $\epsilon_{m0} = 0.004$, $B = 0.001$, $\alpha = 0.002$, $C_1 = 0.003$, and $A = 0.02$. The red, blue and green lines correspond to $A_0 = 0.9$, 1 and 1.1 respectively.

shows a quintom like behaviour i.e., a transition from quintessence to phantom (from $w > -1$ to $w < -1$) and it is decreasing with the evolution of the universe.

4   Dark energy in quintom universe

The study of the GSL of thermodynamics for scalar field models of DE is the focus of this section. The pressure and density may be taken into consideration in light of the logotropic reconstruction of Chaplygin gas model of DE and the generalized holographic fluid. In order to investigate it, we have first shown the initial mathematical calculations of quintessence and phantom models with scalar fields σ and ϕ. Eventually, the validity of the GSL of thermodynamics has been examined in this logotropic reconstruction scheme by re-expressing these scalar fields through the logotropically reconstructed dark fluids.

To describe the new astrophysical data for the transition from quintessence-dominated universe to phantom-dominated universe i.e., from $w > -1$ to $w - 1$, a new model called the quintom model of DE [29] was proposed . For the spatially flat Friedman-Robertson-Walker (FRW) universe, the space-time metric is

$$ds^2 = -dt^2 + a(t)^2(dr^2 + r^2 d\Omega^2). \tag{4.1}$$

The quintom model which is composed of the negative kinetic scalar field ϕ and the normal scalar field σ is charaterized by the following action:

$$S = \int d^4 x \sqrt{-g} \left(\frac{R}{2} - \frac{1}{2} g^{\mu\nu} \partial_\mu \phi \partial_\nu \phi + \frac{1}{2} g^{\mu\nu} \partial_\mu \sigma \partial_\nu \sigma + V(\phi, \sigma) \right), \tag{4.2}$$

where the Lagrangian density of matter field is not taken into account. In the spatially flat FRW universe, the scalar fields' effective pressure p and the effective energy density ρ can be described by

$$p = -\frac{1}{2}\dot{\phi}^2 + \frac{1}{2}\dot{\sigma}^2 - V(\phi, \sigma), \tag{4.3}$$

80

and

$$\rho = -\frac{1}{2}\dot{\phi}^2 + \frac{1}{2}\dot{\sigma}^2 + V(\phi,\sigma). \tag{4.4}$$

Hence, the equation of state can be written as

$$w = \frac{-\dot{\phi}^2 + \dot{\sigma}^2 - 2V(\phi,\sigma)}{-\dot{\phi}^2 + \dot{\sigma}^2 + 2V(\phi,\sigma)}. \tag{4.5}$$

From the equation of state, we can conclude that for $\dot{\sigma} > \dot{\phi}$, $w \geq -1$ and for $\dot{\sigma} < \dot{\phi}$, $w < -1$. A potential with no direct coupling has been considered between two scalar fields as in [30]

$$V(\phi,\sigma) = V_\phi(\phi) + V_\sigma(\sigma) = V_{\phi0}e^{-\lambda_\phi\phi} + V_{\sigma0}e^{-\lambda_\sigma\sigma}, \tag{4.6}$$

where $\lambda_\phi > 0$ and $\lambda_\sigma > 0$, which characterizes the slope of the potential for ϕ and σ respectively. So, the evolution equation for two scalar fields in FRW model are

$$\ddot{\sigma} + 3H\dot{\sigma} - \frac{dV_\sigma(\sigma)}{d\sigma} = 0, \tag{4.7}$$

and

$$\ddot{\phi} + 3H\dot{\phi} - \frac{dV_\phi(\phi)}{d\phi} = 0. \tag{4.8}$$

In the next section, GSL of thermodynamics have been demonstrated for quintom scenario dominated universe.

5   GSL of Thernodynamics

Using the first law of thermodynamics, the expression for normal entropy has been derived to study the GSL [31] in the quintom scenario dominated universe and it is given by

$$TdS = dE + PdV = (P + \rho)dV + Vd\rho. \tag{5.1}$$

We get from Eqs. (4.4) and (4.3)

$$p + \rho = -\dot{\phi}^2 + \dot{\sigma}^2 \tag{5.2}$$

and the Friedmann equation comes out to be

$$H^2 = \frac{1}{3}\left(\frac{-\dot{\phi}^2}{2} + V_\phi + \frac{\dot{\sigma}^2}{2} + V_\sigma\right). \tag{5.3}$$

Hence, from Eqs. (5.2) and (5.3), we have

$$\dot{H} = \frac{1}{2}(\dot{\phi}^2 - \dot{\sigma}^2) = -\frac{1}{2}(p + \rho). \tag{5.4}$$

The left-hand side of Eq.(5.2) might be taken from the reconstructed dark fluids in the following phase of the study. As a result, $\dot{H}$ might be obtained. If $\dot{\phi}^2 < \dot{\sigma}^2$ then $\dot{H} < 0$ for the quintessence dominated universe, and if $\dot{\phi}^2 > \dot{\sigma}^2$ then $\dot{H} > 0$ for the phantom dominated universe. Rewriting the first law of thermodynamics in terms of the aforementioned relationships and applying the relation $V = \frac{4}{3}\pi R_h^3$, in which R_h represents the cosmic horizon's radius, results in

$$TdS = -2\dot{H}dV + Vd\rho = -8\pi R_h^2 \dot{H}dR_h + 8\pi R_h^3 HdH, \tag{5.5}$$

where T is the temperature of the fluid. The time derivative of normal entropy is

$$\dot{S} = \frac{8\pi \dot{H} R_h^2}{T}(HR_h - \dot{R}_h). \tag{5.6}$$

Quintom refers to the conjunction of phantom scalar field and normal scalar field, i.e. quintessence. The corresponding radii of Hubble horizon, apparent horizon, particle horizon and event horizon are:

$$R_H = \frac{1}{H}; \ R_A = \frac{1}{\sqrt{H^2 + \frac{k}{a(t)^2}}}; \ R_P = a(t)\int_0^t \frac{dt}{a(t)}; \ R_E = a(t)\int_t^{t_s} \frac{dt}{a(t)}, \tag{5.7}$$

respectively. Here κ is the curvature of the space, $\kappa = 0, 1$ and -1 for flat, closed and open universes respectively and $\int_t^{t_s} \frac{dt}{a(t)} < \infty$. For different spacetimes t_s, we have different values for e.g. $t_s = \infty$ for de Sitter spacetime. We can easily obtain the following derivatives from the radii of different enveloping horizons:

$$\dot{R}_H = -\frac{\dot{H}}{H^2}; \ \dot{R}_A = -HR_A^3(\dot{H} - \frac{k}{a^2}); \ \dot{R}_P = HR_P + 1; \ \dot{R}_E = HR_E - 1. \tag{5.8}$$

For each scalar field independently, the given equations hold true and $\dot{R}_h \leq 0$ for phantom dominated universe [32] and $\dot{R}_h \geq 0$ for quintessence dominated universe [33]. We have the time derivative of normal entropy of the quintom fluid by using the radius for event horizon in Eq.(5.6) as

$$\dot{S} = \frac{8\pi \dot{H} R_h^2}{T}. \tag{5.9}$$

The Eq.(5.6) indicates that the sign of $\dot{S}$ depends on the sign of $\dot{H}$, thus in a universe dominated by phantom, $\dot{S} > 0$, and in a universe dominated by quintessence, $\dot{S} < 0$. A black hole's entropy is directly correlated with the area of its event horizon. This horizon is highly valued and has significant physical meaning. The status of entropy is related to a cosmic event horizon, albeit this is not fully shown. This makes sense with some cautions in certain situations, such as the de Sitter horizon example, but generally, this is the topic of current research, as shown in [34, 35]. If we assume that $S_h = \pi R_h^2$ is the horizon entropy, then the GSL asserts that

$$\dot{S} + \dot{S}_h \geq 0. \tag{5.10}$$

Hence, we have

$$\dot{S} + \dot{S}_h = \frac{8\pi \dot{H} R_h^2}{T} + 2\pi R_h \dot{R}_h \geq 0. \tag{5.11}$$

In the next subsection, we have the mathematical preliminaries required for the validation of the GSL for the models HDE with G-O cutoff and MCG with different enveloping horizons of the universe in both quintessence and phantom dominated scenarios.

5.1 GSL of thermodynamics for HDE with G-O cutoff and MCG

By substituting Eqs. (2.8) and (2.9) in Eq.(5.4), the time derivative of H can be derived in the scenario of HDE with G-O cutoff in logotropic equation of state as

$$\dot{H}_{HDE} = \frac{1}{2}\left(-3a^{\frac{2-2\alpha}{\beta}}C_1 + \frac{\epsilon_{m0}}{A^3(-1+\alpha)} - ALog\left[5.81395\times10^{-100}a^{\frac{2-2\alpha}{\beta}}C_1 - \frac{1.93798\times10^{-100}\epsilon_{m0}}{A^3(-1+\alpha)}\right]\right).$$
(5.12)

We have the radius of Hubble horizon for HDE with G-O cutoff in logotropic equation of state as

$$R_{HDE,H} = \left(\sqrt{a^{\frac{2-2\alpha}{\beta}}C_1 - \frac{\epsilon_{m0}}{3A^3(-1+\alpha)}}\right)^{-1},$$
(5.13)

and its corresponding time derivative is

$$\dot{R}_{HDE,H} = \frac{3(-1+\alpha)^2}{\left(-3 - \frac{a^{\frac{2(-1+\alpha)}{\beta}}\epsilon_{m0}}{A^3C_1} + 3\alpha\right)\beta}.$$
(5.14)

From the relation

$$T = \frac{1}{2\pi R_h},$$
(5.15)

we can get the temperature of the HDE considered with Hubble horizon as

$$T_{HDE,H} = \frac{\sqrt{a^{\frac{2-2\alpha}{\beta}}C_1 - \frac{\epsilon_{m0}}{3A^3(-1+\alpha)}}}{2\pi}.$$
(5.16)

We have the radius of event horizon for HDE with G-O cutoff in logotropic equation of state as

$$R_{HDE,E} = aC_2 + \frac{a^{\frac{2(-1+\alpha)}{\beta}}\sqrt{a^{\frac{2-2\alpha}{\beta}}C_1 - \frac{\epsilon_{m0}}{3A^3(-1+\alpha)}}\beta\times{_2}F_1\left[1,1+\frac{\beta}{2-2\alpha},\frac{3-3\alpha+\beta}{2-2\alpha},\frac{a^{\frac{2(-1+\alpha)}{\beta}}\epsilon_{m0}}{3A^3C_1(-1+\alpha)}\right]}{C_1(1-\alpha+\beta)},$$
(5.17)

and its corresponding time derivative is

$$\dot{R}_{HDE,E} = aC_2\sqrt{a^{\frac{2-2\alpha}{\beta}}C_1 - \frac{\epsilon_{m0}}{3A^3(-1+\alpha)}} + \frac{1}{18A^6C_1^2(-1+\alpha)^3}a^{-4/\beta}Gamma\left[\frac{1-\alpha+\beta}{2-2\alpha}\right]$$

$$\left(-3a^{2/\beta}A^3C_1\left(-2a^{\frac{2\alpha}{\beta}}\epsilon_{m0} + 3a^{2/\beta}A^3C_1(-1+\alpha)\right)(-1+\alpha)^2{_2}F_1\left[1,1+\frac{\beta}{2-2\alpha},\frac{3-3\alpha+\beta}{2-2\alpha},\frac{a^{\frac{2(-1+\alpha)}{\beta}}\epsilon_{m0}}{3A^3C_1(-1+\alpha)}\right]\right.$$

$$\left.-a^{\frac{2\alpha}{\beta}}\epsilon_{m0}\left(a^{\frac{2\alpha}{\beta}}\epsilon_{m0} - 3a^{2/\beta}A^3C_1(-1+\alpha)\right)(2-2\alpha+\beta){_2}F_1\left[2,2+\frac{\beta}{2-2\alpha},\frac{5-5\alpha+\beta}{2-2\alpha},\frac{a^{\frac{2(-1+\alpha)}{\beta}}\epsilon_{m0}}{3A^3C_1(-1+\alpha)}\right]\right).$$
(5.18)

We can get the temperature of the HDE considered with event horizon as

$$T_{HDE,E} = \frac{1}{2\pi\left(aC_2 + \frac{a^{\frac{2(-1+\alpha)}{\beta}}\sqrt{a^{\frac{2-2\alpha}{\beta}}C_1-\frac{\epsilon_{m0}}{3A^3(-1+\alpha)}}\beta\times{_2}F_1\left[1,1+\frac{\beta}{2-2\alpha},\frac{3-3\alpha+\beta}{2-2\alpha},\frac{a^{\frac{2(-1+\alpha)}{\beta}}\epsilon_{m0}}{3A^3C_1(-1+\alpha)}\right]}{C_1(1-\alpha+\beta)}\right)}.$$
(5.19)

We have the radius of particle horizon for HDE with G-O cutoff in logotropic equation of state as

$$R_{HDE,P} = aC_2 - \frac{a^{\frac{2(-1+\alpha)}{\beta}}\sqrt{a^{\frac{2-2\alpha}{\beta}}C_1 - \frac{\epsilon_{m0}}{3A^3(-1+\alpha)}}\beta \times_2 F_1\left[1, 1+\frac{\beta}{2-2\alpha}, \frac{3-3\alpha+\beta}{2-2\alpha}, \frac{a^{\frac{2(-1+\alpha)}{\beta}}\epsilon_{m0}}{3A^3C_1(-1+\alpha)}\right]}{C_1(1-\alpha+\beta)},$$

(5.20)

and its corresponding time derivative is

$$\dot{R}_{HDE,P} = aC_2\sqrt{a^{\frac{2-2\alpha}{\beta}}C_1 - \frac{\epsilon_{m0}}{3A^3(-1+\alpha)}} + \frac{1}{18A^6C_1^2(-1+\alpha)^3}a^{-4/\beta}Gamma\left[\frac{1-\alpha+\beta}{2-2\alpha}\right]$$

$$\left(3a^{2/\beta}A^3C_1\left(-2a^{\frac{2\alpha}{\beta}}\epsilon_{m0} + 3a^{2/\beta}A^3C_1(-1+\alpha)\right)(-1+\alpha)^2\,_2F_1\left[1, 1+\frac{\beta}{2-2\alpha}, \frac{3-3\alpha+\beta}{2-2\alpha}, \frac{a^{\frac{2(-1+\alpha)}{\beta}}\epsilon_{m0}}{3A^3C_1(-1+\alpha)}\right]\right.$$

$$\left.+a^{\frac{2\alpha}{\beta}}\epsilon_{m0}\left(a^{\frac{2\alpha}{\beta}}\epsilon_{m0} - 3a^{2/\beta}A^3C_1(-1+\alpha)\right)(2-2\alpha+\beta)\,_2F_1\left[2, 2+\frac{\beta}{2-2\alpha}, \frac{5-5\alpha+\beta}{2-2\alpha}, \frac{a^{\frac{2(-1+\alpha)}{\beta}}\epsilon_{m0}}{3A^3C_1(-1+\alpha)}\right]\right).$$

(5.21)

So, the temperature of the HDE considered with particle horizon is

$$T_{HDE,P} = \frac{1}{2\pi\left(aC_2 - \frac{a^{\frac{2(-1+\alpha)}{\beta}}\sqrt{a^{\frac{2-2\alpha}{\beta}}C_1 - \frac{\epsilon_{m0}}{3A^3(-1+\alpha)}}\beta \times_2 F_1\left[1, 1+\frac{\beta}{2-2\alpha}, \frac{3-3\alpha+\beta}{2-2\alpha}, \frac{a^{\frac{2(-1+\alpha)}{\beta}}\epsilon_{m0}}{3A^3C_1(-1+\alpha)}\right]}{C_1(1-\alpha+\beta)}\right)}.$$

(5.22)

So, by using Eqs. (3.2) and (3.4) in Eq.(5.4), we get the time derivative of H for MCG as

$$\dot{H}_{MCG} = -\left(A^3\left(\frac{B+a^{-3(1+A_0)(1+\alpha)}e^{(1+A_0)C_1(1+\alpha)}}{1+A_0}\right)^{\frac{1}{1+\alpha}} + \epsilon_{m0}+\right.$$

$$\left. A^4 Log\left[1.93798\times 10^{-100}\left(\left(\frac{B+a^{-3(1+A_0)(1+\alpha)}e^{(1+A_0)C_1(1+\alpha)}}{1+A_0}\right)^{\frac{1}{1+\alpha}} + \frac{\epsilon_{m0}}{A^3}\right)\right]\right)2^{-1}A^{-3}.$$

(5.23)

We can express the radius of Hubble horizon for MCG as

$$R_{MCG,H} = \frac{\sqrt{3}}{\sqrt{\left(\frac{B+a^{-3(1+A_0)(1+\alpha)}e^{(1+A_0)C_1(1+\alpha)}}{1+A_0}\right)^{\frac{1}{1+\alpha}} + \frac{\epsilon_{m0}}{A^3}}},$$

(5.24)

and its corresponding time derivative is

$$\dot{R}_{MCG,H} = \frac{3a^{-3(1+A_0)(1+\alpha)}e^{(1+A_0)C_1(1+\alpha)}\left(\frac{B+a^{-3(1+A_0)(1+\alpha)}e^{(1+A_0)C_1(1+\alpha)}}{1+A_0}\right)^{-1+\frac{1}{1+\alpha}}}{2\left(\left(\frac{B+a^{-3(1+A_0)(1+\alpha)}e^{(1+A_0)C_1(1+\alpha)}}{1+A_0}\right)^{\frac{1}{1+\alpha}} + \frac{\epsilon_{m0}}{A^3}\right)}.$$

(5.25)

Hence, we have the temperature of MCG with Hubble horizon as

$$T_{MCG,H} = \frac{\sqrt{\left(\frac{B+a^{-3(1+A_0)(1+\alpha)}e^{(1+A_0)C_1(1+\alpha)}}{1+A_0}\right)^{\frac{1}{1+\alpha}} + \frac{\epsilon_{m0}}{A^3}}}{2\pi\sqrt{3}}.$$

(5.26)

The radius of event horizon for MCG is

$$R_{MCG,E} = aC_2 + \frac{1}{A^3 B^{\frac{1}{1+\alpha}} e^{(1+A_0)C_1} + (1+A_0)^{\frac{1}{1+\alpha}} \epsilon_{m0}} (1+A_0)^{\frac{1}{1+\alpha}} A^3$$

$$\sqrt{\frac{3\epsilon_{m0}}{A^3} + \frac{3a^{-3(1+A_0)(1+\alpha)}(1+A_0)^{-\frac{1}{1+\alpha}} B^{-1+\frac{1}{1+\alpha}} e^{(1+A_0)C_1} \left(1+a^{3(1+A_0)(1+\alpha)} B(1+\alpha)\right)}{1+\alpha}}$$

$$_2F_1\left[1, \frac{1}{2} + \frac{1}{3+3A_0+3\alpha+3A_0\alpha}, 1 + \frac{1}{3+3A_0+3\alpha+3A_0\alpha}, -\frac{a^{-3(1+A_0)(1+\alpha)}}{B\left(1+\frac{(1+A_0)^{\frac{1}{1+\alpha}} B^{-\frac{1}{1+\alpha}} e^{-(1+A_0)C_1} \epsilon_{m0}}{A^3}\right)(1+\alpha)}\right],$$

(5.27)

and the time derivative is

$$\dot{R}_{MCG,E} = \frac{aC_2 \sqrt{\frac{\epsilon_{m0}}{A^3} + \frac{a^{-3(1+A_0)(1+\alpha)}(1+A_0)^{-\frac{1}{1+\alpha}} B^{-1+\frac{1}{1+\alpha}} e^{(1+A_0)C_1} \left(1+a^{3(1+A_0)(1+\alpha)} B(1+\alpha)\right)}{1+\alpha}}}{\sqrt{3}} +$$

$$\frac{1}{2B\left(A^3 B^{\frac{1}{1+\alpha}} e^{(1+A_0)C_1} + (1+A_0)^{\frac{1}{1+\alpha}} \epsilon_{m0}\right)} 3a^{-3(1+A_0)(1+\alpha)}(1+A_0)$$

$$\left(\left(2a^{3(1+A_0)(1+\alpha)}(1+A_0)^{\frac{1}{1+\alpha}} B\epsilon_{m0}(1+\alpha) + A^3 B^{\frac{1}{1+\alpha}} e^{(1+A_0)C_1} \left(1+2a^{3(1+A_0)(1+\alpha)} B(1+\alpha)\right)\right)\right)$$

$$_2F_1\left[1, \frac{1}{2} + \frac{1}{3+3A_0+3\alpha+3A_0\alpha}, 1 + \frac{1}{3+3A_0+3\alpha+3A_0\alpha}, -\right.$$

$$\left.\frac{a^{-3(1+A_0)(1+\alpha)}}{B\left(1+\frac{(1+A_0)^{\frac{1}{1+\alpha}} B^{-\frac{1}{1+\alpha}} e^{-(1+A_0)C_1} \epsilon_{m0}}{A^3}\right)(1+\alpha)}\right] - 2\left(a^{3(1+A_0)(1+\alpha)}(1+A_0)^{\frac{1}{1+\alpha}} B\epsilon_{m0}(1+\alpha)+\right.$$

$$\left.A^3 B^{\frac{1}{1+\alpha}} e^{(1+A_0)C_1} \left(1+a^{3(1+A_0)(1+\alpha)} B(1+\alpha)\right)\right) \times_2F_1\left[2, \frac{1}{2} + \frac{1}{3+3A_0+3\alpha+3A_0\alpha}, 1+\right.$$

$$\left.\frac{1}{3+3A_0+3\alpha+3A_0\alpha}, -\frac{a^{-3(1+A_0)(1+\alpha)}}{B\left(1+\frac{(1+A_0)^{\frac{1}{1+\alpha}} B^{-\frac{1}{1+\alpha}} e^{-(1+A_0)C_1} \epsilon_{m0}}{A^3}\right)(1+\alpha)}\right]\right).$$

(5.28)

Hence, we can express the temperature of MCG with event horizon as

$$T_{MCG,E} = \left(2aC_2\pi + 2(1+A_0)^{\frac{1}{1+\alpha}} A^3\pi\right.$$

$$\sqrt{\frac{3\epsilon_{m0}}{A^3} + \frac{3a^{-3(1+A_0)(1+\alpha)}(1+A_0)^{-\frac{1}{1+\alpha}} B^{-1+\frac{1}{1+\alpha}} e^{(1+A_0)C_1} \left(1+a^{3(1+A_0)(1+\alpha)} B(1+\alpha)\right)}{1+\alpha}}$$

$$_2F_1\left[1, \frac{1}{2} + \frac{1}{3+3A_0+3\alpha+3A_0\alpha}, 1 + \frac{1}{3+3A_0+3\alpha+3A_0\alpha}, -\frac{a^{-3(1+A_0)(1+\alpha)}}{B\left(1+\frac{(1+A_0)^{\frac{1}{1+\alpha}} B^{-\frac{1}{1+\alpha}} e^{-(1+A_0)C_1} \epsilon_{m0}}{A^3}\right)(1+\alpha)}\right]$$

$$\left.\left(A^3 B^{\frac{1}{1+\alpha}} e^{(1+A_0)C_1} + (1+A_0)^{\frac{1}{1+\alpha}} \epsilon_{m0}\right)^{-1}\right)^{-1}.$$

(5.29)

The radius of particle horizon for MCG is

$$R_{MCG,P} = aC_2 - \cfrac{1}{A^3 B^{\frac{1}{1+\alpha}} e^{(1+A_0)C_1} + (1+A_0)^{\frac{1}{1+\alpha}} \epsilon_{m0}}(1+A_0)^{\frac{1}{1+\alpha}} A^3$$
$$\sqrt{\frac{3\epsilon_{m0}}{A^3} + \frac{3a^{-3(1+A_0)(1+\alpha)}(1+A_0)^{-\frac{1}{1+\alpha}} B^{-1+\frac{1}{1+\alpha}} e^{(1+A_0)C_1}\left(1+a^{3(1+A_0)(1+\alpha)} B(1+\alpha)\right)}{1+\alpha}}$$

$${}_2F_1\left[1, \frac{1}{2} + \frac{1}{3+3A_0+3\alpha+3A_0\alpha}, 1 + \frac{1}{3+3A_0+3\alpha+3A_0\alpha}, -\cfrac{a^{-3(1+A_0)(1+\alpha)}}{B\left(1+\frac{(1+A_0)^{\frac{1}{1+\alpha}} B^{-\frac{1}{1+\alpha}} e^{-(1+A_0)C_1} \epsilon_{m0}}{A^3}\right)(1+\alpha)}\right],$$

$$(5.30)$$

and its time derivative is

$$\dot{R}_{MCG,P} = \cfrac{aC_2\sqrt{\frac{\epsilon_{m0}}{A^3} + \frac{a^{-3(1+A_0)(1+\alpha)}(1+A_0)^{-\frac{1}{1+\alpha}} B^{-1+\frac{1}{1+\alpha}} e^{(1+A_0)C_1}\left(1+a^{3(1+A_0)(1+\alpha)} B(1+\alpha)\right)}{1+\alpha}}}{\sqrt{3}} +$$

$$\cfrac{1}{2B\left(A^3 B^{\frac{1}{1+\alpha}} e^{(1+A_0)C_1} + (1+A_0)^{\frac{1}{1+\alpha}} \epsilon_{m0}\right)} 3a^{-3(1+A_0)(1+\alpha)}(1+A_0)$$

$$\left(-\left(2a^{3(1+A_0)(1+\alpha)}(1+A_0)^{\frac{1}{1+\alpha}} B\epsilon_{m0}(1+\alpha) + A^3 B^{\frac{1}{1+\alpha}} e^{(1+A_0)C_1}\left(1+2a^{3(1+A_0)(1+\alpha)} B(1+\alpha)\right)\right)\right.$$

$${}_2F_1\left[1, \frac{1}{2} + \frac{1}{3+3A_0+3\alpha+3A_0\alpha}, 1 + \frac{1}{3+3A_0+3\alpha+3A_0\alpha}, -\right.$$

$$\left.\cfrac{a^{-3(1+A_0)(1+\alpha)}}{B\left(1+\frac{(1+A_0)^{\frac{1}{1+\alpha}} B^{-\frac{1}{1+\alpha}} e^{-(1+A_0)C_1} \epsilon_{m0}}{A^3}\right)(1+\alpha)}\right] + 2\left(a^{3(1+A_0)(1+\alpha)}(1+A_0)^{\frac{1}{1+\alpha}} B\epsilon_{m0}(1+\alpha)+\right.$$

$$\left.A^3 B^{\frac{1}{1+\alpha}} e^{(1+A_0)C_1}\left(1+a^{3(1+A_0)(1+\alpha)} B(1+\alpha)\right)\right) \times {}_2F_1\left[2, \frac{1}{2} + \frac{1}{3+3A_0+3\alpha+3A_0\alpha}, 1+\right.$$

$$\left.\left.\frac{1}{3+3A_0+3\alpha+3A_0\alpha}, -\cfrac{a^{-3(1+A_0)(1+\alpha)}}{B\left(1+\frac{(1+A_0)^{\frac{1}{1+\alpha}} B^{-\frac{1}{1+\alpha}} e^{-(1+A_0)C_1} \epsilon_{m0}}{A^3}\right)(1+\alpha)}\right]\right).$$

$$(5.31)$$

So, we can express the temperature of MCG with particle horizon as

$$T_{MCG,P} = \left(2aC_2\pi - 2(1+A_0)^{\frac{1}{1+\alpha}} A^3\pi\right.$$

$$\sqrt{\frac{3\epsilon_{m0}}{A^3} + \frac{3a^{-3(1+A_0)(1+\alpha)}(1+A_0)^{-\frac{1}{1+\alpha}} B^{-1+\frac{1}{1+\alpha}} e^{(1+A_0)C_1}\left(1+a^{3(1+A_0)(1+\alpha)} B(1+\alpha)\right)}{1+\alpha}}$$

$${}_2F_1\left[1, \frac{1}{2} + \frac{1}{3+3A_0+3\alpha+3A_0\alpha}, 1 + \frac{1}{3+3A_0+3\alpha+3A_0\alpha}, -\cfrac{a^{-3(1+A_0)(1+\alpha)}}{B\left(1+\frac{(1+A_0)^{\frac{1}{1+\alpha}} B^{-\frac{1}{1+\alpha}} e^{-(1+A_0)C_1} \epsilon_{m0}}{A^3}\right)(1+\alpha)}\right]$$

$$\left.\left(A^3 B^{\frac{1}{1+\alpha}} e^{(1+A_0)C_1} + (1+A_0)^{\frac{1}{1+\alpha}} \epsilon_{m0}\right)^{-1}\right)^{-1}.$$

$$(5.32)$$

Three scenarios have been considered in order to evaluate the validity of Eq.(5.11): the first involves dominating the phantom fluid, the second involves dominating the quintessence fluid, and the third involves a transition from the quintessence to the phantom.

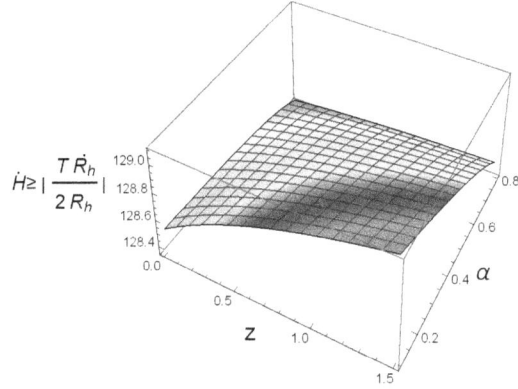

Figure 15: Validity of $\dot{H} \geq \left| \frac{T\dot{R}_h}{2R_h} \right|$ for generalized HDE with Hubble horizon in phantom dominated scenario.

5.2 Phantom Dominated Scenarion

In various pioneering works [36, 37], the phantom era of the universe has been discussed. $\dot{S} + \dot{S}_h < 0$ when $\dot{H} > 0$ and $\dot{R}_h \leq 0$. The condition for validity of GSL when the phantom fluid temperature $T > 0$ is

$$\dot{H} \geq \left| \frac{T\dot{R}_h}{2R_h} \right|. \tag{5.33}$$

If the temperature T is assumed to be proportional to the temperature of de Sitter [33], then

$$T = \frac{bH}{2\pi} \tag{5.34}$$

where b is a parameter. In de Sitter spacetime scenario, the GSL holds when

$$b \leq \frac{4\pi \dot{H} R_h}{H|\dot{R}_h|} \tag{5.35}$$

If $b \leq 1$ when $R_h = \frac{1}{H}$.

GSL for generalized HDE with Hubble horizon in phantom scenario

In this subsection, we have analyzed the validity of GSL for generalized HDE with Hubble horizon in phantom dominated scenario.

We have plotted $\dot{H} \geq \left| \frac{T\dot{R}_h}{2R_h} \right|$ and $b \leq \frac{4\pi \dot{H} R_h}{H|\dot{R}_h|}$ against redshift z and constraint α in Figs. 15 and 16 respectively to pictorially understand the occurrence of inequations (5.33) and (5.35). The parameters chosen are: For the inequation (5.33); $\epsilon_{m0} = 0.002$, $A = 1.1$, $\beta = 1.3$ and $C_1 = 0.005$ and for the inequation (5.35); $\epsilon_{m0} = 0.001$, $A = 0.003$, $\beta = 0.2$ and $C_1 = 0.004$. From the Figs. 15 and 16, we have observed that both the inequations are satisfied for generalized HDE with Hubble horizon in phantom dominated universe.

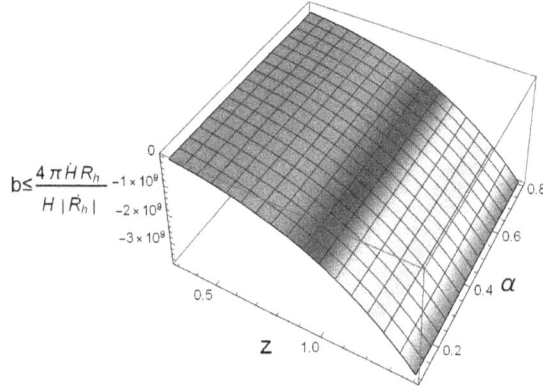

Figure 16: Validity of $b \leq \frac{4\pi \dot{H} R_h}{H|\dot{R}_h|}$ for generalized HDE with Hubble horizon in phantom dominated scenario.

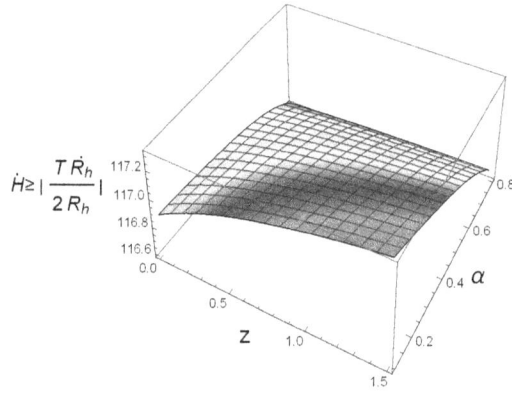

Figure 17: Validity of $\dot{H} \geq \left|\frac{T\dot{R}_h}{2R_h}\right|$ for generalized HDE with event horizon in phantom dominated scenario.

GSL for generalized HDE with event horizon in phantom scenario

In this subsection, we have analyzed the validity of GSL for generalized HDE with event horizon in phantom dominated scenario.

We have plotted $\dot{H} \geq \left|\frac{T\dot{R}_h}{2R_h}\right|$ and $b \leq \frac{4\pi \dot{H} R_h}{H|\dot{R}_h|}$ against redshift z and constraint α in Figs. 17 and 18 respectively to pictorially understand the occurrence of inequations (5.33) and (5.35). The parameters chosen are: For the inequation (5.33); $C_2 = 0.002$, $\epsilon_{m0} = 0.003$, $A = 1$, $\beta = 1.2$ and $C_1 = 0.004$ and for the inequation (5.35); $C_2 = 0.005$, $\epsilon_{m0} = 0.001$, $A = 0.1$, $\beta = 0.002$ and $C_1 = 0.004$. From the Figs. 17 and 18, we have observed that both the inequations are satisfied for generalized HDE with event horizon in phantom dominated universe.

GSL for generalized HDE with particle horizon in phantom scenario

In this subsection, we have analyzed the validity of GSL for generalized HDE with particle horizon in phantom dominated scenario.

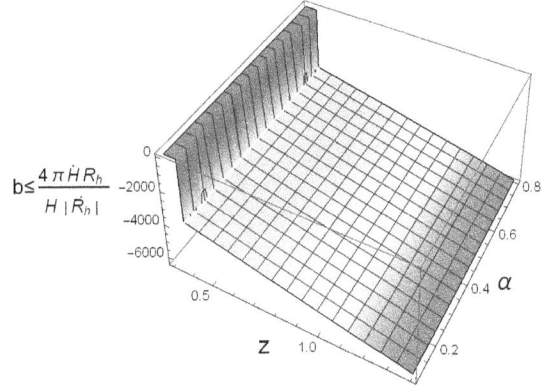

Figure 18: Validity of $b \leq \frac{4\pi \dot{H} R_h}{H |\dot{R}_h|}$ for generalized HDE with event horizon in phantom dominated scenario.

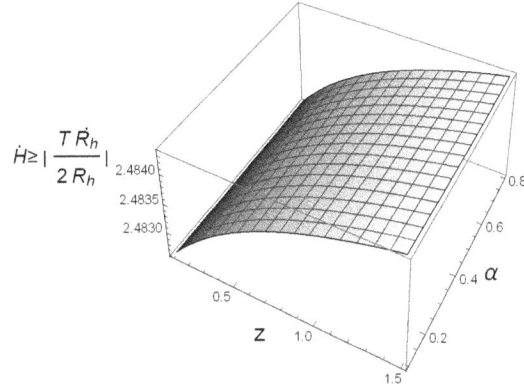

Figure 19: Validity of $\dot{H} \geq \left| \frac{T \dot{R}_h}{2 R_h} \right|$ for generalized HDE with particle horizon in phantom dominated scenario.

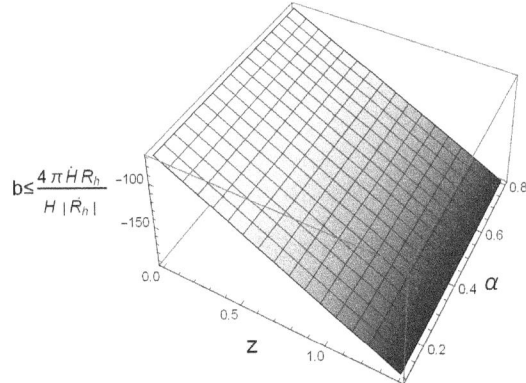

Figure 20: Validity of $b \leq \frac{4\pi \dot{H} R_h}{H |\dot{R}_h|}$ for generalized HDE with particle horizon in phantom dominated scenario.

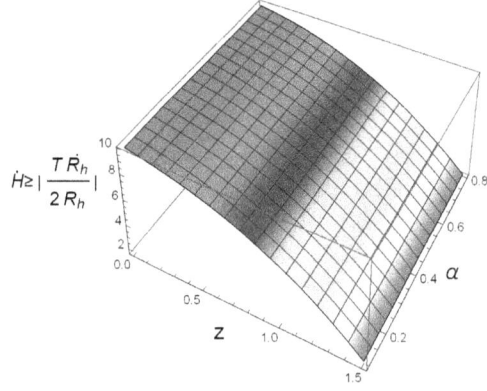

Figure 21: Validity of $\dot{H} \geq \left| \frac{T\dot{R}_h}{2R_h} \right|$ for MCG with Hubble horizon in phantom dominated scenario.

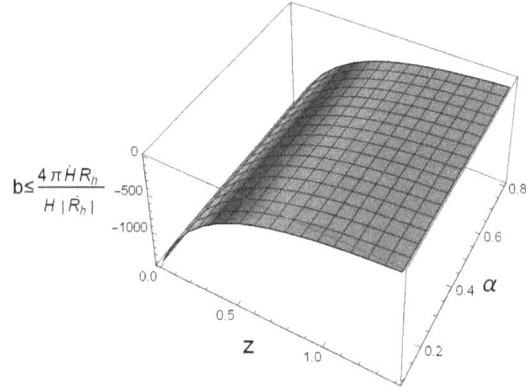

Figure 22: Validity of $b \leq \frac{4\pi\dot{H}R_h}{H|\dot{R}_h|}$ for MCG with Hubble horizon in phantom dominated scenario.

We have plotted $\dot{H} \geq \left| \frac{T\dot{R}_h}{2R_h} \right|$ and $b \leq \frac{4\pi\dot{H}R_h}{H|\dot{R}_h|}$ against redshift z and constraint α in Figs. 19 and 20 respectively to pictorially understand the occurrence of inequations (5.33) and (5.35). The parameters chosen are: For the inequation (5.33); $C_2 = 0.006$, $\epsilon_{m0} = 0.003$, $A = 0.07$, $\beta = 0.005$ and $C_1 = 0.001$ and for the inequation (5.35); $C_2 = 0.001$, $\epsilon_{m0} = 0.004$, $A = 0.005$, $\beta = 0.2$ and $C_1 = 0.002$. From the Figs. 19 and 20, we have observed that both the inequations are satisfied for generalized HDE with particle horizon in phantom dominated universe.

GSL for MCG with Hubble horizon in phantom scenario

In this subsection, we have investigated the validity of GSL for MCG with Hubble horizon in phantom dominated scenario.

We have plotted $\dot{H} \geq \left| \frac{T\dot{R}_h}{2R_h} \right|$ and $b \leq \frac{4\pi\dot{H}R_h}{H|\dot{R}_h|}$ against redshift z and constraint α in Figs. 21 and 22 respectively to pictorially understand the occurrence of inequations (5.33) and (5.35). The parameters chosen are: For the inequation (5.33); $A = 0.1$, $B = 0.003$, $\epsilon_{m0} = 0.002$, $A_0 = 0.007$, $\beta = 0.005$ and $C_1 = 0.005$ and for the inequation (5.35); $A = 0.5$, $B = 0.003$, $\epsilon_{m0} = 0.002$, $A_0 = 0.003$, $\beta = 0.005$ and $C_1 = 0.001$. From the Figs. 21 and 22, we have observed that both the

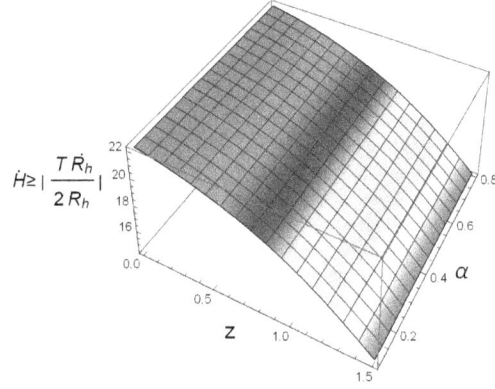

Figure 23: Validity of $\dot{H} \geq \left|\frac{T\dot{R_h}}{2R_h}\right|$ for MCG with event horizon in phantom dominated scenario.

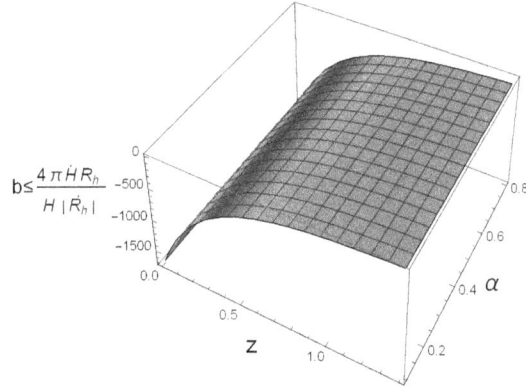

Figure 24: Validity of $b \leq \frac{4\pi\dot{H}R_h}{H|\dot{R_h}|}$ for MCG with event horizon in phantom dominated scenario.

inequations are satisfied for MCG with Hubble horizon in phantom dominated universe.

GSL for MCG with event horizon in phantom scenario

In this subsection, we have investigated the validity of GSL for MCG with event horizon in phantom dominated scenario.

We have plotted $\dot{H} \geq \left|\frac{T\dot{R_h}}{2R_h}\right|$ and $b \leq \frac{4\pi\dot{H}R_h}{H|\dot{R_h}|}$ against redshift z and constraint α in Figs. 23 and 24 respectively to pictorially understand the occurrence of inequations (5.33) and (5.35). The parameters chosen are: For the inequation (5.33); $A = 0.2$, $B = 0.001$, $C_2 = 0.001$, $\epsilon_{m0} = 0.004$, $A_0 = 0.005$, $\beta = 0.005$ and $C_1 = 0.003$ and for the inequation (5.35); $A = 0.2$, $B = 0.001$, $C_2 = 0.001$, $\epsilon_{m0} = 0.004$, $A_0 = 0.005$, $\beta = 0.005$ and $C_1 = 0.003$. From the Figs. 23 and 24, we have observed that both the inequations are satisfied for MCG with event horizon in phantom dominated universe.

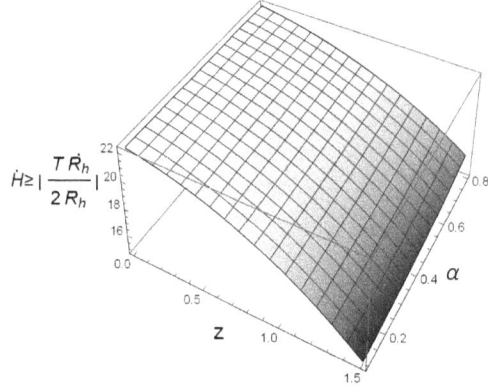

Figure 25: Validity of $\dot{H} \geq \left|\frac{T\dot{R}_h}{2R_h}\right|$ for MCG with particle horizon in phantom dominated scenario.

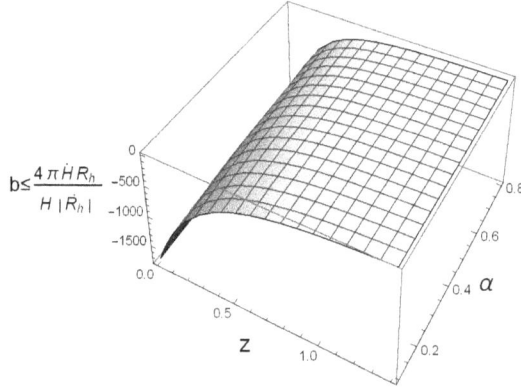

Figure 26: Validity of $b \leq \frac{4\pi\dot{H}R_h}{H|\dot{R}_h|}$ for MCG with particle horizon in phantom dominated scenario.

GSL for MCG with particle horizon in phantom scenario

In this subsection, we have investigated the validity of GSL for MCG with particle horizon in phantom dominated scenario.

We have plotted $\dot{H} \geq \left|\frac{T\dot{R}_h}{2R_h}\right|$ and $b \leq \frac{4\pi\dot{H}R_h}{H|\dot{R}_h|}$ against redshift z and constraint α in Figs. 25 and 26 respectively to pictorially understand the occurrence of inequations (5.33) and (5.35). The parameters chosen are: For the inequation (5.33); $A = 0.2$, $B = 0.3$, $C_2 = 0.002$, $\epsilon_{m0} = 0.003$, $A_0 = 0.004$, $\beta = 0.005$ and $C_1 = 0.003$ and for the inequation (5.35); $A = 0.2$, $B = 0.001$, $C_2 = 0.001$, $\epsilon_{m0} = 0.004$, $A_0 = 0.005$, $\beta = 0.005$ and $C_1 = 0.003$. From the Figs. 25 and 26, we have observed that both the inequations are satisfied for MCG with particle horizon in phantom dominated universe.

5.3 Quintessence Dominated Scenario

$\dot{S} + \dot{S}_h > 0$ when $\dot{R}_h \geq 0$ and $\dot{H} < 0$. The condition for validity of GSL when $T > 0$ is

$$|\dot{H}| \leq \frac{T\dot{R}_h}{2R_h}. \tag{5.36}$$

92

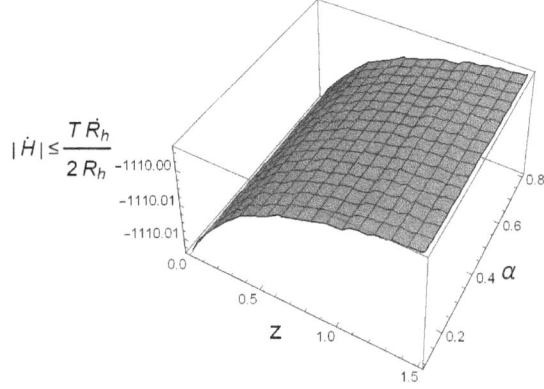

Figure 27: Validity of Validity of $|\dot{H}| \le \frac{T\dot{R}_h}{2R_h}$ for generalized HDE with Hubble horizon in quintessence dominated scenario.

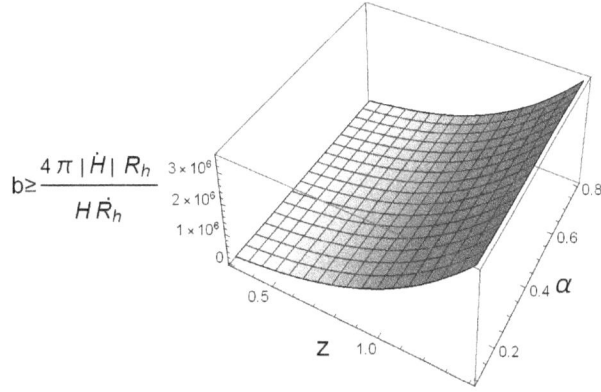

Figure 28: Validity of $b \ge \frac{4\pi|\dot{H}|R_h}{H\dot{R}_h}$ for generalized HDE with Hubble horizon in quintessence dominated scenario.

From Eq.(5.34), we have

$$b \ge \frac{4\pi|\dot{H}|R_h}{H\dot{R}_h}. \tag{5.37}$$

GSL for generalized HDE with Hubble horizon in quintessence scenario

In the following subsection, we have examined the validity of GSL for generalized HDE with Hubble horizon in quintessence dominated scenario.

We have plotted $|\dot{H}| \le \frac{T\dot{R}_h}{2R_h}$ and $b \ge \frac{4\pi|\dot{H}|R_h}{H\dot{R}_h}$ against redshift z and constraint α in Figs. 27 and 28 respectively to pictorially understand the occurrence of inequations (5.36) and (5.37),. The parameters chosen are: For the inequation (5.36); $\beta = 0.9$, $\epsilon_{m0} = 0.002$, $A = 0.01$ and $C_1 = 0.005$ and for the inequation (5.37); $\beta = 0.2$, $\epsilon_{m0} = 0.001$, $A = 0.3$ and $C_1 = 0.004$. From the Figs. 27 and 28, we have observed that both the inequations are satisfied for generalized HDE with Hubble horizon in quintessence dominated universe.

93

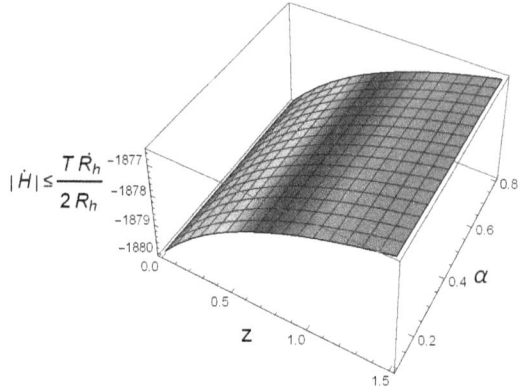

Figure 29: Validity of Validity of $|\dot{H}| \leq \frac{T\dot{R}_h}{2R_h}$ for generalized HDE with event horizon in quintessence dominated scenario.

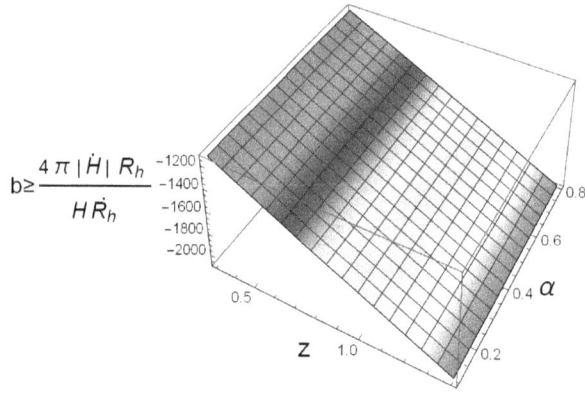

Figure 30: Validity of $b \geq \frac{4\pi|\dot{H}|R_h}{H\dot{R}_h}$ for generalized HDE with event horizon in quintessence dominated scenario.

GSL for generalized HDE with event horizon in quintessence scenario

In the following subsection, we have examined the validity of GSL for generalized HDE with event horizon in quintessence dominated scenario.

We have plotted $|\dot{H}| \leq \frac{T\dot{R}_h}{2R_h}$ and $b \geq \frac{4\pi|\dot{H}|R_h}{H\dot{R}_h}$ against redshift z and constraint α in Figs. 29 and 30 respectively to pictorially understand the occurrence of inequations (5.36) and (5.37),. The parameters chosen are: For the inequation (5.36); $C_2 = 0.002$, $\beta = 1.2$, $\epsilon_{m0} = 0.003$, $A = 0.01$ and $C_1 = 0.004$ and for the inequation (5.37); $C_2 = 0.005$, $\beta = 0.002$, $\epsilon_{m0} = 0.003$, $A = 0.1$ and $C_1 = 0.001$. From the Figs. 29 and 30, we have observed that both the inequations are satisfied for generalized HDE with event horizon in quintessence dominated universe.

GSL for generalized HDE with particle horizon in quintessence scenario

In the following subsection, we have examined the validity of GSL for HDE with particle horizon in quintessence dominated scenario.

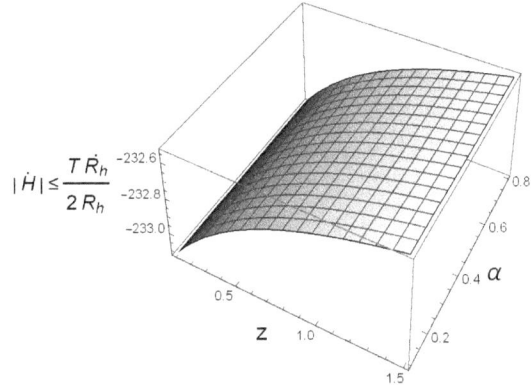

Figure 31: Validity of Validity of $|\dot{H}| \leq \frac{T\dot{R}_h}{2R_h}$ for generalized HDE with particle horizon in quintessence dominated scenario.

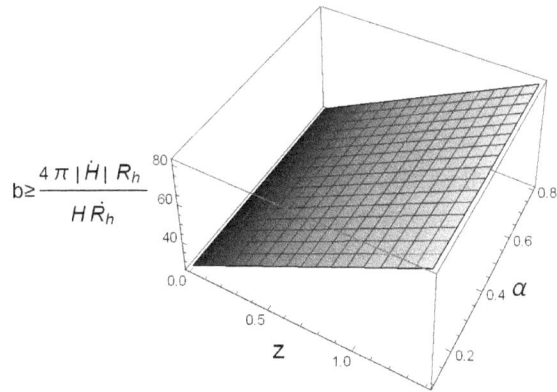

Figure 32: Validity of $b \geq \frac{4\pi|\dot{H}|R_h}{H\dot{R}_h}$ for generalized HDE with particle horizon in quintessence dominated scenario.

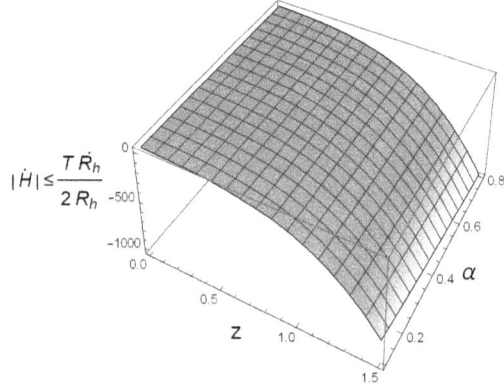

Figure 33: Validity of Validity of $|\dot{H}| \leq \frac{T\dot{R}_h}{2R_h}$ for MCG with Hubble horizon in quintessence dominated scenario.

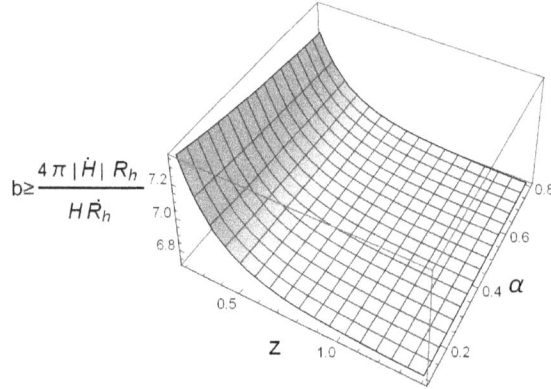

Figure 34: Validity of $b \geq \frac{4\pi|\dot{H}|R_h}{H\dot{R}_h}$ for MCG with Hubble horizon in quintessence dominated scenario.

We have plotted $|\dot{H}| \leq \frac{T\dot{R}_h}{2R_h}$ and $b \geq \frac{4\pi|\dot{H}|R_h}{H\dot{R}_h}$ against redshift z and constraint α in Figs. 31 and 32 respectively to pictorially understand the occurrence of inequations (5.36) and (5.37),. The parameters chosen are: For the inequation (5.36); $C_2 = 0.006$, $\beta = 0.005$, $\epsilon_{m0} = 0.003$, $A = 0.02$ and $C_1 = 0.001$ and for the inequation (5.37); $C_2 = 0.001$, $\beta = 0.2$, $\epsilon_{m0} = 0.004$, $A = 0.08$ and $C_1 = 0.002$. From the Figs. 31 and 32, we have observed that both the inequations are satisfied for generalized HDE with particle horizon in quintessence dominated universe.

GSL for MCG with Hubble horizon in quintessence scenario

In this subsection, we have studied the validity of GSL for MCG with Hubble horizon in quintessence dominated scenario.

We have plotted $|\dot{H}| \leq \frac{T\dot{R}_h}{2R_h}$ and $b \geq \frac{4\pi|\dot{H}|R_h}{H\dot{R}_h}$ against redshift z and constraint α in Figs. 33 and 34 respectively to pictorially understand the occurrence of inequations (5.36) and (5.37),. The parameters chosen are: For the inequation (5.36); $B = 0.002$, $\epsilon_{m0} = 0.005$, $A_0 = 0.9$, $A = 0.3$ and $C_1 = 1.4$ and for the inequation (5.37); $B = 0.001$, $\epsilon_{m0} = 0.4$, $A_0 = 1.2$, $A = 0.2$ and $C_1 = 1.5$.

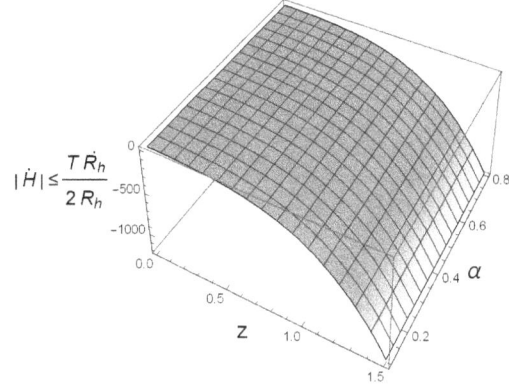

Figure 35: Validity of Validity of $|\dot{H}| \leq \frac{T\dot{R}_h}{2R_h}$ for MCG with event horizon in quintessence dominated scenario.

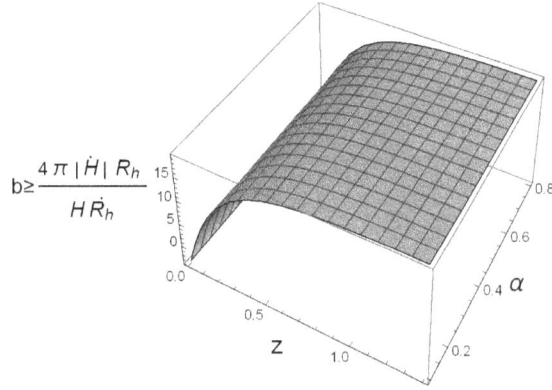

Figure 36: Validity of $b \geq \frac{4\pi|\dot{H}|R_h}{H\dot{R}_h}$ for MCG with event horizon in quintessence dominated scenario.

From the Figs. 33 and 34, we have observed that both the inequations are satisfied for MCG with Hubble horizon in quintessence dominated universe.

GSL for MCG with event horizon in quintessence scenario

In this subsection, we have studied the validity of GSL for MCG with event horizon in quintessence dominated scenario.

We have plotted $|\dot{H}| \leq \frac{T\dot{R}_h}{2R_h}$ and $b \geq \frac{4\pi|\dot{H}|R_h}{H\dot{R}_h}$ against redshift z and constraint α in Figs. 35 and 36 respectively to pictorially understand the occurrence of inequations (5.36) and (5.37),. The parameters chosen are: For the inequation (5.36); $C_2 = 0.001$, $B = 0.001$, $\epsilon_{m0} = 0.004$, $A_0 = 1$, $A = 0.2$ and $C_1 = 1.5$ and for the inequation (5.37); $C_2 = 0.001$, $B = 0.001$, $\epsilon_{m0} = 0.004$, $A_0 = 0.5$, $A = 0.2$ and $C_1 = 2.7$. From the Figs. 35 and 36, we have observed that both the inequations are satisfied for MCG with event horizon in quintessence dominated universe.

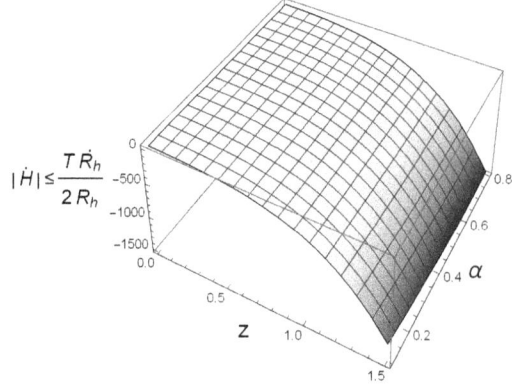

Figure 37: Validity of Validity of $|\dot{H}| \leq \frac{T\dot{R}_h}{2R_h}$ for MCG with particle horizon in quintessence dominated scenario.

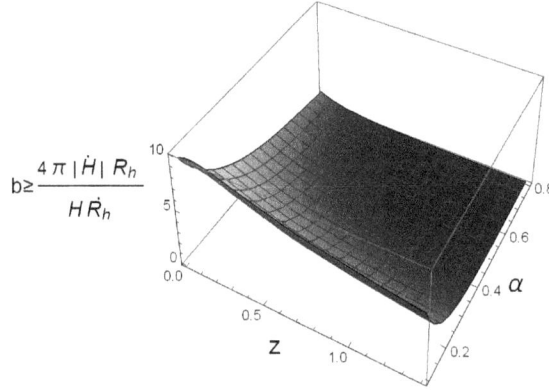

Figure 38: Validity of $b \geq \frac{4\pi|\dot{H}|R_h}{H\dot{R}_h}$ for MCG with particle horizon in quintessence dominated scenario.

GSL for MCG with particle horizon in quintessence scenario

In this subsection, we have examined the validity of GSL for MCG with particle horizon in quintessence dominated scenario.

We have plotted $|\dot{H}| \leq \frac{T\dot{R}_h}{2R_h}$ and $b \geq \frac{4\pi|\dot{H}|R_h}{H\dot{R}_h}$ against redshift z and constraint α in Figs. 37 and 38 respectively to pictorially understand the occurrence of inequations (5.36) and (5.37),. The parameters chosen are: For the inequation (5.36); $C_2 = 0.001$, $B = 0.001$, $\epsilon_{m0} = 0.004$, $A_0 = 1$, $A = 0.2$ and $C_1 = 1.5$ and for the inequation (5.37); $C_2 = 0.001$, $B = 0.001$, $\epsilon_{m0} = 0.004$, $A_0 = 1.5$, $A = 0.2$ and $C_1 = 2.5$. From the Figs. 37 and 38, we have observed that both the inequations are satisfied for MCG with particle horizon in quintessence dominated universe.

5.4   Quinton Dominated Scenario

One may determine By considering how the quintom dominated universe affects the GSL [33], one can conclude that the time derivative of the enveloping horizons and the entropy must be zero at the transition time . They have examined the requirements and restrictions placed on the Hubble parameter H, cosmological future horizon R_h and temperature T in a phantom dominated universe [38]. The GSL of gravitational thermodynamics is shown satisfied in [39], except for the first stage of Chaplygin gas dominated expansion of the universe, and the temperature of the phantom fluid remains non-negative.

We know that $\dot{R}_h \leq 0$ in the phantom dominated scenario and $\dot{R}_h \geq 0$ in the quintessence dominated scenario, and assuming that R_h varies continuously, the transition from quintessence to phantom should yield $\dot{R}_h = 0$. Therefore, S_h and $\dot{H}$ will be zero throughout the transition time, and we may use Eq.(5.9) to derive $\dot{S} = 0$. Consequently, we deduce that the total entropy is both differentiable and continuous during the transition time.

6   Concluding Remarks

This paper suggests a novel technique for reconstructing generalized HDE and analyzing its cosmological consequences based on logotropic equation of state. For the universe's overall energy density one can determine a scale factor, which is provided by baryons, DM, and DE density, by assuming that the DE density is brought by HDE with G–O cutoff. The EoS parameter w_{HDE} for HDE with G–O cutoff in logotropic equation of state in an interacting scenario shows a quintessence like behaviour and also increasing with the evolution of the universe which is consistent with the observations of the models depicting late time accelerating expansion of the universe. The EoS parameter $w_{HDE,log}$ for HDE with G–O cutoff in logotropic equation of state with logarithmic internal energy in an interacting scenario shows a phantom like behaviour and also increasing with the evolution of the universe but it is not crossing the phantom boundary -1.

The EoS parameter w_{MCG} for MCG with logotropic equation of state in an interacting scenario depicts a phantom like behaviour and it is decreasing with the evolution of the universe. The EoS parameter $w_{MCG,logarithmic}$ for MCG in Logotropic equation of state with logarithmic internal energy in an interacting scenario depicts a quintom like behaviour i.e., a transition from quintessence to phantom (from $w > -1$ to $w < -1$) and it is decreasing with the evolution of the universe.

We have observed in the case of inequations $\dot{H} \geq \left| \frac{T\dot{R}_h}{2R_h} \right|$ and $b \leq \frac{4\pi \dot{H} R_h}{H|\dot{R}_h|}$ that the difference is staying in the positive and negative region respectively throughout the evolution of the universe. Hence the inequations are satisfied for the models HDE with G-O cutoff and MCG with different enveloping horizons of the universe in phantom dominated universe. We have observed in the case of inequations $|\dot{H}| \leq \frac{T\dot{R}_h}{2R_h}$ and $b \geq \frac{4\pi |\dot{H}| R_h}{H\dot{R}_h}$ that the difference is staying in the negative and positive region respectively throughout the evolution of the universe. Hence the inequations are satisfied for the models HDE with G-O cutoff and MCG with different enveloping horizons of the universe in quintessence dominated universe.

As a continuation of the current research, let's discuss its future studies before we conclude. We propose to investigate further the unification of the temporal acceleration and early inflation utilising more widely based holographic fluids, as proposed in [40] and further shown in [41]. In keeping with [42, 43], we suggest expanding the study to take into account other versions of Chaplygin gas with polytropic equation of state in modified gravity frameworks as an additional extension of the current work. Let's wrap up by discussing the results of the above-discussed results in light of some pertinent literature. Ref. [44] showed that a wide class of dark energy models, including the entropic dark energy models, are comparable to the generalized HDE. They demonstrated that the proper cut-offs could be determined using either the future horizon and its derivatives or the particle horizon and its derivatives. We might note that in the context of [44], our research has focused on a particular type of IR cutoff, which is an example of a more widespread holographic dark energy that was documented in [45].

7 Acknowledgement

The authors acknowledge the International travel support by the Science and Engineering Research Board (SERB), Government of India to participate in the Third Hermann Minkowski Meeting on the Foundations of Spacetime Physics, held in the resort Albena in Bulgaria on 11-14 September 2023.

References

[1] Riess, A.G., Filippenko, A.V., Challis, P., Clocchiatti, A., Diercks, A., Garnavich, P.M., Gilliland, R.L., Hogan, C.J., Jha, S., Kirshner, R.P. and Leibundgut, B.R.U.N.O., 1998. Observational evidence from supernovae for an accelerating universe and a cosmological constant. The astronomical journal, $116(3)$, p.1009.

[2] Perlmutter, S., Aldering, G., Goldhaber, G., Knop, R.A., Nugent, P., Castro, P.G., Deustua, S., Fabbro, S., Goobar, A., Groom, D.E. and Hook, I.M., 1999. Measurements of Ω and Λ from 42 high-redshift supernovae. The Astrophysical Journal, $517(2)$, p.565.

[3] Hooft, G.T., 1993. Dimensional reduction in quantum gravity. arXiv preprint gr-qc/9310026.

[4] Susskind, L., 1995. The world as a hologram. Journal of Mathematical Physics, $36(11)$, pp.6377-6396.

[5] Bousso, R., 1999. A covariant entropy conjecture. Journal of High Energy Physics, $1999(07)$, p.004.

[6] Cohen, A.G., Kaplan, D.B. and Nelson, A.E., 1999. Effective field theory, black holes, and the cosmological constant. Physical Review Letters, $82(25)$, p.4971.

[7] Wang, S., Wang, Y. and Li, M., 2017. Holographic dark energy. Physics reports, 696, pp.1-57.

[8] Nojiri, S.I. and Odintsov, S.D., 2006. Unifying phantom inflation with late-time acceleration: Scalar phantom–non-phantom transition model and generalized holographic dark energy. General Relativity and Gravitation, 38, pp.1285-1304.

[9] Nojiri, S.I. and Odintsov, S.D., 2017. Covariant generalized holographic dark energy and accelerating universe. The European Physical Journal C, 77, pp.1-8.

[10] Nojiri, S.I., Odintsov, S.D. and Paul, T., 2021. Different faces of generalized holographic dark energy. Symmetry, $13(6)$, p.928.

[11] Nojiri, S.I., Odintsov, S.D., Oikonomou, V.K. and Paul, T., 2020. Unifying holographic inflation with holographic dark energy: A covariant approach. Physical Review D, $102(2)$, p.023540.

[12] Li, M., 2004. A model of holographic dark energy. Physics Letters B, $603(1\text{-}2)$, pp.1-5.

[13] Karami, K. and Fehri, J., 2010. Holographic dark energy in a non-flat universe with Granda-Oliveros cut-off. International Journal of Theoretical Physics, 49, pp.1118-1126.

[14] Granda, L.N. and Oliveros, A., 2008. Infrared cut-off proposal for the holographic density. Physics Letters B, **669**(5), pp.275-277.

[15] Granda, L.N. and Oliveros, A., 2009. New infrared cut-off for the holographic scalar fields models of dark energy. Physics Letters B, **671**(2), pp.199-202.

[16] Nojiri, S.I. and Odintsov, S.D., 2011. Unified cosmic history in modified gravity: from $F(R)$ theory to Lorentz non-invariant models. Physics Reports, **505**(2-4), pp.59-144.

[17] Benaoum, H., 2022. Accelerated universe from modified Chaplygin gas and tachyonic fluid. Universe, **8**(7), p.340.

[18] Sultana, S., Chakraborty, G. and Chattopadhyay, S., 2022. Study of the effect of thermal radiation on singularities in presence of viscous holographic Ricci dark energy and modified Chaplygin gas. International Journal of Geometric Methods in Modern Physics, **19**(11), p.2250179.

[19] Setare, M.R., 2007. Holographic Chaplygin gas model. Physics Letters B, **648**(5-6), pp.329-332.

[20] Sultana, S. and Chattopadhyay, S., 2023. Cosmology of a Chaplygin Gas Model Under f (T) Gravity and Evolution of Primordial Perturbations. Research in Astronomy and Astrophysics, **23**(6), p.065016.

[21] Nojiri, S.I. and Odintsov, S.D., 2004. Final state and thermodynamics of a dark energy universe. Physical Review D, **70**(10), p.103522.

[22] Sultana, S. and Chattopadhyay, S., 2023. An examination of the logotropic equation of state for Chaplygin gas and a generalized holographic model, as well as thermodynamic analysis utilizing scalar fields. Annals of Physics, **455**, p.169392.

[23] Bekenstein, J.D., 1973. Black holes and entropy. Physical Review D, **7**(8), p.2333.

[24] Davies, P.C.W., 1987. Class. Quantum Grav. 4 L225 IOPscience Google Scholar Davies PCW 1988. Class. Quantum Grav, **5**(1349), p.2006.

[25] Chavanis, P.H., 2016. The Logotropic Dark Fluid as a unification of dark matter and dark energy. Physics Letters B, **758**, pp.59-66.

[26] Chavanis, P.H., 2015. Is the Universe logotropic?. The European Physical Journal Plus, **130**, pp.1-44.

[27] Sen, A. and Raja Sekhar, T., 2021. The limiting behavior of the Riemann solution to the isentropic Euler system for logarithmic equation of state with a source term. Mathematical Methods in the Applied Sciences, **44**(8), pp.7207-7227.

[28] Sultana, S. and Chattopadhyay, S., 2023. Cosmology of a Chaplygin Gas Model Under f (T) Gravity and Evolution of Primordial Perturbations. Research in Astronomy and Astrophysics, **23**(6), p.065016.

[29] Feng, B., Wang, X. and Zhang, X., 2005. Dark energy constraints from the cosmic age and supernova. Physics Letters B, **607**(1-2), pp.35-41.

[30] Guo, Z.K., Piao, Y.S., Zhang, X. and Zhang, Y.Z., 2005. Cosmological evolution of a quintom model of dark energy. Physics Letters B, **608**(3-4), pp.177-182.

[31] Setare, M.R., 2006. Generalized second law of thermodynamics in quintom dominated universe. Physics Letters B, **641**(2), pp.130-133.

[32] Sadjadi, H.M., 2006. Generalized second law in a phantom-dominated universe. Physical Review D, **73**(6), p.063525.

[33] Davies, P.C.W., 1987. Class. Quantum Grav. 4 L225 IOPscience Google Scholar Davies PCW

1988. Class. Quantum Grav, **5**(1349), p.2006.

[34] Sadjadi, H.M., 2006. Generalized second law in a phantom-dominated universe. Physical Review D, **73**(6), p.063525.

[35] Davies, P.C.W., 1988. Class. Quantum Grav. 5 1349 Crossref Google Scholar Davies PCW 1987. Class. Quantum Grav, **4**, p.L255.

[36] Elizalde, E., Nojiri, S.I. and Odintsov, S.D., 2004. Late-time cosmology in a (phantom) scalar-tensor theory: dark energy and the cosmic speed-up. Physical Review D, **70**(4), p.043539.

[37] Nojiri, S.I. and Odintsov, S.D., 2003. Quantum de Sitter cosmology and phantom matter. Physics Letters B, **562**(3-4), pp.147-152.

[38] Sadjadi, H.M., 2006. Generalized second law in a phantom-dominated universe. Physical Review D, **73**(6), p.063525.

[39] Izquierdo, G. and Pavón, D., 2006. Dark energy and the generalized second law. Physics Letters B, **633**(4-5), pp.420-426.

[40] Nojiri, S.I. and Odintsov, S.D., 2006. Unifying phantom inflation with late-time acceleration: Scalar phantom–non-phantom transition model and generalized holographic dark energy. General Relativity and Gravitation, **38**, pp.1285-1304.

[41] Nojiri, S.I., Odintsov, S.D. and Paul, T., 2021. Different faces of generalized holographic dark energy. Symmetry, **13**(6), p.928.

[42] Gadbail, G.N., Arora, S. and Sahoo, P.K., 2022. Generalized Chaplygin gas and accelerating universe in $f(Q, T)$ gravity. Physics of the Dark Universe, **37**, p.101074.

[43] Barreiro, T. and Sen, A.A., 2004. Generalized Chaplygin gas in a modified gravity approach. Physical Review D, **70**(12), p.124013.

[44] Nojiri, S.I., Odintsov, S.D. and Paul, T., 2021. Different faces of generalized holographic dark energy. Symmetry, **13**(6), p.928.

[45] Nojiri, S.I. and Odintsov, S.D., 2006. Unifying phantom inflation with late-time acceleration: Scalar phantom–non-phantom transition model and generalized holographic dark energy. General Relativity and Gravitation, **38**, pp.1285-1304.

Part III

Gravitational and Quantum Theories

Eric Ling and Annachiara Piubello (Eds), SPACETIME 1908-2023. Selected peer-reviewed papers presented at the *Third Hermann Minkowski Meeting on the Foundations of Spacetime Physics*, 11-14 September 2023, Albena, Bulgaria (Minkowski Institute Press, Montreal 2024). ISBN 978-1-998902-25-5 (softcover), ISBN 978-1-998902-26-2 (ebook).

8 Wide Binaries, Retardation and the External Field Effect

Asher Yahalom

Abstract Recently a low acceleration gravitational anomaly was reported for wide binaries, that is a system of two binary stars which are separated by more than five kilo astronomical units (kau). The increase in gravitation force was reported to be about 1.37 greater than Newtonian gravity. At the same time binaries which are not wide that is of separation less than one kau where shown to obey standard Newtonian gravity. A possible explanation for this was given in the frame work of MOND correction to gravity which is applicable for low acceleration. However, it was noticed that the explanation is only adequate in the frame work of Milgromian AQUAL theory which takes into account the "external field effect" that is the effect of the gravitational field of the rest of the galaxy on the binary system. Recently it was shown that many "anomalous gravity" effects can be explained in the frame work of general relativity and its weak field approximation. It was shown that anomalous galaxy rotation curves, anomalous gravitation lensing and the Tully-Fisher relations can be understood in terms of retarded gravity. Moreover, the anomalous mass of galaxy clusters derived from the virial theorem was also shown to be a retarded gravity effect. On the other hand it was shown that retarded effects are not extremely important in the solar system and the needed correction to the anomalous shift of Mercury's perihelion are connected with the motion of the solar center of mass with respect to the sun and not to retardation. It is thus desirable to investigate if the wide binary acceleration gravitational anomaly can be also explained in the frame-work of Einstein-Newton general relativity. We show that the scale of five kau separation arises naturally within such a theory if the effect of the galaxy on the wide binary is considered (a Newtonian external field effect) suggesting the anomaly found can be explained without the need to invoke theoretical modifications of the accepted theory.

Keywords: Wide Binaries, Retardation, External Field Effect

1 Introduction

It was suggested [1] that the behaviour of gravity at low accelerations can be directly probed by wide binaries (widely separated stars and thus long-period, but still gravitationally bound binary stars) since any imaginable dark matter in the Milky Way may not deform their internal dynamics (e.g., [23, 4, 30, 2, 31, 22, 15, 13, 21, 32, 19]). Thus using the recent data provided by the Gaia satellite multiple statistical analyses have been reported with the purpose of detecting gravity effects which differ from Newtonian predictions ([7, 3, 9, 24]) based on the Gaia data release 3 [37].

Chae [7] has suggested an algorithm [8] that calculates the probability distribution of a kinematic acceleration assuming it is of the form $g = v^2/r$ with respect to the Newtonian acceleration g_N between the two stars where v is the relative velocity and r is the separation in the three-dimensional real space, and compares it with a naive Newtonian prediction. One key aspect of this algorithm is that the occurrence rate (f_{multi}) of multiplicity higher than two (i.e. harboring hidden additional

Eric Ling and Annachiara Piubello (Eds), Spacetime 1908-2023. Selected peer-reviewed papers presented at the *Third Hermann Minkowski Meeting on the Foundations of Spacetime Physics*, 11-14 September 2023, Albena, Bulgaria (Minkowski Institute Press, Montreal 2024). ISBN 978-1-998902-25-5 (softcover), ISBN 978-1-998902-26-2 (ebook).

components) can be self-calibrated at a high enough acceleration and checked at another high acceleration. Through this algorithm, to be referred to as the "acceleration-plane analysis", Chae [7] found that the observed acceleration started to get boosted from the Newtonian prediction for $g_N \leq 10^{-9}$ m s^{-2} with a boost factor of ≈ 1.4 for $g_N \leq 10^{-10}$ m s^{-2}, at an extremely high ($> 5\sigma$) statistical significance. Chae [7] considered various samples by varying selection criteria and noted that the low-acceleration gravitational anomaly persisted.

Chae [9] further considered a sample of statistically pure binaries (i.e. the limiting case of $f_{\mathrm{multi}} = 0$) for an independent test. For this sample Chae [9] employed another algorithm to be referred to as "stacked velocity profile analysis" in addition to the acceleration-plane analysis. The stacked velocity profile analysis compares the observed distribution of the sky-projected relative velocities against the sky-projected separations with the corresponding Newtonian prediction. Chae [9] found that the results for the pure binary sample through the two independent algorithms agreed well with each other and the [7] results for general or "impure" samples. It follows that the gravitational modifications at a small acceleration does not depend even on a significant variation of the sample selection criteria, that is between $f_{\mathrm{multi}} \sim 0.5$ [7] and $f_{\mathrm{multi}} = 0$ [9].

Hernandez [24] also performed a statistical analysis of the distribution of velocities in a pure binary sample that was selected by the same author [19]. Hernandez defined the normalized velocity on the sky plane [30, 4] by:

$$\tilde{v} \equiv \frac{v_p}{v_c(s)}, \tag{1.1}$$

v_p is the observed sky-projected relative velocity between the pair and $v_c(s)$ is the theoretical Newtonian circular velocity at the sky-projected separation s. The Hernandez [24] algorithm and sample are completely independent of Chae's [7, 9] although the data is from the same Gaia DR3 database. Hernandez [24] obtained a gravitational anomaly that was well consistent with those obtained by Chae [7, 9].

The papers of Chae and Hernandez [7, 9, 24] show that the gravitational acceleration is increased by $\approx 1.3 - 1.5$ and the relative velocity is boosted by a factor of ≈ 1.2 for $s \geq 5$ kau. This was interpreted as connected to the MONDian framework since in this case $g_N \leq 10^{-10}$ m s^{-2} and the MONDian acceleration figure of merit is: $a_0 \simeq 1.2 \ 10^{-10}$ m s^{-2}. Despite these repeating results from different samples of wide binary star systems with different methods, Banik [3] claimed an opposite conclusion based on another statistical method and doubted the data quality control in Chae's [7] samples.

However, Hernandez [20] criticised the claims of Banik. The main fault according to Hernandez is the exclusion by Banik of the Newtonian-regime ($g_N > 10^{-9}$ m s^{-2} or $s < 2$ kau) binaries that are essential for an accurate determination of f_{multi}. Then, Banik employed a distribution fitting (number count) of $\tilde{v}$ (Equation 1.1) in cells in an attempt to simultaneously constrain gravity, f_{multi}, and additional parameters but without the Newtonian-regime data. In such a method they improperly defined cells smaller than the errors of $\tilde{v}$ although Banik relies on number counts in cells [20].

The concern of data quality [3] due to uncertainty of $\tilde{v}$ has no consequence for the analysis of Chae [7] because $\tilde{v}$ is not used, but probability distributions of g and g_N are directly derived on the acceleration plane with sufficiently precise projected relative velocities, and considering ranges of parameters such as eccentricity, inclination, and orbital phase in deprojection to the three-dimensional space. The pure binary sample of Chae [9] admits high signal-to-noise ratios ($S/N \geq 10$) for $\tilde{v}$. The acceleration-plane analysis returns similar results for the gravitational anomaly for $g_N \leq 10^{-9}$ m s^{-2}.

From an empirical perspective general relativity (GR) is known to be verified by many different types of observations [44, 43, 67, 53, 54]. However, currently Einstein's general relativity is at a rather difficult position. It has much support from observational evidence while also having serious challenges. The observational verifications it has gained in both cosmology and astrophysics are in doubt due to the fact that it needs to include unconfirmed ingredients, dark matter and energy, in order to achieve successes on the larger scales of galaxies, clusters of galaxies and universe as a whole. In most cases the unconfirmed ingredient is used while at the same time practitioners neglect a major ingredient of general relativity, the phenomenon of retardation, that negates Newtonian action at a distance.

Indeed, the dark matter enigma has not only been with the astronomical community since the 1930s (or perhaps even since the 1920s when it was known as the question of missing mass), but it has become prevalent as more dark matter (and a more serious neglect of retardation) has had to be postulated on larger scales as those scales were scrutinized. A very detailed and costly forty-years underground and accelerator search failed to prove its existence. The dark matter enigma has become even more problematic in recent years as the Large Hadron Collider failed to find any super symmetric particles, not only the astro-particle community's preferred form of dark matter, but an essential ingredient that string theory needs, and it is that same string theory that is expected to quantize gravity.

As early as 1933, Zwicky noted that a group of galaxies within the Comma Cluster have velocities that are significantly higher than that predicted by the virial calculations based on Newtonian theory [83, 42]. He calculated that the amount of matter required to account for the velocities could be 400 times greater than that of visible matter (this was later mitigated somewhat). Of course, if Zwicky would have use the retarded gravity version of the virial theory [81] no significant problem would arise. In 1959, on a smaller galactic scale, Volders observed that stars in the outer rims of a nearby spiral galaxy (M33) do not move as they should [64]. That is to say that the velocities do not decrease as $1/\sqrt{r}$. This discrepancy was further established in later years. During the seventies Rubin and Ford [57, 58] have shown that for a rather large sample of spiral galaxies the velocities at the outer rim of galaxies do not decrease. Rather, in the general case, they attain a plateau (or continue increasing) at some velocity different for each galaxy. In previous works it was shown that such velocity curves can be deduced directly from GR if retardation is not neglected. The derivation of the retardation effect is described in previous publications [72, 70, 71, 65, 66, 73, 75, 74]. The mechanism is strongly connected to the dynamics of the density of matter inside the galaxy, or more specifically to the densities' second derivative. The density can change due to the depletion of gas in the galaxies surrounding (in which case the second derivative of the galaxies' total mass is negative [73]) but can also be affected by dynamical processes involving star formation and supernovae explosions [65, 66]. It was determined that all possible processes can be captured by three different length scales: the typical length of the density gradient, the typical length of the velocity field gradient and the dynamical length scale. It is the shortest among those length scales that determine the significance of retardation [76].

The famous relation of Tully and Fisher [62] connecting the baryonic mass of a galaxy to the fourth power of its rim velocity can also be deduced from retarded gravity [77].

Retarded gravity does not affect only slowly moving particles but also photons. Although the mathematical analysis is slightly different in both cases [76, 79], it is concluded that the apparent "dark mass" must be the same as in the galactic rotation curves.

While the standard dark matter paradigm may still prevail, the current situation is alarming enough to contemplate the possibility that this prevailing paradigm might at least be reconsidered. Thus there is indeed room for the present suggestion. Unlike other theories that suggest modifying general relativity such as Milgrom's MOND [51], Mannheim's Conformal Gravity [48, 49] or Moffat's MOG [52], the current retarded gravity approach does not do so. We adhere strictly to Occam's razor, as suggested by Newton and Einstein. Retarded gravity replaces dark matter with effects within the standard General Relativity. Notice, however, that the connection between retardation and MOND was recently elucidated [82], showing in what sense low acceleration MOND criteria can be derived from retardation theory and how MOND interpolation function can be a good approximation to retarded gravity.

We emphasize that appreciable retardation effects do not require that velocities of matter in the galaxy are high (although this may help), in fact the vast majority of galactic bodies (stars, gas) are slow with respect to the speed of light. In other words, the ratio of $\frac{v}{c} \ll 1$. Typical velocities in galaxies are 100 km/s, which makes this ratio 0.001 or smaller. To obtain appreciable retardation effects what is needed is a small typical gradient scale with respect to the size of the system [76]. It was shown that retardation effects may become significant even at low speeds, provided that the distance over a typical length scale is large enough.

Within the solar system retardation effects are not appreciable [78, 80]. However, galaxies' velocity curves indicate that the retardation effects cannot be neglected beyond a certain distance [65, 66, 73]. The purpose of this study is to establish the situation in interim gravitational system that is wide binaries, in which distances are larger than the typical distances in the solar system but smaller than the typical size of a galaxy.

2   Beyond the Newtonian Approximation

Retarded gravity can be obtained from the weak field approximation to general relativity [73]. The metric perturbation h_{00} can be given in term of a retarded potential ϕ as follows [73, 76]:

$$\phi = -G \int \frac{\rho(\vec{x}', t - \frac{R}{c})}{R} d^3 x', \quad \phi \equiv \frac{c^2}{2} h_{00}, \quad h_{00} = \frac{2}{c^2} \phi \tag{2.1}$$

In the above G is the gravitational universal constant, $\vec{x}$ is where the potential is measured, $\vec{x}'$ is the location of the mass element generating the potential, $\vec{R} \equiv \vec{x} - \vec{x}'$, $R \equiv |\vec{R}|$, and ρ is the mass density. The duration $\frac{R}{c}$ for galaxies may be a few tens of thousands of years, but can be considered short in comparison to the time taken for the galactic density to change significantly. Similarly for clusters of galaxies the duration $\frac{R}{c}$ for galaxies may be a few tens of millions of years, but can be considered short in comparison to the time taken for the galactic cluster density to change significantly. Thus, we can write a Taylor series for the density:

$$\rho(\vec{x}', t - \frac{R}{c}) = \sum_{n=0}^{\infty} \frac{1}{n!} \rho^{(n)}(\vec{x}', t)(-\frac{R}{c})^n, \quad \rho^{(n)} \equiv \frac{\partial^n \rho}{\partial t^n}. \tag{2.2}$$

108

By inserting Equations (2.2) into Equation (2.1) and keeping the first three terms, we will obtain:

$$\phi = -G \int \frac{\rho(\vec{x}',t)}{R} d^3x' + \frac{G}{c} \int \rho^{(1)}(\vec{x}',t) d^3x' - \frac{G}{2c^2} \int R\rho^{(2)}(\vec{x}',t) d^3x' \qquad (2.3)$$

The Newtonian potential is the first term:

$$\phi_N = -G \int \frac{\rho(\vec{x}',t)}{R} d^3x' \qquad (2.4)$$

the second term does not contribute to the force affecting subluminal particles as its gradient is null and the third term is the lower order correction to the Newtonian potential:

$$\phi_r = -\frac{G}{2c^2} \int R\rho^{(2)}(\vec{x}',t) d^3x' \qquad (2.5)$$

The geodesic equation for a any "slow" test particle moving under the above space-time metric can be approximated [73] using the force per unit mass as follows:

$$\frac{d\vec{v}}{dt} = \vec{g}, \qquad \vec{g} \equiv -\vec{\nabla}\phi \qquad (2.6)$$

The total force per unit mass is thus:

$$\begin{aligned} \vec{g} &= \vec{g}_N + \vec{g}_r \\ \vec{g}_N &\equiv -\vec{\nabla}\phi_N = -G \int \frac{\rho(\vec{x}',t)}{R^2} \hat{R} d^3x', \quad \hat{R} \equiv \frac{\vec{R}}{R}, \\ \vec{g}_r &\equiv -\vec{\nabla}\phi_r = -\frac{G}{2c^2} \int \rho^{(2)}(\vec{x}',t) \hat{R} d^3x' \end{aligned} \qquad (2.7)$$

Now consider a point particle which has a mass m_j and is located at $\vec{r}_j(t)$, such a particle will have a mass density of:

$$\rho_j = m_j \delta^{(3)}(\vec{x}' - \vec{r}_j(t)) \qquad (2.8)$$

in which $\delta^{(3)}$ is a three dimensional Dirac delta function. This particle will cause a Newtonian potential:

$$\phi_{Nj} = -G \frac{m_j}{R_j(t)}, \qquad \vec{R}_j(t) = \vec{x} - \vec{r}_j(t), \qquad R_j(t) = |\vec{R}_j(t)| \qquad (2.9)$$

and a retardation potential of the form:

$$\begin{aligned} \phi_{rj} &= -\frac{Gm_j}{2c^2} \frac{\partial^2}{\partial t^2} R_j(t) = \frac{Gm_j}{2c^2} \left(\hat{R}_j \cdot \vec{a}_j - \frac{\vec{v}_j^2 - (\vec{v}_j \cdot \hat{R}_j)^2}{R_j(t)} \right), \\ \hat{R}_j &\equiv \frac{\vec{R}_j}{R_j}, \; \vec{v}_j \equiv \frac{d\vec{r}_j}{dt}, \; \vec{a}_j \equiv \frac{d\vec{v}_j}{dt}. \end{aligned} \qquad (2.10)$$

Thus any point particle moving at the vicinity of particle j will be accelerated as follows:

$$\begin{aligned} \vec{g}_j &= \vec{g}_{Nj} + \vec{g}_{rj} \\ \vec{g}_{Nj} &= -\vec{\nabla}\phi_{Nj} = -G \frac{m_j}{R_j^2} \hat{R}_j, \\ \vec{g}_{rj} &= -\vec{\nabla}\phi_r = \frac{Gm_j}{2R_j^2 c^2} \left(R_j \vec{a}_{\perp j} + \hat{R}_j \vec{v}_{\perp j}^2 - 2(\vec{v}_j \cdot \hat{R}_j)\vec{v}_{\perp j} \right) \\ \vec{a}_{\perp j} &\equiv \vec{a}_j - (\vec{a}_j \cdot \hat{R}_j)\hat{R}_j, \qquad \vec{v}_{\perp j} \equiv \vec{v}_j - (\vec{v}_j \cdot \hat{R}_j)\hat{R}_j. \end{aligned} \qquad (2.11)$$

Notice the different notation used for the test particle acceleration $\vec{g}_j$ and the gravity source acceleration $\vec{a}_j$.

Now consider a point particle of mass m_k which is located at $\vec{r}_k(t)$, this particle will feel the force:

$$
\begin{aligned}
\vec{F}_{j,k} &= \vec{F}_{Nj,k} + \vec{F}_{rj,k} \\
\vec{F}_{Nj,k} &= -G\frac{m_j m_k}{R_{k,j}^2}\hat{R}_{k,j}, \qquad \vec{R}_{k,j} \equiv \vec{r}_k - \vec{r}_j, \\
R_{k,j} &\equiv |\vec{R}_{k,j}(t)|, \ \hat{R}_{k,j} \equiv \frac{\vec{R}_{k,j}}{R_{k,j}}, \\
\vec{F}_{rj,k} &= \frac{Gm_j m_k}{2R_{k,j}^2 c^2}\left(R_{k,j}\vec{a}_{\perp j,k} + \hat{R}_{k,j}\vec{v}_{\perp j,k}^2 - 2(\vec{v}_j \cdot \hat{R}_{k,j})\vec{v}_{\perp j,k} \right) \\
\vec{a}_{\perp j,k} &\equiv \vec{a}_j - (\vec{a}_j \cdot \hat{R}_{k,j})\hat{R}_{k,j}, \\
\vec{v}_{\perp j,k} &\equiv \vec{v}_j - (\vec{v}_j \cdot \hat{R}_{k,j})\hat{R}_{k,j}. \tag{2.12}
\end{aligned}
$$

We notice once again (see [75]) that while Newtonian forces are prominent at "short" distances the retardation forces are the most significant at large distances in which it drops as $\frac{1}{R}$ and this fact is not related to the Taylor series approximation that we have used here. Now let us consider the gravitational effect of particle k on particle j, this is easily calculated by exchanging the indices j and k in the above expression. As $R_{j,k} = R_{k,j}$ but $\hat{R}_{j,k} = -\hat{R}_{k,j}$ it follows that the Newtonian force satisfies Newton's third law: $\vec{F}_{Nk,j} = -\vec{F}_{Nj,k}$, however, since there is no simple relation between the velocity and acceleration of the particle j and k it follows that generally speaking $\vec{F}_{rk,j} \neq -\vec{F}_{rj,k}$ and thus both the retardation force and the total gravitational force do not satisfy Newton's third law. This is well known in electromagnetism and discussed in a series of papers [84, 85, 86, 87, 88, 89].

3 Retardation Corrections

Let us establish the conditions under which the retardation correction is important. We shall write the retardation force in equation (2.12) as:

$$
\begin{aligned}
\vec{F}_{rj,k} &= \frac{1}{2}F_{Nj,k}\left(\frac{R_{k,j}\vec{a}_{\perp j,k}}{c^2} + \hat{R}_{k,j}\frac{v_{\perp j,k}^2}{c^2} - 2\frac{(\vec{v}_j \cdot \hat{R}_{k,j})\vec{v}_{\perp j,k}}{c^2} \right), \\
F_{Nj,k} &\equiv |\vec{F}_{Nj,k}| \tag{3.1}
\end{aligned}
$$

If the velocity of particle j is non relativistic that is much smaller than the speed of light we have:

$$
\vec{F}_{rj,k} \simeq \frac{1}{2}F_{Nj,k}\left(\frac{R_{k,j}\vec{a}_{\perp j,k}}{c^2} \right). \tag{3.2}
$$

It thus follows that this is the correct expression for most gravitational systems including the wide binary stars discussed by Chae [1]. Thus this retarded force can be neglected if:

$$
\frac{1}{2}\left(\frac{R_{k,j}a_{\perp j,k}}{c^2} \right) \ll 1 \quad \Rightarrow \quad R_{k,j} \ll R_c \equiv \frac{2c^2}{a_j} < \frac{2c^2}{a_{\perp j,k}} \tag{3.3}
$$

The critical distance R_c is acceleration dependent it will be relatively small for highly accelerated bodies and much larger for non accelerating bodies. Of course to know what the acceleration is one must well define the inertial system otherwise this notion is arbitrary. The same statement is of course also correct for MOND type theories which also rely on acceleration to calculate the force, and even more so as in the MONDian case a slight acceleration can make a huge difference in force calculations. The other extreme:

$$R_{k,j} \geq R_c \qquad \text{Or} \qquad R_{k,j} \simeq R_c \qquad (3.4)$$

signifies the regime in which retardation forces are important. Thus the gravitational interaction between a galactic central star moving at high acceleration and a distant gas atom at the outskirts of a galaxy is more likely to be affected by retardation than the gravitational interaction of a low acceleration binary star system even if widely separated. We will put this assertion in quantitative terms in later sections.

4 Binary Systems

Let us consider a system of N particles, each particle will feel a force generated by all other particles. $\vec{F}_k$ is the total force acting on particle k by all the other particles:

$$\vec{F}_k \equiv \sum_{j=1, j \neq k}^{N} \vec{F}_{j,k} \qquad (4.1)$$

Without the loss of generality we shall separate a binary subsystem from the system of N particles, made out of the particle 1 and 2. We thus write the forces acting on particle 1 and 2 as:

$$\vec{F}_1 = \vec{F}_{2,1} + \vec{F}_{e1}, \qquad \vec{F}_{e1} \equiv \sum_{j=3}^{N} \vec{F}_{j,1}$$

$$\vec{F}_2 = \vec{F}_{1,2} + \vec{F}_{e2}, \qquad \vec{F}_{e2} \equiv \sum_{j=3}^{N} \vec{F}_{j,2} \qquad (4.2)$$

If the particles are non-relativistic we may write the following equations of motion:

$$m_1 \vec{a}_1 = \vec{F}_1 = \vec{F}_{2,1} + \vec{F}_{e1},$$
$$m_2 \vec{a}_2 = \vec{F}_2 = \vec{F}_{1,2} + \vec{F}_{e2}. \qquad (4.3)$$

We shall now define the center of mass and the relative position of the binary particles (stars) in the customary way:

$$\vec{R}_{cm} \equiv \frac{m_1 \vec{r}_1 + m_2 \vec{r}_2}{M}, \qquad \vec{r} \equiv \vec{R}_{2,1} = \vec{r}_2 - \vec{r}_1, \qquad M \equiv m_1 + m_2. \qquad (4.4)$$

Thus:

$$\vec{r}_1 = \vec{R}_{cm} + (\frac{m_1}{M} - 1)\vec{r} = \vec{R}_{cm} - \frac{m_2}{M}\vec{r},$$
$$\vec{r}_2 = \vec{R}_{cm} + (1 - \frac{m_2}{M})\vec{r} = \vec{R}_{cm} + \frac{m_1}{M}\vec{r}. \qquad (4.5)$$

The acceleration of each star in the binary system can be partitioned to a contribution from the center of mass acceleration and a relative acceleration as follows:

$$\vec{a}_1 = \ddot{\vec{r}}_1 = \ddot{\vec{R}}_{cm} - \frac{m_2}{M}\ddot{\vec{r}} = \vec{a}_{cm} - \frac{m_2}{M}\vec{a}, \qquad \vec{a}_{cm} \equiv \ddot{\vec{R}}_{cm}, \quad \vec{a} \equiv \ddot{\vec{r}}$$

$$\vec{a}_2 = \ddot{\vec{r}}_2 = \ddot{\vec{R}}_{cm} + \frac{m_1}{M}\ddot{\vec{r}} = \vec{a}_{cm} + \frac{m_1}{M}\vec{a}. \tag{4.6}$$

On the other hand we can calculate the center of mass acceleration as:

$$\vec{a}_{cm} = \frac{m_1\vec{a}_1 + m_2\vec{a}_2}{M} = \frac{1}{M}[\vec{F}_{1,2} + \vec{F}_{2,1} + \vec{F}_{e1} + \vec{F}_{e2}]$$

$$= \frac{1}{M}[\vec{F}_{r1,2} + \vec{F}_{r2,1} + \vec{F}_{e1} + \vec{F}_{e2}]. \tag{4.7}$$

In which the last equation sign follows from Newton's third law: $\vec{F}_{N1,2} = -\vec{F}_{N1,2}$. We can also calculate the relative acceleration as follows:

$$\vec{a} = \vec{a}_2 - \vec{a}_1 = \vec{g}_{1,2} - \vec{g}_{2,1} + \vec{g}_{e2} - \vec{g}_{e1}, \qquad \vec{g}_{j,k} \equiv \frac{\vec{F}_{j,k}}{m_k}, \quad \vec{g}_{ek} \equiv \frac{\vec{F}_{ek}}{m_k}. \tag{4.8}$$

5 Orders of Magnitude

We shall assume that the center of the galaxy is the same with the origin of axis of the inertial system. As in Chae [7, 9], we consider a sample of binaries within 200 pc from the Sun included in the El-Badry [16] database of 1.8 million binary candidates that are free of clusters, background pairs, and resolved (1″) triples or higher-order multiples. The sample of 81,880 binary candidates defined by Chae [7] has the range of relative distances of $r_{min} = 0.05$ kau $< r = |\vec{r}| < r_{max} = 50$ kau. Now $R_{cm} = |\vec{R}_{cm}|$ for each of the binary systems is the same order of magnitude as the distance of the sun from the galactic center which is about $R_{sun} \simeq 26,660$ ly $\simeq 8,174$ pc $\simeq 2.5 \ 10^{20}$ m. The relative distance of the stars in the binary system is much smaller as $r_{max} = 50$ kau $\simeq 0.24$ pc and the relative distance r is smaller than this number even for wide binaries. Thus for each binary system we have:

$$r \ll R_{cm}, r_1, r_2, \qquad \epsilon \equiv \frac{r}{R_{cm}} < \frac{r_{max}}{R_{cm}} \simeq 3 \ 10^{-5} \ll 1. \tag{5.1}$$

This leads to the picture depicted in figure 39. Now the sun circulates the center of the galaxy with a velocity of: $v_{sun} \simeq 251 \ 10^3$ m/s leading to a center of mass acceleration of about:

$$a_{cm} \simeq a_{sun} \equiv \frac{v_{sun}^2}{R_{sun}} \simeq 2.5 \ 10^{-10} \ \text{m/s}^2. \tag{5.2}$$

A number not far from the MONDian acceleration scale of $a_0 \simeq 1.2 \ 10^{-10}$ m/s². The relative acceleration will of course depend on the distance r between the stars in the binary system, thus for order of magnitude estimation only:

$$a_{min} < a < a_{max},$$
$$a_{max} \equiv \frac{GM_\odot}{r_{min}^2} \simeq 2.4 \ 10^{-6} \ \text{m/s}^2,$$
$$a_{min} \equiv \frac{GM_\odot}{r_{max}^2} \simeq 2.4 \ 10^{-12} \ \text{m/s}^2 \tag{5.3}$$

112

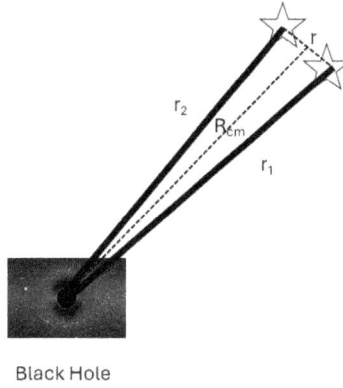

Figure 39: A schematic binary system. The distance r is much exaggerated with respect to R_{cm}.

in which we took as typical mass the solar mass $M_\odot \simeq 2 \; 10^{30}$ kg. It follows that according to equation (4.6) there is a difference between close binaries in which the relative acceleration of the system a dominates over a_{cm} and wide binaries in which the opposite may be true. Curiously enough the border line is at about the distance above which Chae [9] claimed gravitational anomalies which is $r_{gr\ anomaly} \simeq 5$ kau.

$$a_{gr\ anomaly} \equiv \frac{GM_\odot}{r_{gr\ anomaly}^2} \simeq 2.4 \; 10^{-10} \text{ m/s}^2. \tag{5.4}$$

The question of the importance of retardation within the binary system is resolved by equation (3.3) and depends on the relation between the typical size of the system and the typical acceleration. For close binaries the critical radius is:

$$R_{c\ max} \simeq \frac{2c^2}{a_{max}} \simeq 5 \; 10^8 \text{ kau} \tag{5.5}$$

which is clearly much bigger than a typical size of a close binary system and thus retardation can be neglected for such systems. For wide binaries the importance shifts according to equation (4.6) from internal accelerations to center of mass accelerations. Leading to a much higher critical radius:

$$R_{c\ wide} \simeq \frac{2c^2}{a_{cm}} \simeq 4.8 \; 10^{12} \text{ kau} \tag{5.6}$$

which is of course much bigger than a wide binary system making the gravitational retardation interaction of all binaries considered negligible. This does mean that gravitational retardation effects are not important for such systems, but what this means is that they are encapsulated in the $\vec{F}_e$ terms which take into account the gravity of distance stars far away from the binary system. With this in mind we may rewrite equation (4.7) for the center of mass acceleration as:

$$\vec{a}_{cm} = \frac{1}{M}[\vec{F}_{r1,2} + \vec{F}_{r2,1} + \vec{F}_{e1} + \vec{F}_{e2}] \simeq \frac{1}{M}[\vec{F}_{e1} + \vec{F}_{e2}]. \tag{5.7}$$

And also the relative acceleration as:

$$\vec{a} = \vec{g}_{1,2} - \vec{g}_{2,1} + \vec{g}_{e2} - \vec{g}_{e1} \simeq \vec{g}_{N1,2} - \vec{g}_{N2,1} + \vec{g}_{e2} - \vec{g}_{e1},$$

$$\vec{g}_{Nj,k} \equiv \frac{\vec{F}_{Nj,k}}{m_k}. \tag{5.8}$$

113

Now:

$$g_{N1,2} - \vec{g}_{N2,1} = \frac{\vec{F}_{N1,2}}{m_2} - \frac{\vec{F}_{N2,1}}{m_1} = \frac{\vec{F}_{N1,2}}{\mu} = -\frac{GM}{r^2}\hat{r},$$

$$\frac{1}{\mu} \equiv \frac{1}{m_1} + \frac{1}{m_2}, \quad \hat{r} \equiv \frac{\vec{r}}{r}. \tag{5.9}$$

It follows that we can write the relative acceleration as:

$$\vec{a} \simeq -\frac{GM}{r^2}\hat{r} + \vec{g}_{e2} - \vec{g}_{e1}. \tag{5.10}$$

6   Slowly Varying External Gravitational Fields

Let us estimate the external gravitational acceleration $\vec{g}_{e1}$ using equation (4.5), it follows that:

$$\vec{g}_{e1} = \vec{g}_e(\vec{r}_1) = \vec{g}_e(\vec{r}_1) = \vec{g}_e(\vec{R}_{cm} - \frac{m_2}{M}\vec{r}). \tag{6.1}$$

Now according to equation (5.1) $r << R_{cm}$ so if $\vec{g}_e$ is slowly varying we may approximate:

$$\vec{g}_{e1} \simeq \vec{g}_e(\vec{R}_{cm}) - \frac{m_2}{M}\vec{r} \cdot \vec{\nabla}\vec{g}_e(\vec{x})|_{\vec{R}_{cm}}. \tag{6.2}$$

Similarly:

$$\vec{g}_{e2} \simeq \vec{g}_e(\vec{R}_{cm}) + \frac{m_1}{M}\vec{r} \cdot \vec{\nabla}\vec{g}_e(\vec{x})|_{\vec{R}_{cm}}. \tag{6.3}$$

Inserting equation (6.2) and equation (6.3) into equation (5.7) we arrive at the simple form:

$$\vec{a}_{cm} \simeq \frac{1}{M}[\vec{F}_{e1} + \vec{F}_{e2}] = \frac{1}{M}[m_1\vec{g}_{e1} + m_2\vec{g}_{e2}] \simeq \vec{g}_e(\vec{R}_{cm}) \equiv \vec{g}_{e\ cm}. \tag{6.4}$$

Moreover:

$$\vec{g}_{e2} - \vec{g}_{e1} \simeq \vec{r} \cdot \vec{\nabla}\vec{g}_e(\vec{x})|_{\vec{R}_{cm}}. \tag{6.5}$$

Inserting equation (6.5) into equation (5.10) leads to:

$$\vec{a} \simeq -\frac{GM}{r^2}\hat{r} + \vec{g}_{e2} - \vec{g}_{e1} \simeq -\frac{GM}{r^2}\hat{r} + \vec{r} \cdot \vec{\nabla}\vec{g}_e(\vec{x})|_{\vec{R}_{cm}}. \tag{6.6}$$

We are now at a position to calculate the individual acceleration of each star in the binary system by plugging equation (6.4) and equation (6.6) into equation (4.6)

$$\vec{a}_1 = \vec{a}_{cm} - \frac{m_2}{M}\vec{a} \simeq \vec{g}_{e\ cm} + \frac{Gm_2}{r^2}\hat{r} - \frac{m_2}{M}\vec{r} \cdot \vec{\nabla}\vec{g}_e(\vec{x})|_{\vec{R}_{cm}},$$

$$\vec{a}_2 = \vec{a}_{cm} + \frac{m_1}{M}\vec{a} \simeq \vec{g}_{e\ cm} - \frac{Gm_1}{r^2}\hat{r} + \frac{m_1}{M}\vec{r} \cdot \vec{\nabla}\vec{g}_e(\vec{x})|_{\vec{R}_{cm}}. \tag{6.7}$$

Although it is quite clear that:

$$g_{e\ cm} \gg |\vec{r} \cdot \vec{\nabla}\vec{g}_e(\vec{x})|_{\vec{R}_{cm}}|, \tag{6.8}$$

let us study a simple model to show this quantitatively. Assume that the mass of the entire galaxy (excluding the binary system) is located at the origin, that is it is a giant black hole of mass M_G which is depicted in figure 39, in this case the external acceleration is:

$$\vec{g}_e(\vec{x}) = -\frac{GM_G}{x^3}\vec{x}, \tag{6.9}$$

It thus follows that:

$$\frac{\partial g_{e\ i}}{\partial x_j} = -\frac{GM_G}{x^3}(\delta_{ij} - 3\frac{x_i x_j}{x^2}) = -\frac{g_e}{x}(\delta_{ij} - 3\frac{x_i x_j}{x^2}). \tag{6.10}$$

in which δ_{ij} is a Kronecker delta. We can then easily calculate:

$$|\vec{r} \cdot \vec{\nabla}\vec{g}_e(\vec{x})|_{\vec{R}_{cm}}| = g_{e\ cm}\frac{r}{R_{cm}}|\hat{r} - 3(\hat{r} \cdot \hat{R}_{cm})\hat{R}_{cm}|$$
$$= g_{e\ cm}\epsilon|\hat{r} - 3(\hat{r} \cdot \hat{R}_{cm})\hat{R}_{cm}|. \tag{6.11}$$

were we used the definition of ϵ given in equation (5.1). We notice that:

$$|\hat{r} - 3(\hat{r} \cdot \hat{R}_{cm})\hat{R}_{cm}| = \sqrt{1 + 3(\hat{r} \cdot \hat{R}_{cm})^2} \quad \Rightarrow$$
$$1 \le |\hat{r} - 3(\hat{r} \cdot \hat{R}_{cm})\hat{R}_{cm}| \le 2. \tag{6.12}$$

Thus:

$$|\vec{r} \cdot \vec{\nabla}\vec{g}_e(\vec{x})|_{\vec{R}_{cm}}| \le 2g_{e\ cm}\epsilon \ll g_{e\ cm}, \tag{6.13}$$

as expected. We can thus simplify equation (6.7) as follows:

$$\vec{a}_1 \simeq \vec{g}_{e\ cm} + \frac{Gm_2}{r^2}\hat{r},$$
$$\vec{a}_2 \simeq \vec{g}_{e\ cm} - \frac{Gm_1}{r^2}\hat{r}. \tag{6.14}$$

This indicates that there is another critical distance (unrelated to retardation per se) below in which one may neglect all external effects in the system and consider it as "stand alone" in the universe.

$$g_{e\ cm} \ll \frac{Gm_i}{r^2}, \quad \Rightarrow \quad r \ll r_{ci} = \sqrt{\frac{Gm_i}{g_{e\ cm}}} \tag{6.15}$$

For a typical star with the sun mass this takes the value:

$$r_{c\odot} = \sqrt{\frac{Gm_\odot}{g_{e\ cm}}} \simeq 5 \text{ kau} \simeq r_{gr\ anomaly}. \tag{6.16}$$

Thus we see again that the scale in which gravitational anomaly is announced is also the scale in which the galaxy "interferes" in the affairs of specific binary systems which are wide enough to react to such interference.

7 Conclusion

The main results of the current paper are twofold, first we establish a criterion by which slow star systems may suffer retarded gravity effects. This is given by an "uncertainty relation" type formula:

$$R\, a > 2c^2 \tag{7.1}$$

demanding that a star j may affect a star k by retarded gravity only if their distance R times the acceleration of either of them denoted a is larger than twice the speed of light in vacuum square. This may provide an additional mechanism to retarded gravity in stable galaxies beyond the gas depletion mechanism described in [73] and the galactic wind mechanism described in [65, 66]. Hence we derived the uncertainty relation of retarded gravity. Second we have shown that within binary systems, retarded gravity is not important and thus may affect the binary star system only externally (that is by matter which is very far away from the binary stars). We have shown that anomalous gravity is reported only in binary systems in which the external gravitational pull is about the size of the internal gravitational pull. As we do not know the precise coordinates of each binary star but only the projection of its trajectory on the sky this requires careful consideration.

References

[1] Kyu-Hyun Chae, "Measurements of the Low-Acceleration Gravitational Anomaly from the Normalized Velocity Profile of Gaia Wide Binary Stars and Statistical Testing of Newtonian and Milgromian Theories" arXiv:2402.05720 [astro-ph.GA]

[2] Banik, I., Kroupa, P. 2019, Monthly Notices of the Royal Astronomical Society, 487, 1653

[3] Banik, I., Pittordis, C., Sutherland, W., Famaey, B., Ibata, R., Mieske, S., Zhao, H. 2024, MNRAR, in press (arXiv:2311.03436)

[4] Banik, I., Zhao, H. 2018, Monthly Notices of the Royal Astronomical Society, 480, 2660

[5] Bekenstein, J., Milgrom, M. 1984, Astrophysical Journal, 286, 7

[6] Chae, K.-H., Lelli, F., Desmond, H., McGaugh, S. S., Schombert, J. M. 2022, Physical Review D, 106, 103025

[7] Chae, K.-H. 2023a, Astrophysical Journal, 952, 128

[8] Chae, K.-H. 2023b, Python scripts to test gravity with the dynamics of wide binary stars v5, Zenodo

[9] Chae, K.-H. 2024, Astrophysical Journal, 960, 114

[10] Chae, K.-H., Desmond, H., Lelli, F., McGaugh, S. S., Schombert, J. M. 2021, ApJ, 921, 104

[11] Chae, K.-H., Lelli, F., Desmond, H., McGaugh, S. S., Li, P., Schombert, J. M. 2020, ApJ, 904, 51

[12] Chae, K.-H., Milgrom, M. 2022, Astrophysical Journal, 928, 24

[13] Clarke, C. J. 2020, Monthly Notices of the Royal Astronomical Society, 491, L72

[14] Einstein, A. 1916, Annalen der Physik, 354, 769

[15] El-Badry, K. 2019, Monthly Notices of the Royal Astronomical Society, 482, 5018

[16] El-Badry, K., Rix, H.-W., Heintz, T. M. 2021, Monthly Notices of the Royal Astronomical Society, 506, 2269

[17] Famaey, B., Binney, J. 2005, Monthly Notices of the Royal Astronomical Society, 363, 603

[18] Foreman-Mackey, D., Hogg, D. W., Lang, D., Goodman, J. 2013, Publications of the Astrophysical Society of Japan, 125, 306

[19] Hernandez, X. 2023, Monthly Notices of the Royal Astronomical Society, 525, 1401

[20] Hernandez, X., Chae, K.-H. 2023, arXiv:2312.03126

[21] Hernandez, X., Cookson, S., Cortés, R. A. M. 2022, Monthly Notices of the Royal Astronomical Society, 509, 2304

[22] Hernandez, X., Cortés, R. A. M., Allen, C., Scarpa, R. 2019, IJMPD, 28, 1950101

[23] Hernandez, X., Jiménez, M. A., Allen, C. 2012, EPJC, 72, 1884

[24] Hernandez, X., Verteletskyi, V., Nasser, L., Aguayo-Ortiz, A. 2024, MNRAS, in press (arXiv:2309.10995)

[25] Hwang, H.-C., Ting, Y.-S., Zakamska, N. L. 2022, Monthly Notices of the Royal Astronomical Society, 512, 3383

[26] McGaugh, S. S., Lelli, F., Schombert, J. M. 2016, Physical Review Letters, 117, 201101

[27] Milgrom, M. 1983, Astrophysical Journal, 270, 365

[28] Milgrom, M. 2010, Monthly Notices of the Royal Astronomical Society, 403, 886

[29] Pecaut, M. J., Mamajek, E. E. 2013, Astrophysical Journal Supplement, 208, 9

[30] Pittordis, C., Sutherland, W. 2018, Monthly Notices of the Royal Astronomical Society, 480, 1778

[31] Pittordis, C., Sutherland, W. 2019, Monthly Notices of the Royal Astronomical Society, 488, 4740

[32] Pittordis, C., Sutherland, W. 2023, OJAp, 6, 4

[33] Shaya, E. J., Olling, R. P. 2011, Astrophysical Journal Supplement, 192, 2

[34] Tokovinin, A. 2014, Astronomical Journal, 147, 87

[35] Tokovinin, A. 2022, Astrophysical Journal, 926, 1

[36] Tokovinin, A., Kiyaeva, O. 2016, Monthly Notices of the Royal Astronomical Society, 456, 2070

[37] Vallenari, A., Brown, A. G. A., Prusti, T., et al. (Gaia Collaboration) 2023, Astronomy and Astrophysics, 674, A1

[38] Wall, J. V., Jenkins, C. R. 2012, Practical Statistics for Astronomers (Cambridge University Press, 2nd ed.), Section 5.3.1

[39] Babcock H. W., 1939, Lick Observatory Bulletin 19, 41

[40] Abbott B. P., Abbott R., Abbott T. D. et al., 2016, Phys. Rev.

[41] Castelvecchi D., Witze W., Nature News 2016, doi:10.1038/nature.2016.19361.

[42] de Swart J. G., Bertone G., van Dongen J., Nature Astronomy, 2017, 1, 0059 Macmillan Publishers Limited https://doi.org/10.1038/s41550-017-0059

[43] Eddington A. S. , "The mathematical theory of relativity" Cambridge University Press (1923)

[44] Einstein A., Sitzungsberichte der Königlich Preussischen Akademie der Wissenschaften Berlin; Part 1; 1916; pp. 688–696. The Prusssian Academy of Sciences, Berlin, Germany.

[45] Feynman R. P., Leighton R. B., Sands M. L., Feynman Lectures on Physics, Basic Books; revised 50th anniversary edition (2011).

[46] Jackson J. D., Classical Electrodynamics, Third Edition. Wiley: New York, (1999).

[47] Landau L. D., 1975, The classical theory of fields, 4th edn. (Pergamon)

[48] Mannheim P. D., 1993, The Astrophysical Journal, 419, 150

[49] Mannheim P. D., 1996, Foundations of Physics, 26, 1683

[50] McGaugh S., McGaugh's Data Pages. N.p., n.d. 2017. Available online: http://astroweb.case.edu/ssm/data/ (accessed on 22 January 2017).

[51] Milgrom M., 1983, The Astrophysical Journal, 270, 371

[52] Moffat J. W., (2006). Journal of Cosmology and Astroparticle Physics. 2006 (3): 4. arXiv:gr-qc/0506021. doi:10.1088/1475-7516/2006/03/004.

[53] Misner C. W., Thorne K.S., Wheeler J.A., "Gravitation" W.H. Freeman & Company (1973)

[54] Narlikar, J. V. (1993). Introduction to Cosmology, Second Edition. Cambridge University Press.

[55] Navarro J. F., Frenk C. S., White S. D. M., (May 10, 1996). The Astrophysical Journal. 462: 563-575. arXiv:astro-ph/9508025. doi:10.1086/177173.

[56] Nobel Prize, A. Press Release The Royal Swedish Academy of Sciences; **1993**.The Royal Swedish Academy of Sciences, Stockholm, Sweden.

[57] Rubin V.C., Ford W.K. Jr., *Astrophys. J.*, vol. 159, 379, 1970.

[58] Rubin V.C., Ford W.K. Jr., Thonnard N., *Astrophysical Journal*, vol. 238, 471, 1980.

[59] Sancisi R., proceedings of IAU Symposium 220, "Dark Matter in Galaxies", eds. S. Ryder, D.J. Pisano, M. Walker and K. Freeman, Publ. Astron. Soc. Pac arXiv:astro-ph/0311348

[60] Sanders R., McGaugh S., Annu. Rev. Astron. Astrophys. 2002, 40, 217.

[61] Schwinger J., Lester L., DeRaad Jr K. W., 1998, Classical Electrodynamics, Advanced Book Program (Reading, Massacusetts: Perseus Books).

[62] Tully R. B., Fisher J. R., (1977). Astronomy and Astrophysics. 54 (3): 661–673.

[63] van Dokkum P., Danieli S., Cohen Y., Merritt A., Romanowsky A.J., Abraham R., Brodie J., Conroy C., Lokhorst D., Mowla L., et al. Nature **2018**, 555, 629–632, doi:10.1038/nature25767.

[64] Volders L.M.J.S., *Bull. astr. Inst. Netherl.*, vol. 14, 323, 1959. Rubin V.C., Ford Jr. W.K., Thonnard N., and Roberts M.S., *Astrophys. J.*, vol. 81, 687 and 719, 1976.

[65] Wagman M., Retardation Theory in Galaxies. Ph.D. Thesis, Senate of Ariel University, Samria, Israel, 23 September 2019.

[66] Wagman M., Horwitz L. P., Yahalom A., 2023 J. Phys.: Conf. Ser. 2482 012005. Proceedings of the 13th Biennial Conference on Classical and Quantum Relativistic Dynamics of Particles and Fields (IARD 2022), 05/06/2022 - 09/06/2022 Prague, Czechia. DOI 10.1088/1742-6596/2482/1/012005.

[67] Weinberg S., "Gravitation and Cosmology: Principles and Applications of the General Theory of Relativity" John Wiley & Sons, Inc. (1972)

[68] Yahalom A., Foundations of Physics, Volume 38, Number 6, Pages 489-497 (June 2008).

[69] Yahalom A., (2009). International Journal of Modern Physics D, Vol. 18, Issue: 14, pp. 2155-2158.

[70] Yahalom A., Retardation Effects in Electromagnetism and Gravitation. In Proceedings of the Material Technologies and Modeling the Tenth International Conference, Ariel University, Ariel, Israel, 20–24 August 2018. (arXiv:1507.02897v2)

[71] Yahalom A., Dark Matter: Reality or a Relativistic Illusion? In Proceedings of Eighteenth Israeli-Russian Bi-National Workshop 2019, Ein Bokek, Israel, 17–22 February 2019.

[72] Yahalom A., J. Phys.: Conf. Ser. 1239 (2019) 012006.

[73] Yahalom A., Symmetry 2020, 12(10), 1693; https://doi.org/10.3390/sym12101693

[74] Yahalom A., Proceedings of IARD 2020. 2021 J. Phys.: Conf. Ser. 1956 012002

[75] Yahalom A., Universe. 2021; 7(7):207. https://doi.org/10.3390/universe7070207. https://arxiv.org/abs/2108.08246

[76] Yahalom A., Symmetry 2021, 13, 1062. https://doi.org/10.3390/sym13061062. https://arxiv.org/abs/2108.04683.

[77] Yahalom A., International Journal of Modern Physics D, (2021), Volume No. 30, Issue No. 14, Article No. 2142008 (8 pages). ©World Scientific Publishing Company.https://doi.org/10.1142/S0218271822420184

[78] Yahalom A., Proceedings of the International Conference: COSMOLOGY ON SMALL SCALES 2022 Dark Energy and the Local Hubble Expansion Problem, Prague, September 21-24, 2022. Edited by Michal Krizek and Yuri V. Dumin, Institute of Mathematics, Czech Academy of Sciences.

[79] Yahalom A., IJMPD Vol. 31, No. 14, 2242018 (10 pages), received 23 May 2022, Accepted 31 August 2022, published online 30 September 2022.

[80] Yahalom A., Symmetry 2023, 15, 39. https://doi.org/10.3390/sym15010039

[81] Yahalom A., International Journal of Modern Physics D, 2342013, Received 12 May 2023, Accepted 3 July 2023, Published: 28 July 2023. https://doi.org/10.1142/S0218271823420130, https://www.worldscientific.com/doi/abs/10.1142/S0218271823420130

[82] Yahalom A., accepted by Bulgarian Journal of Physics vol. 50 (2023) 1–16.

[83] Zwicky F., In: Proc. Natl. Acad. Sci. U S A., May 1937, vol. 23(5), pp. 251–256.

[84] Tuval, M.; Yahalom, A. Newton's Third Law in the Framework of Special Relativity. Eur. Phys. J. Plus **2014**, <u>129</u>, 240, doi:10.1140/epjp/i2014-14240-x.

[85] Tuval, M.; Yahalom, A. Momentum Conservation in a Relativistic Engine. Eur. Phys. J. Plus **2016**, <u>131</u>, 374, doi:10.1140/epjp/i2016-16374-1.

[86] Yahalom, A. Retardation in Special Relativity and the Design of a Relativistic Motor. Acta Phys. Pol. A **2017**, <u>131</u>, 1285–1288.

[87] Shailendra Rajput, Asher Yahalom & Hong Qin "Lorentz Symmetry Group, Retardation and Energy Transformations in a Relativistic Engine" Symmetry 2021, 13, 420. https://doi.org/10.3390/sym13030420.

[88] Rajput, Shailendra, and Asher Yahalom. 2021. "Newton's Third Law in the Framework of Special Relativity for Charged Bodies" Symmetry 13, no. 7: 1250. https://doi.org/10.3390/sym13071250

[89] Yahalom, Asher. 2022. "Newton's Third Law in the Framework of Special Relativity for Charged Bodies Part 2: Preliminary Analysis of a Nano Relativistic Motor" Symmetry 14, no. 1: 94. https://doi.org/10.3390/sym14010094.

9

Torsion and Chern-Simons gravity in 4D space-times from a Geometrodynamical four-form

Patrick Das Gupta

Abstract In Hermann Minkowski's pioneering mathematical formulation of special relativity, the space-time geometry in any inertial frame is described by the line-element $ds^2 = \eta_{\mu\nu}dx^\mu dx^\nu$. It is interesting to note that not only the Minkowski metric $\eta_{\mu\nu}$ is invariant under proper Lorentz transformations, the totally antisymmetric Levi-Civita tensor $e_{\mu\nu\alpha\beta}$ too is.

In Einstein's general relativity (GR), $\eta_{\mu\nu}$ of the flat space-time gets generalized to a dynamical, space-time dependent metric tensor $g_{\mu\nu}$ that characterizes a curved space-time geometry. In the present study, it is put forward that the flat space-time Levi-Civita tensor gets elevated to a dynamical four-form field $\tilde{w}$ in curved space-time manifolds, i.e. $e_{\mu\nu\alpha\beta} \to w_{\mu\nu\alpha\beta}(x) = \phi(x)\, e_{\mu\nu\alpha\beta}$, so that $\tilde{w} = \frac{1}{4!}\, w_{\mu\nu\rho\sigma}\, \tilde{d}x^\mu \wedge \tilde{d}x^\nu \wedge \tilde{d}x^\rho \wedge \tilde{d}x^\sigma$. It is shown that this geometrodynamical four-form field extends GR by leading naturally to a torsion in the theory as well as to a Chern-Simons gravity. Furthermore, it is demonstrated that the scalar-density $\phi(x)$ associated with the geometrodynamical four-form $\tilde{w}$ may be used to construct a generalized exterior derivative that converts a p-form density to a (p+1)-form density of identical weight.

In order to relate the hypothesized four-form field $\tilde{w}$ to observational evidence, we first argue that the associated scalar-density $\phi(x)$ corresponds to an axion-like pseudo-scalar field in the Minkowski space-time, and that it can also masquerade as dark matter. Thereafter, we provide a simple semi-classical analysis in which a self-gravitating Bose-Einstein condensate of such ultra-light pseudo-scalars leads to the formation of a supermassive black hole. A brief analysis of propagation of weak gravitational waves in the presence of $\tilde{w}$ is also considered in this article.

Keywords: Torsion, Chern-Simons gravity, Dynamical four-form, Ultra-light Dark Matter, Supermassive Black holes, Bose-Einstein Condensation

1   Introduction

By envisaging a four dimensional manifold's space-time geometry be described by a line-element, $ds^2 = \eta_{\mu\nu}dx^\mu dx^\nu$, Hermann Minkowski had overturned the idea of treating space and time disjointly. Furthermore, the cause and effect relation, in the realm of classical physics, attained a significant elucidation in the Minkowskian framework: infinitesimally separated events (assuming a signature $(+ - - -)$) for which $ds^2 < 0$ $(ds^2 \geq 0)$ are causally disconnected (causally connected). Of course, the underlying physics involved in it is that an effect essentially follows a cause due to transfer of energy from the latter to the former, and that the rate of flow of energy is limited by the speed c. What is usually not stressed is that, because of this causality criteria, the imaginary numbers entered physics in a fundamental manner even before the advent of quantum mechanics, i.e. the proper distance ds between any two causally disconnected events is always an imaginary number.

Eric Ling and Annachiara Piubello (Eds), SPACETIME 1908-2023. Selected peer-reviewed papers presented at the *Third Hermann Minkowski Meeting on the Foundations of Spacetime Physics*, 11-14 September 2023, Albena, Bulgaria (Minkowski Institute Press, Montreal 2024). ISBN 978-1-998902-25-5 (softcover), ISBN 978-1-998902-26-2 (ebook).

Although space and time coordinates get mixed up while transiting from one inertial frame to another, the Minkowski metric $\eta_{\mu\nu}$ itself remains invariant. This is utilized in the Minkowskian framework to define an arbitrary Lorentz transformation $x^\mu \to x'^\mu = \Lambda^\mu{}_\nu x^\nu$ by the condition,

$$\eta_{\mu\nu} = \Lambda^\alpha{}_\mu \Lambda^\beta{}_\nu \, \eta_{\alpha\beta} \tag{1.1}$$

It must be pointed out that there is another tensor in the flat space-time that is invariant under proper Lorentz transformations: the totally antisymmetric Levi-Civita tensor $\epsilon_{\mu\nu\rho\sigma}$. Since, under $x^\mu \to \Lambda^\mu{}_\nu x^\nu$,

$$\epsilon_{\mu\nu\alpha\beta} \quad \to \quad \Lambda^\sigma{}_\mu \Lambda^\tau{}_\nu \Lambda^\gamma{}_\alpha \Lambda^\kappa{}_\beta \, \epsilon_{\sigma\tau\gamma\kappa} \tag{1.2}$$

$$= det(\Lambda) \, \epsilon_{\mu\nu\alpha\beta} \quad , \tag{1.3}$$

and $det(\Lambda) = 1$ for proper Lorentz transformations, it is clear that the flat space-time Levi-Civita tensor does not change under arbitrary boosts and rotations of inertial frames.

Given this special status that the Minkowski metric and the Levi-Civita tensor enjoy in the Minkowskian space-time, as well as the fact that $\eta_{\mu\nu}$ metamorphoses into a dynamical space-time metric $g_{\mu\nu}(x)$ in GR, raises a pertinent point of elevating $\epsilon_{\mu\nu\alpha\beta}$ to a dynamical field. In the present article, we explore such a possibility by positing a geometrodynamical field, $w_{\mu\nu\rho\sigma}(x)$, totally antisymmetric in its indices, that is independent of the metric tensor and that influences space-time physics. Then, the interesting feature that $\eta_{\mu\nu}$ and $\epsilon_{\mu\nu\alpha\beta}$ both share in the flat space-time becomes relevant also in arbitrary curved space-times [1].

We propose that this geometrodynamical field represents a new physical degree of freedom that couples universally to all matter. In the following section, we discuss some of the differential geometric aspects of $w_{\mu\nu\rho\sigma}(x)$ that may be significant in the context of space-time physics.

2 Geometrodynamical four-form field and Torsion

Now, in any 4D space-time manifold, because of its complete antisymmetry, the four-form field $w_{\mu\nu\rho\sigma}(x)$ has only one algebraically independent component and hence, renders itself to be expressed as,

$$w_{\mu\nu\rho\sigma}(x) = \phi(x) \, \epsilon_{\mu\nu\rho\sigma} \quad , \tag{2.1}$$

where $\phi(x)$ is a scalar-density of weight $+1$, with $\phi(x) \to \phi'(x') = \phi(x)/J(x,x')$ under a general coordinate transformation $x^\mu \to x'^\mu$, $J(x,x')$ being the corresponding Jacobian.

In a coordinate basis, this four-form field may be expressed as,

$$\tilde{w} = \frac{1}{4!} \, w_{\mu\nu\rho\sigma}(x) \, \tilde{d}x^\mu \wedge \tilde{d}x^\nu \wedge \tilde{d}x^\rho \wedge \tilde{d}x^\sigma \tag{2.2}$$

$$= \frac{1}{4!} \, \phi(x) \, \epsilon_{\mu\nu\rho\sigma} \, \tilde{d}x^\mu \wedge \tilde{d}x^\nu \wedge \tilde{d}x^\rho \wedge \tilde{d}x^\sigma \tag{2.3}$$

Eq.(2.3) reflects the equivalence between four-forms and 0-forms in the case of four dimensional manifolds.

Since we demand that this new dynamical field be independent of the metric tensor, instead of raising the indices of $w_{\mu\nu\rho\sigma}$ using the metric tensor, the totally antisymmetric components $w^{\mu\nu\rho\sigma}(x)$

122

corresponding to the four-form field $\tilde{w}$ are obtained from the associated p-vector $\mathbf{w}$, so that one obtains the relation, $w^{\mu\nu\rho\sigma}\ w_{\mu\nu\rho\sigma} = -4!$ [2]. This implies,

$$w^{\mu\nu\rho\sigma}(x) = \epsilon^{\mu\nu\rho\sigma}/\phi(x) \ . \tag{2.4}$$

We assume throughout that $\phi(x) = w_{0123}(x)$ has no physical dimension. As the canonical volume-form in GR is given by,

$$\tilde{V} = \frac{1}{4!}\sqrt{-g}\ \epsilon_{\mu\nu\rho\sigma}\ \tilde{d}x^{\mu} \wedge \tilde{d}x^{\nu} \wedge \tilde{d}x^{\rho} \wedge \tilde{d}x^{\sigma}\ , \tag{2.5}$$

the geometrodynamical four-form field of eq.(2.3) may be expressed in terms of a scalar field $\chi(x)$,

$$\chi(x) \equiv \frac{\phi(x}{\sqrt{-g}} \quad \Rightarrow \quad \tilde{w} = \chi(x)\tilde{V} \tag{2.6}$$

A fundamental four-form field that varies with space and time may be utilized to extend GR by having a dynamical torsion field in the theory. Introducing an asymmetric affine connection (e.g. see [3, 4]),

$$\bar{\Gamma}^{\mu}_{\alpha\beta} \equiv \Gamma^{\mu}_{\alpha\beta} - a_T\ g^{\mu\nu}g^{\gamma\lambda}\frac{\partial\chi}{\partial x^{\lambda}}\ w_{\nu\gamma\alpha\beta} = \Gamma^{\mu}_{\alpha\beta} - a_T\ g^{\mu\nu}\chi^{;\gamma}\ w_{\nu\gamma\alpha\beta}\ , \tag{2.7}$$

where $\Gamma^{\mu}_{\alpha\beta}$ is the standard Christoffel-Levi Civita connection while a_T is a dimensionless constant that is a measure of the coupling between the metric tensor and the four-form, we define a torsion field,

$$S_{\alpha\beta}{}^{\mu}(x) = \frac{1}{2}\left(\bar{\Gamma}^{\mu}_{\alpha\beta} - \bar{\Gamma}^{\mu}_{\beta\alpha}\right) = -a_T\ g^{\mu\nu}\chi^{;\gamma}\ w_{\nu\gamma\alpha\beta} = -a_T\sqrt{-g}\ g^{\mu\nu}g^{\gamma\sigma}\ \chi\ \frac{\partial\chi}{\partial x^{\sigma}}\ \epsilon_{\nu\gamma\alpha\beta} \tag{2.8}$$

In the language of differential geometry, since torsion is a vector valued 2-form, the above expression for it can be seen as arising from an inner contraction (e.g. see [5, 2]) of $\tilde{w}$ with the vector field $\chi^{;\gamma}\frac{\partial}{\partial x^{\gamma}}$ followed by inner contractions with four vector fields $-g^{\mu\nu}\frac{\partial}{\partial x^{\nu}}$, $\mu = 0, 1, ..3$.

Covariant derivatives of tensor fields are obtained in the usual way,

$$D_{\nu}A^{\mu} = \partial_{\nu}A^{\mu} + \bar{\Gamma}^{\mu}_{\nu\beta}A^{\beta} = A^{\mu}{}_{;\nu} - a_T\ g^{\mu\alpha}\chi^{;\gamma}w_{\alpha\gamma\nu\beta}A^{\beta}\ , \tag{2.9}$$

$$D_{\nu}B_{\mu} = \partial_{\nu}B_{\mu} - \bar{\Gamma}^{\beta}_{\mu\nu}B_{\beta} = B_{\mu;\ \nu} + a_T g^{\beta\sigma}\chi^{;\gamma}w_{\sigma\gamma\mu\nu}B_{\beta}\ , \tag{2.10}$$

and,

$$D_{\nu}C_{\mu\tau} = \partial_{\nu}C_{\mu\tau} - \bar{\Gamma}^{\beta}_{\mu\nu}C_{\beta\tau} - \bar{\Gamma}^{\beta}_{\tau\nu}C_{\mu\beta} = C_{\mu\tau;\ \nu} + a_T g^{\beta\sigma}\chi^{;\gamma}[w_{\sigma\gamma\mu\nu}C_{\beta\tau} + w_{\sigma\gamma\tau\nu}C_{\mu\beta}]\ , \tag{2.11}$$

with $C_{\mu\tau;\ \nu}$ being the covariant derivative of $C_{\mu\tau}$ in standard GR when there is no torsion (i.e. $a_T = 0$ case). Using eq.(2.11), the metric compatibility of $S_{\alpha\beta}{}^{\mu}$ can easily be proved by showing that $D_{\alpha}g_{\mu\nu} = 0$. Similarly, for any vector field $A^{\alpha}(x)$, eq.(2.7) entails that,

$$\bar{\Gamma}^{\mu}_{\alpha\beta}A^{\alpha}A^{\beta} = \Gamma^{\mu}_{\alpha\beta}A^{\alpha}A^{\beta}$$

implying that the torsion field given by eq.(2.8) is autoparallel compatible too. In our case, the contortion tensor $K_{\alpha\beta}{}^{\mu}$ and $S_{\alpha\beta\gamma}$ have the expressions,

$$K_{\alpha\beta}{}^{\mu} \equiv -S_{\alpha\beta}{}^{\mu} + S_{\beta}{}^{\mu}{}_{\alpha} - S^{\mu}{}_{\alpha\beta} = -S_{\alpha\beta}{}^{\mu} \tag{2.12}$$

123

and,

$$S_{\alpha\beta\gamma} = g_{\gamma\mu}S_{\alpha\beta}{}^\mu = a_T \; \chi^{;\sigma} w_{\sigma\alpha\beta\gamma} = a_T \sqrt{-g} \; g^{\gamma\sigma}\chi \, \frac{\partial\chi}{\partial x^\gamma}\epsilon_{\sigma\alpha\beta\gamma} \tag{2.13}$$

which is simply an inner contraction of $\tilde{w}$ with the vector field $\chi^{;\sigma}\frac{\partial}{\partial x^\sigma}$.

The important point to note is that the torsion vanishes if either $a_T = 0$ or if the scalar field $\chi(x)$ that corresponds to the geometrodynamical four form is independent of both space as well as time. **For the sake of simplicity, in the remaining portions of the article we will concentrate on the case $a_T = 0$ while assuming $\frac{\partial\chi}{\partial x^\alpha}$ to be non-vanishing.** It is straightforward to extend the analysis to the $a_T \neq 0$ case in order to study the effects of torsion. For the rest of the sections, we have chosen a unit system in which $\hbar = c = 1$ and the metric signature: $(+\text{ - - -})$.

3 Action, Field Equations and Chern-Simons extensions

3.1 Dynamical equations

As ϕ is a scalar-density of weight $+1$,

$$\phi_{;\beta} = \phi_{,\beta} - \Gamma^\alpha_{\alpha\beta}\,\phi \;, \tag{3.1}$$

where we follow the standard notation,

$$f_{,\beta} \equiv \partial_\beta f \quad . \tag{3.2}$$

Using the above equations along with the relation $w^{\mu\nu\rho\sigma}\,w_{\mu\nu\rho\sigma} = -\,4!$, the covariant derivatives of the four-form field and its corresponding p-vector are easily shown to be,

$$w_{\mu\nu\alpha\beta\;;\lambda} = [(\ln\phi)_{,\lambda} - \Gamma^\sigma_{\sigma\lambda}]w_{\mu\nu\alpha\beta} \;, \tag{3.3}$$

and,

$$w^{\mu\nu\alpha\beta}{}_{;\lambda} = -[(\ln\phi)_{,\lambda} - \Gamma^\sigma_{\sigma\lambda}]w^{\mu\nu\alpha\beta} \;, \tag{3.4}$$

respectively.

The action $\mathcal{A}$ for the standard matter, geometry as well as the dynamical four-form that is invariant under general coordinate transformations may be expressed as,

$$\mathcal{A} = -\frac{m_{Pl}^2}{16\pi}\int R\sqrt{-g}\;d^4x + \int L\sqrt{-g}\;d^4x + $$

$$+ \frac{a_1}{4!}\int \phi\; w^{\mu\nu\alpha\beta}\,w_{\mu\nu\alpha\beta}{}^{;\lambda}\;d^4x + a_2\int\phi(x)d^4x \;, \tag{3.5}$$

where $m_{Pl} \equiv \sqrt{\hbar c/G}$ and L are the Planck mass and the Lagrangian density of the matter fields, respectively, with a_1 and a_2 being real valued constants of the theory having physical dimensions $(\text{mass})^2$ and $(\text{mass})^4$, respectively. The part of the action in eq.(3.5) that pertains to the four-form is by no means unique. For instance, a term $\propto \int R\,\phi\,d^4x$ could be added to the above action, but we limit ourselves to gravitational minimal coupling, at present. Later, we shall discuss supplementing the above action with Chern-Simons terms induced by the geometrodynamical four-form.

By extremising $\mathcal{A}$ with respect to $g_{\mu\nu}$ and ϕ, respectively, we obtain the following equations of motion,

$$R_{\mu\nu} - \frac{1}{2}g_{\mu\nu}R = \frac{8\pi}{m_{Pl}^2}[T_{\mu\nu} + \Theta_{\mu\nu}] \,, \tag{3.6}$$

$$\chi^{;\alpha}_{\ ;\alpha} \equiv \frac{1}{\sqrt{-g}}(\sqrt{-g}g^{\alpha\beta}\chi_{,\alpha})_{,\beta} = \frac{1}{2}\left[g^{\mu\nu}\frac{\chi_{,\mu}\chi_{,\nu}}{\chi} + \frac{a_2}{a_1}\chi\right] \,, \tag{3.7}$$

where the scalar field $\chi(x)$ is defined, as earlier, to be $\chi \equiv \frac{\phi}{\sqrt{-g}}$, while $T_{\mu\nu}$ and $\Theta_{\mu\nu}$ are the energy-momentum tensors for the standard matter and the geometrodynamical four-form, respectively. The expression for the latter is given by,

$$\Theta_{\mu\nu} = 2a_1\left[\frac{\chi_{,\mu}\chi_{,\nu}}{\chi} - g_{\mu\nu}\chi^{;\alpha}_{\ ;\alpha}\right] \,. \tag{3.8}$$

When χ satisfies the equation of motion given by eq.(3.7), the above energy-momentum tensor gets simplified to,

$$\Theta_{\mu\nu} = 2a_1\left[\frac{\chi_{,\mu}\chi_{,\nu}}{\psi} - \frac{g_{\mu\nu}}{2}\left(g^{\alpha\beta}\frac{\chi_{,\alpha}\chi_{,\beta}}{\chi} + \frac{a_2}{a_1}\chi\right)\right] \,. \tag{3.9}$$

3.2 Chern-Simons extensions

In this sub-section, we construct Chern-Simons (CS) terms in (3+1)-dimensional space-time. Our CS extensions of GR follows closely the seminal paper of Jackiw and Pi [6], except for a crucial difference: they had introduced an external fixed four-vector v_μ instead of a dynamical $\phi_{;\mu}$ in their formulation. Because of this, their model had a manifest violation of Lorentz invariance. In our case, both general as well as Lorentz covariances are maintained throughout [1].

First, we consider a Chern-Simons (CS) term that couples the electromagnetic field to the four-form $\tilde{w}$,

$$\mathcal{A}_{CS} = \text{J}\int w^{\mu\nu\alpha\beta}F_{\mu\nu}A_\alpha\phi_{;\beta}\ d^4x$$

$$= \text{J}\int \epsilon^{\mu\nu\alpha\beta}F_{\mu\nu}A_\alpha(\ln\chi)_{,\beta}\ d^4x \,, \tag{3.10}$$

where J is a dimensionless constant and $F_{\mu\nu} = A_{\nu,\mu} - A_{\mu,\nu}$. The action expressed in eq.(3.10) is invariant under diffeomorphism as well as gauge transformations, $A_\mu \to A_\mu + \partial_\mu\xi(x)$.

The upshot of adding $\mathcal{A}_{CS}$ to the standard action $\mathcal{A}_{EM}$ for an electromagnetic field interacting with charge particles in the presence of gravitation is that, upon extremising the full action $\mathcal{A}_{EM} + \mathcal{A}_{CS}$ with respect to A_μ, one obtains the following modified Einstein-Maxwell equation,

$$F^{\alpha\beta}_{\ ;\beta} = -4\pi j^\alpha + 8\pi\text{J}\ w^{\mu\nu\alpha\beta}F_{\mu\nu}\ \chi_{,\beta} \,, \tag{3.11}$$

with j^α being the 4-current density associated with charge particles.

Use of the covariant derivative to the above equation with respect to x^α leads to the standard charge density continuity equation that entails conservation of electric charge Q,

$$(\sqrt{-g}j^\alpha)_{,\alpha} = 0 \quad \Rightarrow \quad Q = \int j^0 \sqrt{-g}\, d^3x \ . \tag{3.12}$$

Application of the above CS formulation to study its effects on magnetohydrodynamics as well as the $\vec{E}.\vec{B} \neq 0$ regions of a pulsar magnetosphere acting as a source of propagating $\tilde{w}$ have been studied earlier [7].

We continue adopting the procedure laid out in the paper by Jackiw and Pi [6] to construct a Chern-Simons action for gravity in the 3+1-dimensional space-time. Instead of their fixed Lorentz vector v_μ, we use the covariant derivative of $\phi(x)$ so that,

$$\mathcal{A}_{GCS} = \mathrm{H} \int w^{\mu\nu\alpha\beta}[\Gamma^\sigma_{\nu\tau}\partial_\alpha\Gamma^\tau_{\beta\sigma} + \frac{2}{3}\Gamma^\sigma_{\nu\tau}\Gamma^\tau_{\alpha\eta}\Gamma^\eta_{\beta\sigma}]\phi_{;\mu}\, d^4x$$

$$= \mathrm{H} \int \epsilon^{\mu\nu\alpha\beta}[\Gamma^\sigma_{\nu\tau}\partial_\alpha\Gamma^\tau_{\beta\sigma} + \frac{2}{3}\Gamma^\sigma_{\nu\tau}\Gamma^\tau_{\alpha\eta}\Gamma^\eta_{\beta\sigma}](\ln\chi)_{,\mu}\, d^4x \ , \tag{3.13}$$

where H is a dimensionless, real constant. After integrating by parts once, eq.(3.13) can be expressed in terms of the Riemann tensor as,

$$\mathcal{A}_{GCS} = -\frac{\mathrm{H}}{2} \int \ln\chi \ ^*RR \, d^4x \ , \tag{3.14}$$

with,

$$^*RR \equiv \frac{1}{2}\epsilon^{\mu\nu\alpha\beta}R^\tau_{\ \sigma\alpha\beta}R^\sigma_{\ \tau\mu\nu} = 8\left[R^{\tau\sigma}_{\ \ 01}R_{\sigma\tau23} + R^{\tau\sigma}_{\ \ 12}R_{\sigma\tau03} + R^{\tau\sigma}_{\ \ 13}R_{\sigma\tau20}\right] \tag{3.15}$$

and,

$$^*R^{\tau\rho\mu\nu} \equiv \frac{1}{2}\epsilon^{\mu\nu\alpha\beta}R^{\tau\rho}_{\ \ \alpha\beta} \ . \tag{3.16}$$

We have used the following conventions pertaining to the Riemann curvature and Ricci tensors,

$$R^\tau_{\ \sigma\alpha\beta} = \partial_\alpha\Gamma^\tau_{\sigma\beta} - \partial_\beta\Gamma^\tau_{\sigma\alpha} + \Gamma^\tau_{\alpha\eta}\Gamma^\eta_{\beta\sigma} - \Gamma^\tau_{\beta\eta}\Gamma^\eta_{\alpha\sigma}$$

and $R_{\alpha\beta} = R^\tau_{\ \alpha\tau\beta}$. (We note that $\ln\chi$ in eq.(3.13) acts like the parameter θ in Jackiw and Pi's paper [6] in which the external vector v_μ is set as $\theta_{,\mu}$.)

Adding $\mathcal{A}_{CS} + \mathcal{A}_{GCS}$ to $\mathcal{A}$ of eq.(3.5) and then extremising the full action with respect to ϕ and $g_{\mu\nu}$ leads to,

$$\chi^{;\alpha}_{\ ;\alpha} = \frac{1}{2}\left[g^{\mu\nu}\frac{\chi_{,\mu}\chi_{,\nu}}{\chi} + \frac{a_2}{a_1}\chi + \frac{J}{2a_1}\chi\, w^{\mu\nu\alpha\beta}F_{\mu\nu}F_{\alpha\beta} - \frac{H}{4a_1}\chi\, w^{\mu\nu\alpha\beta}R^\tau_{\ \sigma\alpha\beta}R^\sigma_{\ \tau\mu\nu}\right], \tag{3.17}$$

$$R_{\mu\nu} - \frac{1}{2}g_{\mu\nu}R = \frac{8\pi}{m_{Pl}^2}[T_{\mu\nu} + \Theta_{\mu\nu} + C_{\mu\nu}] \ , \tag{3.18}$$

where the modified Cotton tensor $C^{\mu\nu}$ is defined as,

$$C^{\mu\nu} \equiv -2\frac{\mathrm{H}}{\sqrt{-g}}\left[\frac{1}{4}\ ^*RR g^{\mu\nu} - (\ln\chi)_{;\alpha;\beta}\left(^*R^{\beta\mu\alpha\nu} + {}^*R^{\beta\nu\alpha\mu}\right) + (\ln\chi)_{,\alpha}\left(\epsilon^{\alpha\mu\sigma\tau}R^\nu_{\ \sigma;\tau} + \epsilon^{\alpha\nu\sigma\tau}R^\mu_{\ \sigma;\tau}\right)\right]. \tag{3.19}$$

The first term in the RHS of Eq.(3.19) is new and is not present in the expression for the Cotton tensor as delineated in [6]. It appears in our work because under an infinitesimal variation $g_{\mu\nu} \to g_{\mu\nu} + \delta g_{\mu\nu}$, the induced change in $(\ln\chi)_{,\mu}$ occurring in eq.(3.13) (while considering the ensuing variation of $\mathcal{A}_{GCS}$) is given by,

$$\delta(\ln\chi)_{,\mu} = \delta(\phi_{;\mu}/\phi) = -\delta\Gamma^{\alpha}_{\alpha\mu} = -\frac{1}{2}\delta(g^{\alpha\beta}g_{\alpha\beta,\mu}) \ . \tag{3.20}$$

However, when the equation of motion for the dynamical four-form given by eq.(3.17) is substituted in eq.(3.8), its energy-momentum tensor in the presence of CS-term takes the form,

$$\Theta_{\mu\nu} = 2a_1\left[\frac{\chi_{,\mu}\chi_{,\nu}}{\chi} - \frac{g_{\mu\nu}}{2}\left(g^{\alpha\beta}\frac{\chi_{,\alpha}\chi_{,\beta}}{\chi} + \frac{a_2}{a_1}\chi\right)\right] + \frac{H}{2\sqrt{-g}}{}^*RRg_{\mu\nu} \tag{3.21}$$

so that when eq.(3.21) is substituted in eq.(3.18) the first term in the RHS of eq.(3.19) cancels with the last term in the RHS of eq.(3.21). In other words, the term ${}^*RRg_{\mu\nu}$ does not appear in or contribute to the Einstein equations given by eq.(3.18).

Assuming that the geometrodynamical four-form is a dark energy candidate, the above dynamical equations were studied in an earlier work to address the observed late time acceleration of the expansion rate of the universe ([1], and the references therein). Our objective, in this article, is somewhat different since we propose here that the particle associated with $\tilde{w}$ is a pseudo-scalar which acts as a cold dark matter (CDM) candidate aiding in the formation of supermassive black holes when the universe was young. This is the subject of Section IV.

3.3 Generalized Exterior Derivative

Apart from the possibility of torsion and Chern-Simons extensions ensuing from the four-form field $\tilde{w}$, there may be another differential geometric significance of this field. The scalar-density ϕ can also be used to have an antiderivation acting on antisymmetric tensor-densities of arbitrary weights.

Suppose $\tilde{\alpha}$ is a p-form density with weight w such that its components $\alpha_{\nu_1\nu_2..\nu_p}$ transform to $J^{-w}(x,x')\,\alpha_{\nu_1\nu_2..\nu_p}$, under a general coordinate transformation $x \to x'$, $J(x,x')$ being the corresponding Jacobian. We define a generalized exterior derivative $\tilde{d}_w$ acting on $\tilde{\alpha}$ in the following manner:

$$\tilde{d}_w\tilde{\alpha} \equiv \frac{1}{p!}\,\partial_\mu\alpha_{\nu_1\nu_2..\nu_p}\,\tilde{d}x^\mu \wedge \tilde{d}x^{\nu_1} \wedge ... \wedge \tilde{d}x^{\nu_p} -$$

$$-w\,\partial_\mu(\ln\phi)\,\tilde{d}x^\mu \wedge \tilde{\alpha} = \tilde{d}\tilde{\alpha} - w\,\tilde{d}\,(\ln\phi)\,\wedge\tilde{\alpha} \ . \tag{3.22}$$

It is easy to verify that $\tilde{d}_w\tilde{\alpha}$ is a (p+1)-form density of weight w.

Suppose α_1 and α_2 are scalar-densities of weights w_1 and w_2, respectively. Then, from eq.(3.22), we have,

$$\tilde{d}_w\alpha_i = \partial_\mu\alpha_i\,\tilde{d}x^\mu - w_i\alpha_i\,\partial_\mu(\ln\phi)\,\tilde{d}x^\mu, \quad i=1,2 \tag{3.23}$$

which are one-form densities and furthermore, one can show that,

$$\tilde{d}_w(\alpha_1\tilde{d}_w\alpha_2) = \tilde{d}_w\alpha_1 \wedge \tilde{d}_w\alpha_2 \ . \tag{3.24}$$

127

The generalized exterior derivative defined by eq.(3.22) also satisfies (a) $\tilde{d}_w \tilde{d}_w = 0$ and (b) $\tilde{d}_w(\tilde{\alpha} \wedge \tilde{\beta}) = \tilde{d}_w \tilde{\alpha} \wedge \tilde{\beta} + (-1)^p \tilde{\alpha} \wedge \tilde{d}_w \tilde{\beta}$, where $\tilde{\alpha}$ and $\tilde{\beta}$ are p- and q-form densities of weights w_1 and w_2, respectively. These properties are sufficient to qualify $\tilde{d}_w$ to be a well defined anti-derivation on differential form-densities [5].

From eq.(3.23) it follows that,

$$\tilde{d}_w \sqrt{-g} = -\sqrt{-g} \; \tilde{d} \ln(\frac{\phi}{\sqrt{-g}}) \; , \tag{3.25}$$

since $\sqrt{-g}$ is a scalar-density of weight $+1$.

There are other physically meaningful tensor densities e.g. Dirac delta function (which is a scalar-density) and anti-symmetric tensor-densities, e.g. $\widetilde{F}^{\mu\nu} = $ dual of $F^{\mu\nu}$, on which $\tilde{d}_w$ can act. In passing, we note that $\tilde{d}_w \phi = 0$, which is analogous to $g_{\mu\nu;\lambda} = 0$.

4 Dark bosons, Bose-Einstein Condensates and Formation of Supermassive Black Holes

The wave-particle duality of quantum mechanics transcends to field-particle duality in quantum field theory (QFT). Particles associated with quantum fields are intimately linked with their Poincare' algebra and special relativistic covariance in the standard framework of QFT. In the context of the four-form field $\tilde{w}$, or equivalently the scalar-density $\phi(x)$, we may ask what kind of particle is associated with it?

Now, $\phi(x)$ in the flat Minkowski space-time transforms to,

$$\phi \to \phi' = J(t, \vec{r}; t, -\vec{r}) \; \phi = -\phi \; , \tag{4.1}$$

under a space reversal $\vec{r} \to -\vec{r}$ as the Jacobian $J(t, \vec{r}; t, -\vec{r}) = -1$. Since ϕ changes its sign under a parity transformation, the quantum particle associated with the field $\tilde{w}$ is a pseudo scalar, and thus is a boson.

Because such a pseudo-scalar is likely to interact with matter with a strength at the most comparable with that of gravity, we propose here that $\tilde{w}$ is the candidate for ultra-light cold dark matter (CDM), although such CDM particles have been conventionally linked with axions in the literature (e.g. see [8], and the references therein). Can the dark bosons associated with the geometrodynamical four-form sort out the intriguing problem of the very early formation of supermassive black holes (SMBHs)?

So far, around a million quasars have been observed, each powered by the infall of matter into the deep gravitational potential of SMBHs weighing $\gtrsim 10^7 M_\odot$, and thereby getting overheated to turn into hot plasma [9, 10, 11, 12]. Quasar J0100+2802 found at a redshift, $z = 6.33$, is associated with a SMBH of mass $\approx 10^{10} \; M_\odot$, while another (J2157–3602) at a redshift, $z = 4.7$, has a black hole weighing $\cong 2.4 \times 10^{10} \; M_\odot$. Similarly, quasars J1120+0641 and J1342+0928 located at redshifts $z = 7.09$ and $z = 7.54$, respectively, are associated with SMBHs with mass $\gtrsim 10^8 \; M_\odot$. These findings pose a severe challenge to the existing models that invoke seed black holes (BHs) growing via accretion, since one needs extremely large seed BHs with mass $\gtrsim 10^3 \; M_\odot$ at $z \gtrsim 40$ [12, 13].

128

Recently, using the JWST observations, a very distant SMBH with mass $\sim 10^6~M_\odot$ has been discovered at a redshift of 10.6 [14]. Of course, even heavier SMBHs at lower redshifts, like with mass $\cong 4 \times 10^{10}~M_\odot$ at the centre of Holm 15A galaxy belonging to the galaxy cluster Abell 85 have been detected [15]. At a redshift of 3.96, a SMBH weighing $\cong 1.7 \times 10^{10}~M_\odot$, accreting matter at a rate one solar mass per day has been seen [16].

In order to address the formation of SMBHs when the universe was $\lesssim 10^9$ yrs old, we had earlier put forward a scenario using the framework of Gross-Pitaevskii equation in which ultra heavy BHs are created because of gravitational contraction of Bose-Einstein condensates (BECs) of ultra-light bosonic CDM [17, 18]. In this section, we provide a simple semi-classical analysis that captures the essential physics of the problem.

Considering a galactic DM halo of size R_h that is constituted of these gravitationally bound dark bosons of mass m, a significant fraction of the bosons that have sufficiently low momenta p would form a BEC if their de Broglie wavelength,

$$\lambda_{DB} \sim \frac{h}{p} \gtrsim \left(\frac{3N}{4\pi R_h^3}\right)^{-1/3} = R_h \left(\frac{3N}{4\pi}\right)^{-1/3} = R_h \left(\frac{3M}{4\pi m}\right)^{-1/3} \; , \tag{4.2}$$

where M and N are the total mass of dark pseudo-scalars and their number, respectively, making up the BEC. Eq.(4.2) represents the condition for BEC formation as it entails that the de Broglie wavelength of the ultra-light pseudo-scalars be larger than the mean separation between the neighbouring dark bosons.

In the context of a ΛCDM scenario with a positive cosmological constant present, the energy E_b of a typical dark boson is given by [19],

$$E_b \sim \frac{p^2}{2m} - \frac{GMm}{R_h} + \frac{\Lambda\, r^2}{6} mc^2 < 0 \; . \tag{4.3}$$

The condition $E_b < 0$ follows from the fact that the DM halo is a gravitationally bound structure. Hence,

$$p^2 < \frac{2GMm^2}{R_h} - \frac{\Lambda\, r^2}{3} m^2 c^2 \; . \tag{4.4}$$

Since a very weakly interacting dark boson can be anywhere within the DM halo, application of the Heisenberg's uncertainty principle implies that,

$$\Delta p \sim p \gtrsim \frac{\hbar}{2R_h} \; . \tag{4.5}$$

From eqs.(4.2) and (4.5), we have an inequality,

$$\frac{h}{4\pi p} \lesssim R_h \lesssim \frac{h}{p} \left(\frac{3N}{4\pi}\right)^{1/3} \tag{4.6}$$

that is self-consistent since $N \gg 1$.

Using the result $p \sim \frac{\hbar}{2R_h}$ from eq.(4.5) in eq.(4.3), we obtain,

$$E_b(R_h) \sim \frac{\hbar^2}{8mR_h^2} - \frac{GMm}{R_h} + \frac{\Lambda\, R_h^2}{6} mc^2 < 0 \; , \tag{4.7}$$

that leads to a condition,

$$R_h \gtrsim \frac{\hbar^2}{8GMm^2}\left(1 - \frac{R_h^3 \Lambda c^2}{6GM}\right)^{-1} . \tag{4.8}$$

In the context of the ΛCDM model, the numerical value of the cosmological constant that ensues from the cosmological constant density parameter, $\Omega_{\Lambda,0} \cong 0.7$ and the Hubble parameter, $H_0 \cong 68$ km/s/Mpc, is $\Lambda \cong 10^{-56}$ cm^{-2}.

Even with a low dark matter halo mass $M \cong 10^6$ $M_\odot$, one finds $\lambda_M \equiv \Lambda c^2/6GM \cong 10^{-68}$ cm^{-3}. Hence, on galactic scales $R_h \lesssim 10^{22}$ cm, $R_h^3 \Lambda c^2/6GM \ll 1$ so that eq.(4.8) can be approximated to,

$$R_h \gtrsim \frac{\hbar^2}{8GMm^2}\left(1 + \frac{R_h^3 \Lambda c^2}{6GM}\right) \approx \frac{\hbar^2}{8GMm^2} . \tag{4.9}$$

In a BEC, most of the identical bosons occupy the ground state. Therefore, minimizing E_b (given by eq.(4.7)) with respect to R_h, one obtains,

$$\frac{\partial E_b}{\partial R_h} = \frac{\Lambda R_h}{3}mc^2 + \frac{GMm}{R_h^2} - \frac{\hbar^2}{4mR_h^3} = 0 , \tag{4.10}$$

which leads to a quartic equation that needs to be solved in order to estimate the BEC size,

$$\Lambda R_h^4 + \frac{3GM}{c^2}R_h - \frac{3\hbar^2}{4m^2c^2} = 0 . \tag{4.11}$$

The above equation can be solved exactly and, out of the four roots, only one happens to be real and positive. However, given the smallness of the quantity λ_M, this real and positive solution to eq.(4.11), even up to λ_M^3 orders, is of the form,

$$R_{bec} \cong \frac{\hbar^2}{4GMm^2} \cong 22\left(\frac{10^9 M_\odot}{M}\right)\left(\frac{10^{-22} \text{ eV}}{m}\right)^2 \text{ pc} , \tag{4.12}$$

that corresponds to a single boson energy,

$$E_{min} = E_b(R_{bec}) \cong -\frac{GMm}{2R_{bec}} = -2 \, mc^2\left(\frac{m_{Pl}^2}{m \, M}\right)^{-2} . \tag{4.13}$$

Eqs.(4.12) and (4.13) are of course exact solutions for the $\Lambda = 0$ case, ensuing trivially from eq.(4.11).

The preceding equations entail that for larger BEC mass M or larger rest mass m of the pseudo-scalar not only the BEC size is smaller, the condensate is also more tightly bound since its energy is more negative. The physical repercussion of these features is that for a sufficiently large mass M or m, R_{bec} can be very close to the corresponding Schwarzschild radius $R_s \equiv 2GM/c^2$ so that even a slight perturbation would cause an irreversible gravitational collapse leading to the formation of a BH.

The BEC would form a BH if its size R_{bec} shrinks below the Schwarzschild radius R_S limit,

$$R_S = \frac{2GM}{c^2} = \frac{2M}{m_{Pl}^2} , \tag{4.14}$$

where m_{Pl} represents the Planck mass.

Using the criteria $R_{bec} \lesssim R_S$ along with eqs.(4.12) and (4.14), the condition for the BEC to implode into a BH is given by the inequality,

$$\left(\frac{m_{Pl}^2}{4M\,m}\right)^2 \lesssim 1 \,, \tag{4.15}$$

$$\Rightarrow\ m\,M \gtrsim\ 0.25\ m_{Pl}^2 = 3 \times 10^9 \left(\frac{m}{10^{-20}\ \mathrm{eV}}\right)^{-1} M_\odot \,. \tag{4.16}$$

A more rigorous analysis carried out earlier, for the $\Lambda = 0$ case, by employing a Gross-Pitaevskii equation framework had shown that the dynamical evolution of ultra-light dark bosons in the BEC phase leads to the formation of a BH on time scales $\sim 10^8$ years with a similar constraint given by eq.(4.16) [17, 18].

Now, the Hawking temperature T_{BH} of a BH of mass M is given by [20],

$$k_B T_{BH} = \frac{m_{Pl}^2}{8\pi M} \,. \tag{4.17}$$

So, it is interesting to note that when eq.(4.16) is substituted in eq.(4.17), one gets,

$$k_B T_{BH} \lesssim \frac{m}{2\pi} \,. \tag{4.18}$$

In other words, a SMBH created from the collapse of a BEC of ultra-light dark bosons would be predominantly radiating, via Hawking evaporation, massless particles as well as these pseudo-scalars associated with $\tilde{w}$.

5 Propagation of Gravitational Waves in the presence of the geometrodynamical four-form

We take up in this section the effect of $\tilde{w}$ on the propagation of gravitational waves (GWs) assuming that, except for a weak GW and the geometrodynamical four form, there is no other matter present.

By making the weak field approximation $g_{\mu\nu} = \eta_{\mu\nu} + h_{\mu\nu}$ (with $|h_{\mu\nu}| \ll 1$)) and considering a traceless-transverse (TT)-gauge for a plane GW travelling along the x-axis, we express the GW amplitudes,

$$h_\oplus \equiv h_{22}(t,x) = -h_{33}(t,x) \quad \text{and} \quad h_\otimes \equiv h_{23}(t,x) = h_{32}(t,x) \,. \tag{5.1}$$

Retaining only upto quadratic terms in $h_\oplus$ and $h_\otimes$ for the gravitational part, the action given by eqs.(3.5) and (3.14) reduce to (also, see [21]),

$$\mathcal{A} \approx -\frac{m_{Pl}^2}{8\pi} \int [h_\oplus(\ddot{h}_\oplus - h_\oplus'') + h_\otimes(\ddot{h}_\otimes - h_\otimes'')]d^4x + \mathcal{A}_{GCS} + \mathcal{A}_\phi \tag{5.2}$$

with,

$$\mathcal{A}_{GCS} \approx -H \int \eta^{\tau\lambda} \left(\frac{\partial^2 \ln \chi}{\partial x \partial x^\tau}(h_\oplus h_{\otimes,0\lambda} - h_\otimes h_{\oplus,0\lambda}) + \frac{\partial^2 \ln \chi}{\partial t \partial x^\tau}(h_\otimes h_{\oplus,1\lambda} - h_\oplus h_{\otimes,1\lambda})\right)d^4x$$

131

$$-H \int \frac{\partial \ln \chi}{\partial x} \left(h_\oplus (\ddot{h}_{\otimes,0} - h''_{\otimes,0}) - h_\otimes (\ddot{h}_{\oplus,0} - h''_{\oplus,0}) \right) d^4 x$$

$$+H \int \frac{\partial \ln \chi}{\partial t} \left(h_\oplus (\ddot{h}_{\otimes,1} - h''_{\otimes,1}) - h_\otimes (\ddot{h}_{\oplus,1} - h''_{\oplus,1}) \right) d^4 x \tag{5.3}$$

and,

$$\mathcal{A}_\phi = \int \left[\frac{a_1}{\chi} \eta^{\mu\nu} \chi_{,\mu} \chi_{,\nu} + a_2 \chi - \frac{a_1}{\chi} \left(h_\oplus (\chi_{,2} \chi_{,2} - \chi_{,3} \chi_{,3}) + 2 h_\otimes \chi_{,2} \chi_{,3} \right) \right] d^4 x \quad . \tag{5.4}$$

After defining $\psi(x^\mu) \equiv \ln(\chi(x^\mu))$, the dynamical equations of motion that ensue from variations of $\mathcal{A}$ (eq.(5.2)) with respect to $h_\oplus$, $h_\otimes$ and ψ are given by,

$$\ddot{h}_\oplus - h''_\oplus = -\frac{8\pi}{m_{Pl}^2} \left[H \left\{ \frac{\partial^2 \psi}{\partial t \partial x} (\ddot{h}_\otimes + h''_\otimes) - h_{\otimes,01} \left(\frac{\partial^2 \psi}{\partial t^2} + \frac{\partial^2 \psi}{\partial x^2} \right) + \frac{\partial \psi}{\partial x} (\ddot{h}_{\otimes,0} - h''_{\otimes,0}) - \frac{\partial \psi}{\partial t} (\ddot{h}_{\otimes,1} - h''_{\otimes,1}) \right\} + \right.$$

$$\left. + a_1 \exp(\psi) \left\{ \left(\frac{\partial \psi}{\partial y} \right)^2 - \left(\frac{\partial \psi}{\partial z} \right)^2 \right\} \right] , \tag{5.5}$$

$$\ddot{h}_\otimes - h''_\otimes = \frac{8\pi}{m_{Pl}^2} \left[H \left\{ \frac{\partial^2 \psi}{\partial t \partial x} (\ddot{h}_\oplus + h''_\oplus) - h_{\oplus,01} \left(\frac{\partial^2 \psi}{\partial t^2} + \frac{\partial^2 \psi}{\partial x^2} \right) + \frac{\partial \psi}{\partial x} (\ddot{h}_{\oplus,0} - h''_{\oplus,0}) - \frac{\partial \psi}{\partial t} (\ddot{h}_{\oplus,1} - h''_{\oplus,1}) \right\} + \right.$$

$$\left. + 2 a_1 \exp(\psi) \frac{\partial \psi}{\partial y} \frac{\partial \psi}{\partial z} \right] \tag{5.6}$$

and

$$\partial^\mu \partial_\mu \psi + \frac{1}{2} \eta^{\mu\nu} \frac{\partial \psi}{\partial x^\mu} \frac{\partial \psi}{\partial x^\nu} - \frac{a_2}{4 a_1} = h_\oplus \left[\frac{\partial^2 \psi}{\partial y^2} - \frac{\partial^2 \psi}{\partial z^2} + \frac{1}{2} \left(\frac{\partial \psi}{\partial y} \right)^2 - \frac{1}{2} \left(\frac{\partial \psi}{\partial z} \right)^2 \right] + 2 h_\otimes \left[\frac{\partial^2 \psi}{\partial y \partial z} + \frac{1}{2} \frac{\partial \psi}{\partial y} \frac{\partial \psi}{\partial z} \right] . \tag{5.7}$$

(The derivatives with respect to time t and x-coordinate are denoted by dot and prime, respectively.)

A simplification occurs if we assume ψ to depend only on the x-coordinate and time, like the GW amplitude itself. With $\psi = \psi(t, x)$, the last terms in eqs.(5.5) and (5.6) drop out. Introducing the circularly polarized GW amplitudes, as demonstrated in the reference [21],

$$h_R = \frac{1}{\sqrt{2}} (h_\oplus + i h_\otimes) \quad \text{and} \quad h_L = \frac{1}{\sqrt{2}} (h_\oplus - i h_\otimes) ,$$

one finds that the coupled differential equations given by eqs.(5.5) and (5.6), involving $h_\oplus$ and $h_\otimes$, get mutually separated,

$$\ddot{h}_R - h''_R = -\frac{8\pi H i}{m_{Pl}^2} \left[\dot{\psi} \frac{\partial}{\partial x} (\ddot{h}_R - h''_R) - \psi' \frac{\partial}{\partial t} (\ddot{h}_R - h''_R) + \dot{h}'_R (\ddot{\psi} + \psi'') - \dot{\psi}' (\ddot{h}_R + h''_R) \right] \tag{5.8}$$

and

$$\ddot{h}_L - h''_L = \frac{8\pi H i}{m_{Pl}^2} \left[\dot{\psi} \frac{\partial}{\partial x} (\ddot{h}_L - h''_L) - \psi' \frac{\partial}{\partial t} (\ddot{h}_L - h''_L) + \dot{h}'_L (\ddot{\psi} + \psi'') - \dot{\psi}' (\ddot{h}_L + h''_L) \right] , \tag{5.9}$$

displaying a symmetry under $R \longleftrightarrow L$ in the above equations modulo a negative sign, corresponding to the two independent circular polarizations.

Since $\psi = \psi(t, x)$, the equation of motion for the four-form given by eq.(5.7) simplifies to,

$$\ddot{\psi} - \psi'' + \frac{1}{2}(\dot{\psi}^2 - \psi'^{\,2}) + 2\mu^2 = 0 \tag{5.10}$$

where $\mu \equiv \sqrt{-\frac{a_2}{4a_1}}$ is akin to the rest mass of the four-form field.

Seeking exact solutions of eqs.(5.8) and (5.10), involving monochromatic and circularly polarized GWs as well as ψ, we substitute,

$$h_R(t, x) = h \exp(i(\omega t - kx))$$

in eq.(5.8), leading to a relation,

$$\omega^2 - k^2 = -\beta[k(\omega^2 - k^2)\dot{\psi} + \omega(\omega^2 - k^2)\psi' - i\omega k(\ddot{\psi} + \psi'') - i(\omega^2 + k^2)\dot{\psi}'] \tag{5.11}$$

where $\beta \equiv \frac{8\pi H}{m_{Pl}^2}$.

If $w = \pm k$, eq.(5.11) implies,

$$2\dot{\psi}' \pm (\ddot{\psi} + \psi'') = 0 \; . \tag{5.12}$$

The only self-consistent solution of eqs.(5.10) and (5.12) is,

$$\psi(t, x) = (b_0 - \frac{\mu^2}{b_0})t + (b_0 + \frac{\mu^2}{b_0})x \tag{5.13}$$

where b_0 is an integration constant. But the above solution is unphysical, as it implies an exponentially growing $\chi(t, x)$. On the other hand, when $w^2 > k^2$,

$$k\dot{\psi} + \omega\psi' + \frac{1}{\beta} = 0 \Rightarrow \psi'' = \frac{k^2}{w^2}\ddot{\psi} \tag{5.14}$$

and therefore, $\psi = \psi(x - \frac{\omega}{k}t)$, leading to an exact solution for $b_1 \geq \lambda$,

$$\psi = \ln\left(\sqrt{\frac{b_1}{\lambda}}\cos^2\left(\pm\sqrt{\frac{\lambda}{2}}(t - \frac{k}{\omega}x) + b_2\right)\right) \tag{5.15}$$

where $\lambda \equiv \frac{2\omega^2\mu^2}{\omega^2 - k^2}$, while b_1 and b_2 are integration constants.

Hence,

$$\chi(t - \frac{k}{\omega}x) = \exp(\psi) \propto \cos^2\left(\pm\sqrt{\frac{\lambda}{2}}(t - \frac{k}{\omega}x) + b_2\right) , \tag{5.16}$$

with $|\omega| > |k|$, is physically meaningful and is an acceptable solution.

Indeed, we find that it is possible to have exact solutions corresponding to the four-form and circularly polarized, monochromatic GWs with phase velocity exceeding the speed of light. The caveat, however, is that eq.(5.10) being a nonlinear differential equation, deriving exact solutions corresponding to eqs.(5.8) and (5.10) with superposed GW modes with a physically acceptable group velocity is non-trivial.

6 Conclusions

Based on the invariances of the Minkowski tensor and the flat space-time Levi-Civita tensor under proper Lorentz transformations, the present study explores the possibility of extending Einstein's geometrical formulation of gravitation by including another geometrical field, $\tilde{w}$ - a generalisation of the Levi-Civita symbol, in the theory. The geometrodynamical four-form $\tilde{w}$ leads not only to a dynamical torsion, it also generates Chern-Simon (CS) extensions in the 3+1-dimensional space-times. Torsion, among several other interesting implications, is also important in teleparallel gravity theories [4, 22].

Pursuing closely the seminal work of Jackiw and Pi [6] but avoiding the use of an unphysical Lorentz vector, CS coupling between electromagnetic fields and $\tilde{w}$ comes about naturally, bringing about a modification of Einstein-Maxwell equations. Adopting the formalism of [6], a gravitational CS term is also constructed that leads to a modified Cotton tensor. However, when the dynamical equation of the four-form is used in the ensuing Einstein equation, it is shown that it is the standard Cotton tensor that affects the space-time dynamics.

The scalar-density ϕ associated with $\tilde{w}$ leads to a well-defined exterior derivative that turns a differential p-form density into a (p+1)-form density of same weight. Since the notion of an n-form and exterior derivative in a differential manifold does not require either an affine connection or a metric, further studies are required to investigate the role of $\tilde{w}$ in situations where metric is degenerate as well as its impact on the manifold-orientability.

The geometrodynamical four-form field is shown to correspond to a pseudo-scalar particle. From a simplified semi-classical treatment, it is demonstrated that the Bose-Einstein condensates of such pseudo-scalars can give rise to formation of supermassive black holes through an interplay of self-gravity and quantum mechanics. Therefore, self-gravitating BECs of pseudo-scalar particles associated with the geometrodynamical four-form field may solve the long standing problem concerning the frequent discoveries of tens of billion solar mass SMBHs at epochs when the universe was less than billion years old.

It is also proved that in the presence of the dynamical four-form, equations for the linearly polarized gravitational waves can be decoupled using circularly polarized waveforms as well as exact and self-consistent solutions of the dynamical four-form equation coupled to a monochromatic gravitational wave (albeit with phase velocity $> c$) can be obtained. However, much more work is needed to obtain realistic solutions of gravitational waves, with group velocities $\leq c$, propagating in the background consisting of $\tilde{w}$.

7 Acknowledgements

It is a pleasure to thank Prof. Vesselin Petkov for organising the Third Minkowski Meeting that entailed stimulating discussions on the fundamental aspects of space-time physics.

References

[1] Das Gupta, P. (2009), arXiv:0905.1621

[2] Schutz, B.F., 1980. Geometrical methods of mathematical physics. Cambridge university press.

[3] Hehl, F.W., 1973. Spin and torsion in general relativity: I. Foundations. General relativity and gravitation, 4, pp.333-349.

[4] Hehl, Friedrich W.; von der Heyde, Paul; Kerlick, G. David; Nester, James M., Reviews of Modern Physics, Volume 48, Issue 3, July 1976, pp.393-416

[5] Mukunda, N. (1997), in Geometry, fields and cosmology: techniques and applications (Vol. 88), eds. Iyer, B.R. and Vishveshwara, C.V., Springer Science & Business Media 2013.

[6] Jackiw, R. and Pi, S.-Y., 2003. Chern-Simons modification of general relativity. Physical Review D, 68(10), p.104012.

[7] Das Gupta, P. (2010), On Chern–Simons corrections to magnetohydrodynamics equations. Radiation Effects & Defects in Solids: Incorporating Plasma Science & Plasma Technology, 165(2), 106-113.

[8] Fukuyama, T. (2024). Axion and the SuperMassive Black Holes at high z. International Journal of Modern Physics A, 2350191.

[9] Ho, L. C. (1999), *Observational Evidence for Black Holes in the Universe*, (Springer Netherlands) pp. 157-186

[10] Ferrarese, L., & Merritt, D. (2000), Ap J, 539, L9

[11] Kormendy, J., & Ho, L. C. (2013), Ann. Rev. Astr. & Ap, 51, 511

[12] Volonteri, M., Habouzit, M. and Colpi, M., 2021. The origins of massive black holes. Nature Reviews Physics, 3(11), pp.732-743.

[13] Banados, E., Venemans, B. P., Mazzucchelli, C., Farina, E. P., Walter, F., Wang, F., ... & Winters, J. M. (2018). An 800-million-solar-mass black hole in a significantly neutral Universe at a redshift of 7.5. Nature, 553(7689), 473-476.

[14] Maiolino, R., Scholtz, J., Witstok, J. et al. A small and vigorous black hole in the early Universe. Nature (2024)

[15] Mehrgan, K., Thomas, J., Saglia, R., Mazzalay, X., Erwin, P., Bender, R., Kluge, M. and Fabricius, M., 2019. A 40 billion solar-mass black hole in the extreme core of Holm 15A, the central galaxy of Abell 85. The Astrophysical Journal, 887(2), p.195.

[16] Wolf, Christian, et al. "The accretion of a solar mass per day by a 17-billion solar mass black hole." Nature Astronomy (2024): 1-10.

[17] Das Gupta, P. & Thareja, E. (2017). Supermassive black holes from collapsing dark matter Bose–Einstein condensates. Classical and Quantum Gravity, 34(3), 035006.

[18] Das Gupta, P. , & Rahman, F. (2018). Aspects of Black Hole Physics and Formation of Super-massive Black Holes from Ultra-light Dark Bosons. The Physical Universe eds. S. M. Wagh, S. D. Maharaj & G. Chon (2018, Published by Central India Research Institute, India) arXiv:1801.02559.

[19] Das Gupta, P. (1997). The Cosmological Constant: A Tutorial. in Geometry, Fields and Cosmology: Techniques and Applications (Vol.88), eds. B. R. Iyer & C. V. Vishveshwara, 525-548.

[20] Hawking, S. W. (1974). Nature, 248, 30-31.

[21] Alexander, S. and Martin, J., 2005. Phys. Rev. D 71, 063526.

[22] Bahamonde, S., Dialektopoulos, K.F., Escamilla-Rivera, C., Farrugia, G., Gakis, V., Hendry,

M., Hohmann, M., Said, J.L., Mifsud, J. and Di Valentino, E., 2023. Teleparallel gravity: from theory to cosmology. Reports on Progress in Physics, 86(2), p.026901.

10   Relationship between the metric tensor and the field tensor. The structure of space-time

Piotr Ogonowski

Abstract   The document presents the main conclusions regarding the relationship between the metric tensor and the field tensor. Possible implications for further research of the presented approach are analyzed.

Keywords: Field theory, Quantum mechanics, General Relativity, Lagrangian mechanics

1   Introduction

Since H. Minkowski described flat space-time with a metric tensor [1], and then A. Einstein showed that the geometry of space-time does not have to be flat and is related to the energy–momentum tensor [2], we have been trying to describe and understand what space-time actually is [3].

There are many works that develop Einstein's concepts by incorporating further fields into the General Relativity, e.g. [4], [5], however, most of them do not answer the fundamental question about the essence of space-time itself. In most published articles, space-time is simply the scenery in which physical phenomena take place [6], [7].

In this article, the author would like to present the conclusions from his previous research, which shows a potential solution to this over 100-year-old puzzle. Let the introduction to this article be the question: "Is there space-time without a field?"

Space-time without a field would essentially mean that we are dealing with an inertial frame. But can an inertial frame exist in nature? All frames we know are non-inertial, and even Lagrange and Hamilton considered the inertial frame to be idealized and unreal, but useful for physical considerations. [8]

However, if there is no space-time without a field, what is the relationship between field and space-time? Can the field be related to the metric tensor in some way?

In the following chapters it will be shown, that not only is this possible, but that in fact adopting

Eric Ling and Annachiara Piubello (Eds), SPACETIME 1908-2023. Selected peer-reviewed papers presented at the *Third Hermann Minkowski Meeting on the Foundations of Spacetime Physics*, 11-14 September 2023, Albena, Bulgaria (Minkowski Institute Press, Montreal 2024). ISBN 978-1-998902-25-5 (softcover), ISBN 978-1-998902-26-2 (ebook).

this method of analysis provides very expected results.

The author uses the Einstein summation convention, metric signature $(+,-,-,-)$ and commonly used notations.

2 Main conclusions from the last research

According to [9], stress-energy tensor $T^{\alpha\beta}$ for a system in a given space-time described by a metric tensor $g^{\alpha\beta}$ may be defined as

$$T^{\alpha\beta} = \varrho\, U^{\alpha} U^{\beta} - \left(c^2 \varrho + \Lambda_\rho\right)\left(g^{\alpha\beta} - \xi\, h^{\alpha\beta}\right) \tag{2.1}$$

where ϱ_o is for rest mass density, γ is Lorentz gamma factor and

$$\varrho \equiv \varrho_o \gamma \tag{2.2}$$

$$\frac{1}{\xi} \equiv \frac{1}{4}\, g_{\mu\nu}\, h^{\mu\nu} \tag{2.3}$$

$$\Lambda_\rho \equiv \frac{1}{4\mu_o} \mathbb{F}^{\alpha\mu}\, g_{\mu\gamma}\, \mathbb{F}^{\beta\gamma} g_{\alpha\beta} \tag{2.4}$$

$$h^{\alpha\beta} \equiv 2\,\frac{\mathbb{F}^{\alpha\delta}\, g_{\delta\gamma}\, \mathbb{F}^{\beta\gamma}}{\sqrt{\mathbb{F}^{\alpha\delta}\, g_{\delta\gamma}\, \mathbb{F}^{\beta\gamma}\, g_{\mu\beta}\, \mathbb{F}_{\alpha\eta}\, g^{\eta\xi}\, \mathbb{F}^{\mu}{}_{\xi}}} \tag{2.5}$$

where $\mathbb{F}^{\alpha\beta}$ represents electromagnetic field tensor.

The stress–energy tensor for electromagnetic filed, denoted as $\varUpsilon^{\alpha\beta}$ may be presented as follows

$$\varUpsilon^{\alpha\beta} \equiv \Lambda_\rho\left(g^{\alpha\beta} - \xi\, h^{\alpha\beta}\right) = \Lambda_\rho g^{\alpha\beta} - \frac{1}{\mu_o}\mathbb{F}^{\alpha\delta}\, g_{\delta\gamma}\, \mathbb{F}^{\beta\gamma} \tag{2.6}$$

The pressure p in the system is equal to

$$p \equiv c^2 \varrho + \Lambda_\rho \tag{2.7}$$

so the stress-energy tensor $T^{\alpha\beta}$ for a system may be then denoted as just

$$T^{\alpha\beta} = \varrho\, U^{\alpha} U^{\beta} - \frac{p}{\Lambda_\rho}\varUpsilon^{\alpha\beta} \tag{2.8}$$

It has been shown that the above stress-energy tensor can be extended by other fields without losing properties of the solution.

The described solution requires an amendment to continuum mechanics, which introduces a relationship between density tensors and the curvature of space-time.

$$\partial_\alpha U^\alpha = -\frac{d\gamma}{dt} \quad \rightarrow \quad \partial_\alpha \varrho\, U^\alpha = 0 \tag{2.9}$$

Denoting four-momentum density as $\varrho U^\mu = \varrho_o \gamma U^\mu$, total four-force density f^μ acting in the system is

$$f^\mu \equiv \varrho A^\mu = \partial_\alpha \varrho U^\mu U^\alpha \tag{2.10}$$

Denoting rest charge density in the system as ρ_o and

$$\rho \equiv \rho_o \gamma \tag{2.11}$$

electromagnetic four-current J^α is equal to

$$J^\alpha \equiv \rho U^\alpha = \rho_o \gamma U^\alpha \tag{2.12}$$

In the flat Minkowski space-time, total four-force density f^α acting in the system calculated from $\partial_\beta T^{\alpha\beta} = 0$ is the sum of electromagnetic (f^α_{EM}), gravitational (f^α_{gr}) and other (f^α_{oth}) four-force densities

$$f^\alpha = \begin{cases} f^\alpha_{EM} \equiv \partial_\beta \Upsilon^{\alpha\beta} \quad (electromagnetic) \\ + \\ f^\alpha_{gr} \equiv \left(\eta^{\alpha\beta} - \xi\, h^{\alpha\beta}\right) \partial_\beta p \quad (gravitational) \\ + \\ f^\alpha_{oth} \equiv \frac{\varrho c^2}{\Lambda_\rho} f^\alpha_{EM} \quad (other) \end{cases} \tag{2.13}$$

As was shown in [9], in curved space-time ($g_{\alpha\beta} = h_{\alpha\beta}$) part of the stress-energy tensor $T^{\alpha\beta}$ related to fields vanishes, and presented method reproduces Einstein Field Equations with an accuracy of $\frac{4\pi G}{c^4}$ constant and with cosmological constant Λ dependent on invariant of electromagnetic field tensor $\mathbb{F}^{\alpha\gamma}$

$$\Lambda = -\frac{\pi G}{c^4 \mu_o} \mathbb{F}^{\alpha\mu} h_{\mu\gamma} \mathbb{F}^{\beta\gamma} h_{\alpha\beta} = -\frac{4\pi G}{c^4} \Lambda_\rho \tag{2.14}$$

where $h_{\alpha\beta}$ appears to be metric tensor of the space-time in which all motion occurs along geodesics and where Λ_ρ describes vacuum energy density.

It was also shown in [9], that Einstein tensor describes the space-time curvature related to vanishing in curved space-time four-force densities $f^\alpha_{gr} + f^\alpha_{oth}$ where Ricci Tensor and Einstein tensor are expressed with an accuracy of $\frac{4\pi G}{c^4}$ constant as

$$R^{\alpha\beta} = 2\varrho U^\alpha U^\beta - p\, h^{\alpha\beta} \tag{2.15}$$

$$G_{\alpha\beta} = R_{\alpha\beta} - \frac{1}{2} R\, h_{\alpha\beta} \tag{2.16}$$

therefore

$$G_{\alpha\beta} - \Lambda_\rho h_{\alpha\beta} = 2\, T_{\alpha\beta} \tag{2.17}$$

The presented solution creates a coherent picture in which space-time is in fact a way of perceiving the field (in this case: electromagnetic field). This solution allows for further development, introducing additional fields, different parameterization and simple transformation between Minkowski space-time and curvilinear reference systems.

139

It also shows, that description of motion in curved space-time and its description in flat Minkowski space-time with fields are equivalent, and the transformation between curved space-time and Minkowski space-time is known, because the geometry of curved space-time depends on the field tensor. This allows for a significant simplification of research, because the results obtained in flat Minkowski space-time can be easily transformed into curved space-time. The last missing link seems to be the quantum description.

3 Potential, farther consequences

It is discussed in [10], that by imposing additional condition on normalized stress-energy tensor in flat Minkowski space-time with fields

$$0 = \partial_\beta \left(\frac{T^{\alpha\beta}}{\eta_{\mu\gamma}T^{\mu\gamma}} \right) + \partial^\alpha \ln \left(\eta_{\mu\gamma}T^{\mu\gamma} \right) \tag{3.1}$$

one obtains following results

- Lagrangian density for the systems appears to be equal to $\mathcal{L} = \Lambda_\rho = \frac{1}{4\mu_o}\mathbb{F}^{\alpha\beta}\mathbb{F}_{\alpha\beta}$

- Stress-energy tensor may be simplified to familiar form: $T^{\alpha\beta} = \frac{1}{\mu_o}\mathbb{F}^{\alpha\gamma}\partial^\beta\mathbb{A}_\gamma - \Lambda_\rho\eta^{\alpha\beta}$

- $H^\beta \equiv -\frac{1}{c}\int T^{0\beta}\,d^3x$ acts as cannonical four-momentum for the point-like particle, it includes electromagnetic four-potential and other terms responsible for other fields

- The vanishing four-divergence of the canonical four-momentum H^β turns out to be the consequence of Poynting theorem

- Some gauge of electromagnetic four-potential may be expressed as $\mathbb{A}^\mu = -\frac{\Lambda_\rho}{p}\frac{\varrho_o}{\rho_o}U^\mu$

Gravitational four-force in this solution appears to be equal to

$$f^\alpha_{gr} = \varrho \left(\frac{d\ln(p)}{d\tau}U^\mu - c^2\partial^\mu\ln(p) \right) \tag{3.2}$$

One may also express canonical four-momentum H^μ as

$$H^\mu = P^\mu + V^\mu = -\frac{\gamma L}{c^2}U^\mu + \mathbb{S}^\mu \tag{3.3}$$

where P^μ is four-momentum, L is for Lagrangian, $\mathbb{S}^\mu$ due to its properties, seems to be some description of the spin

$$\mathbb{S}^\beta \equiv \int \frac{\epsilon_o\Lambda_\rho}{\gamma c\rho_o}\,\mathbb{F}^{0\mu}\partial_\mu U^\beta\,d^3x \tag{3.4}$$

and four-vector V^μ describes the transport of energy due to the field

$$V^\mu = q\mathbb{A}^\mu + \frac{\varrho c^2\gamma^2}{p}P^\beta + \frac{\varrho c^2}{p}\mathbb{S}^\mu + Y^\mu \tag{3.5}$$

where $\mathbb{A}^\mu$ is electromagnetic four-potential and Y^μ is the volume integral of the Poyinting four-vector.

If, indeed, in the absence of fields, Lagrangian, Hamiltonian and Action vanish...
Since in the limit of the inertial system one gets $P^\mu X_\mu = mc^2\tau$, therefore, to ensure vanishing Hamilton's principal function in the inertial system, one can expect that

$$V^\mu X_\mu \equiv -mc^2\tau \qquad (3.6)$$

what yields vanishing in the inertial system Lagrangian in form of

$$-\gamma L = F^\mu X_\mu \qquad (3.7)$$

where F^μ is four-force and where one obtains

$$\mathbb{S}^\mu \mathbb{S}_\mu = H^\mu H_\mu - \left(\frac{\gamma L}{c}\right)^2 \qquad (3.8)$$

To ensure compliance with the equations of quantum mechanics it suffices if

$$\mathbb{S}^\mu \mathbb{S}_\mu = m^2 c^2 - \left(\frac{\gamma L}{c}\right)^2 \qquad (3.9)$$

By introducing quantum wave function Ψ

$$\Psi \equiv e^{\pm i K^\mu X_\mu} \qquad (3.10)$$

where K^μ is wave four-vector related to cannonical four-momentum

$$\hbar K^\mu \equiv H^\mu \qquad (3.11)$$

from (3.8) one obtains Klein-Gordon equation

$$\left(\Box + \frac{m^2 c^2}{\hbar^2}\right)\Psi = 0 \qquad (3.12)$$

It seems, that considered method [10] may allow the analysis of the system in the quantum approach, classical approach and the introduction of a field-dependent metric for curved space-time, which may help with connecting previously divergent descriptions of physical systems.

As it was shown in [10], there is also possibility to obtain Hamiltonian density that agrees with the classical Hamiltonian density for electromagnetic field, considered in Quantum Field Theory. Such Hamiltonian density was currently considered mainly for sourceless regions and to consider the system with electromagnetic field only.

According to the presented results, it may appear that, actually, this Hamiltonian density describes the entire physical system, containing all known interactions. For this reason, the discussed method might also greatly simplify Quantum Field Theory equations.

4 Conclusions

As demonstrated in the above article, it is possible to consider space-time as a method of field perception. This not only leads to a dual description in which a physical system can be analyzed in a flat space-time with fields or a curved space-time without fields. All other methods of analysis are also possible in which space-time is partially curved and other phenomena are still described by residual fields.

The use of the described method allows to obtain not only a simple transformation between the GR equations and flat space-time with fields, but also to obtain quantum solutions and analyze the system both in the QM mathematical apparatus and in QFT.

As demonstrated in the source articles, it is also possible to analyze such a physical system with the use of the tools of continuum mechanics, based on the Cauchy momentum equation.
Perhaps a completely new area of research opens up that can be conducted based on the above method.

References

[1] V. Petkov, Minkowski spacetime: a hundred years later, 2010.

[2] O. Darrigol, Relativity principles and theories from Galileo to Einstein, 2021.

[3] G. Musser, What is spacetime?, Nature 557 (2018) S3–S6.

[4] M. L. Ruggiero, A. Ortolan, C. C. Speake, Galactic dynamics in general relativity: the role of gravitomagnetism, Classical and Quantum Gravity 39 (2021).

[5] R. T. Hammond, New fields in general relativity, Contemporary Physics 36 (1995) 103–114.

[6] P. G. LeFloch, T.-C. Nguyen, The seed-to-solution method for the Einstein equations, arXiv: Analysis of PDEs (2019).

[7] M. Hohmann, C. Pfeifer, N. Voicu, Mathematical foundations for field theories on Finsler spacetimes, Journal of Mathematical Physics (2021).

[8] L. B. Sklar, Philosophy and the foundations of dynamics, 2012.

[9] P. Ogonowski, Proposed method of combining continuum mechanics with Einstein Field Equations, International Journal of Modern Physics D 2350010 (2023) 15.

[10] P. Ogonowski, Developed method. Interactions and their quantum picture (2023). arXiv:2306.14906.

11   Time Crystals and Phase-Space Noncommutative Quantum Mechanics

Orfeu Bertolami and A. E. Bernardini

Abstract   We argue that time crystal properties naturally arise from phase-space noncommutative quantum mechanics. In order to exemplify our point we consider the 2-dimensional noncommutative quantum harmonic oscillator and show that it exibihits periodic oscillations that can be identified as time crystals.

Keywords: time crystals, phase-space noncommutative quantum mechanics, 2-dimensional noncommutative quantum harmonic oscillator

Based on talk presented by one of us (O.B.) at the Third Minkowsky Meeting on the Foundations of the Spacetime Physics at Albena, Bulgaria, 11-14 September 2023.

1   Introduction

In this contribution we review the arguments presented in Ref. [1] where it was shown that time crystal features arise in the context of the phase-space noncommutative 2-dimensional quantum harmonic oscillator. As discussed in the following this is yet another new property emerging from phase-space noncommutative quantum mechanics (PSNCQM).

Time crystals are time-periodic self-organized structures that presumably arise due to the spontaneous breaking of time translation symmetry [2, 3]. They are analogous to spatial crystal lattices that form when the spontaneous breaking of space translation symmetry takes place [4]. Time crystal features were claimed to appear in ultra-cold atoms [5, 6] and spin-based solid state systems [7, 8, 9, 10, 11, 12], through which it has been argued that periodically driven systems exhibit a discrete time symmetry [13, 14]. In fact, these experiments suggest that novel phases of matter do exist [10, 11, 12] which exhibit a discrete time translation symmetry hinting the breakdown of the continuous time translation symmetry, $\hat{\mathcal{T}}_H \equiv e^{-i\hat{H}t}$. For a contextualization of time crystals with respect to the research on the physics of time, see, for instance, Ref. [15].

As is well known, if a time-independent system driven by a time-independent Hamiltonian, H, is prepared in an eigenstate $|\psi_n\rangle$, such that $H|\psi_n\rangle = E_n|\psi_n\rangle$, for the energy eigenvalue E_n, in the context of quantum mechanics (QM) the probability density at a fixed position in the configuration

Eric Ling and Annachiara Piubello (Eds), Spacetime 1908-2023. Selected peer-reviewed papers presented at the *Third Hermann Minkowski Meeting on the Foundations of Spacetime Physics*, 11-14 September 2023, Albena, Bulgaria (Minkowski Institute Press, Montreal 2024). ISBN 978-1-998902-25-5 (softcover), ISBN 978-1-998902-26-2 (ebook).

space is also time-independent. Nevertheless, the mentioned experiments suggest that time crystals exist and thus, $[\hat{H}, \rho_n] \equiv [\hat{\mathcal{T}}_H, \rho_n] \neq 0$ for $\rho_n = |\psi_n\rangle\langle\psi_n|$.

As originally argued [2], this would correspond to a spontaneous breakdown of time translation symmetry followed by a non-stationary behaviour of the eigensystem solutions. However, a no-go theorem [16], based on the time-dependent correlation functions of the order parameter, rules out the possibility of time crystals defined in this way for the ground state and for a canonical ensemble of a general Hamiltonian. We argue that the emergence of a non-stationary behaviour, and its connection with time crystal properties can be explained in terms of both position and/or momentum noncommutativity in the phase-space [1], in opposition to the *ab initio* breaking symmetry assumptions proposed in Refs. [2, 3].

2 Phase-Space Noncommutative Quantum Mechnics

We present now some of the main features of PSNCQM. Noncommutativity was firstly considered in the space coordinate domain as a way to regularize quantum field theories [17] and subsequently in string theory [18, 19, 20, 21]. The PSNCQM extension [22, 23, 24, 25, 26, 27, 28], considered here, can be formulated in terms of the Weyl-Wigner-Groenewold-Moyal (WWGM) framework [29, 30, 31], supported by a $2n$-dimensional phase-space deformed Heisenberg-Weyl algebra, where position and momentum operators, $\hat{q}_i$ and $\hat{p}_j$, obey the commutation relations,

$$[\hat{q}_i, \hat{q}_j] = i\theta_{ij}, \qquad [\hat{q}_i, \hat{p}_j] = i\hbar\delta_{ij}, \qquad [\hat{p}_i, \hat{p}_j] = i\eta_{ij}, \tag{2.1}$$

where $i, j = 1, ..., d$, and η_{ij} and θ_{ij} are the entries of invertible antisymmetric real constant $(d \times d)$ matrices, $\boldsymbol{\Theta}$ and $\mathbf{N}$, such that an equally invertible matrix, $\boldsymbol{\Sigma}$, with $\Sigma_{ij} \equiv \delta_{ij} + \hbar^{-2}\theta_{ik}\eta_{kj}$, exists, which demands that $\theta_{ik}\eta_{kj} \neq -\hbar^2\delta_{ij}$. Of course, given that $\eta_{ij} \neq 0$, $\theta_{ij} \neq 0$, the relations from Eq. (2.1) can affect the symmetries related to conserved quantities associated to quantum operators, $\hat{\mathcal{O}}$, for which $d\langle\hat{\mathcal{O}}\rangle/dt = i\hbar^{-1}\langle[\hat{H}, \hat{\mathcal{O}}]\rangle = 0$.

In this context, the key issue is if quantum operators identified by $\hat{\mathcal{O}} \to \hat{\mathcal{O}}(\{\hat{q}_i, \hat{p}_i\})$ do present a time crystal behaviour arising from the breakdown of time translational symmetries, $\langle[\hat{H}, \hat{\mathcal{O}}(\{\hat{q}_i, \hat{p}_i\})]\rangle \neq 0$, in opposition to usual QM. In order to investigate this point, the NC algebra, Eq. (2.1) can be mapped into the Heisenberg-Weyl algebra through the linear Seiberg-Witten (SW) transformation [20],

$$\hat{q}_i = A_{ij}\hat{Q}_j + B_{ij}\hat{\Pi}_j, \qquad \hat{p}_i = C_{ij}\hat{Q}_j + D_{ij}\hat{\Pi}_j, \tag{2.2}$$

where A_{ij}, B_{ij}, C_{ij} and D_{ij} are real entries of constant matrices, $\mathbf{A}, \mathbf{B}, \mathbf{C}$ and $\mathbf{D}$. In this case, one recovers the algebra of ordinary QM,

$$[\hat{Q}_i, \hat{Q}_j] = 0, \quad [\hat{Q}_i, \hat{P}_j] = i\hbar\delta_{ij}, \quad [\hat{P}_i, \hat{P}_j] = 0, \tag{2.3}$$

through the following matrix equation constraints [24], $\mathbf{AD}^T - \mathbf{BC}^T = \mathbf{I}_{d\times d}$, $\mathbf{AB}^T - \mathbf{BA}^T = \hbar^{-1}\boldsymbol{\Theta}$, and $\mathbf{CD}^T - \mathbf{DC}^T = \hbar^{-1}\mathbf{N}$, where the superscript T denotes matrix transposition.

From the WWGM framework [24, 26] for the algebra, Eq. (2.1), it is possible to show that the resulting quantum mechanical extensions have some striking features which include putative violations of the Robertson-Schrödinger uncertainty relation [32, 33], quantum correlations and information collapse in gaussian quantum systems [34, 35, 36, 37, 38], new regularizing features in minisuperspace quantum cosmology models [32, 39] and in black-hole physics [40, 41, 42], putative violations of the Equivalence Principle [43, 44] and, likewise ordinary QM, non-locality properties that can be captured by the Bell operator [45]. In fact, the generalized WWGM star-product, the extended Moyal bracket and the noncommutative (NC) Wigner function framework ensure that observables are independent of any particular choice of the SW map [26].

3 The Noncommutative 2-dimensional Quantum Harmonic Oscillator and the Emergence of Time Crystal Behaviour

Aiming to exemplify the emerging time crystal behaviour we consider the 2-dimensional harmonic oscillator in the PSNCQM [46] with Hamiltonian,

$$\hat{H}_{HO}(\hat{\mathbf{q}}, \hat{\mathbf{p}}) = \frac{\hat{\mathbf{p}}^2}{2m} + \tfrac{1}{2}m\omega^2\hat{\mathbf{q}}^2, \tag{3.1}$$

on the NC "$x - y$" plane, with position and momentum satisfying the NC algebra, Eq. (2.1), now with $i, j = 1, 2$, $\theta_{ij} = \theta\epsilon_{ij}$ and $\eta_{ij} = \eta\epsilon_{ij}$, where ϵ_{ij} is the 2-dimensional Levi-Civita tensor. The map to commutative operators is given by

$$\hat{Q}_i = \mu\left(1 - \frac{\theta\eta}{\hbar^2}\right)^{-1/2}\left(\hat{q}_i + \frac{\theta}{2\lambda\mu\hbar}\epsilon_{ij}\hat{p}_j\right), \quad \hat{\Pi}_i = \lambda\left(1 - \frac{\theta\eta}{\hbar^2}\right)^{-1/2}\left(\hat{p}_i - \frac{\eta}{2\lambda\mu\hbar}\epsilon_{ij}\hat{q}_j\right), \tag{3.2}$$

in terms of the SW map,

$$\hat{q}_i = \lambda\hat{Q}_i - \frac{\theta}{2\lambda\hbar}\epsilon_{ij}\hat{\Pi}_j, \qquad \hat{p}_i = \mu\hat{\Pi}_i + \frac{\eta}{2\mu\hbar}\epsilon_{ij}\hat{Q}_j, \tag{3.3}$$

which is invertible for $\theta\eta \neq \hbar^2$, and the parameters λ and μ satisfying the condition

$$\frac{\theta\eta}{4\hbar^2} = \lambda\mu(1 - \lambda\mu). \tag{3.4}$$

The Hamiltonian in terms of the commutative variables, $\hat{Q}_i$ and $\hat{\Pi}_i$, reads [46]

$$\hat{H}_{HO}(\hat{\mathbf{Q}}, \hat{\mathbf{\Pi}}) = \alpha^2\hat{\mathbf{Q}}^2 + \beta^2\hat{\mathbf{\Pi}}^2 + \gamma\sum_{i,j=1}^{2}\epsilon_{ij}\hat{\Pi}_i\hat{Q}_j, \tag{3.5}$$

where $\alpha^2 \equiv m\omega^2\lambda^2/2 + \eta^2/(8m\hbar^2\mu^2)$, $\beta^2 \equiv \mu^2/(2m) + m\omega^2\theta^2/(8\hbar^2\lambda^2)$, and $\gamma \equiv m\omega^2\theta/(2\hbar) + \eta/(2m\hbar)$, from which one obtains the following set of coupled equations of motion,

$$\begin{aligned} \dot{\Pi}_i &= -\tfrac{i}{\hbar}\langle\left[\hat{\Pi}_i, \hat{H}_{HO}\right]\rangle = -2\alpha^2\,Q_i - \gamma\,\varepsilon_{ji}\Pi_j, \\ \dot{Q}_i &= -\tfrac{i}{\hbar}\langle\left[\hat{Q}_i, \hat{H}_{HO}\right]\rangle = 2\beta^2\,\Pi_i - \gamma\,\varepsilon_{ji}Q_j, \end{aligned} \tag{3.6}$$

with $Q_i \equiv \langle\hat{Q}_i\rangle$ and $\Pi_i \equiv \langle\hat{\Pi}_i\rangle$. In this case, $\mathbf{Q} = (Q_1, Q_2)$ and $\mathbf{\Pi} = (\Pi_1, \Pi_2)$ may be interpreted as the dynamical variables within the WWGM formalism for which the solutions are given by [46]

$$\begin{aligned} Q_1(t) &= x\cos(\Omega t)\cos(\gamma t) + y\cos(\Omega t)\sin(\gamma t) + \tfrac{\beta}{\alpha}\left[\pi_y\sin(\Omega t)\sin(\gamma t) + \pi_x\sin(\Omega t)\cos(\gamma t)\right], \\ Q_2(t) &= y\cos(\Omega t)\cos(\gamma t) - x\cos(\Omega t)\sin(\gamma t) - \tfrac{\beta}{\alpha}\left[\pi_x\sin(\Omega t)\sin(\gamma t) - \pi_y\sin(\Omega t)\cos(\gamma t)\right], \\ \Pi_1(t) &= \pi_x\cos(\Omega t)\cos(\gamma t) + \pi_y\cos(\Omega t)\sin(\gamma t) - \tfrac{\alpha}{\beta}\left[y\sin(\Omega t)\sin(\gamma t) + x\sin(\Omega t)\cos(\gamma t)\right], \\ \Pi_2(t) &= \pi_y\cos(\Omega t)\cos(\gamma t) - \pi_x\cos(\Omega t)\sin(\gamma t) + \tfrac{\alpha}{\beta}\left[x\sin(\Omega t)\sin(\gamma t) - y\sin(\Omega t)\cos(\gamma t)\right], \tag{3.7} \end{aligned}$$

where x, y, π_x, and π_y are arbitrary parameters, and

$$\Omega = 2\alpha\beta = \sqrt{(2\lambda\mu - 1)^2\omega^2 + \gamma^2} = \sqrt{\omega^2 + \gamma^2 - \frac{\theta\eta}{\hbar^2}}, \tag{3.8}$$

with λ and μ being eliminated by the constraint Eq. (3.4). Of course, if one sets $\theta = \eta = 0$, and therefore $\gamma = 0$, one recovers the solutions for the 2-dimensional harmonic oscillator with uncoupled $x - y$ coordinates and $\Omega = \omega$. For θ, $\eta \neq 0$, the above results lead to two decoupled time-invariant quantities,

$$\sum_{i=1}^{2} \left(\frac{\alpha}{\beta} Q_i(t)^2 + \frac{\beta}{\alpha} \Pi_i(t)^2 \right) = \frac{\alpha}{\beta}(x^2 + y^2) + \frac{\beta}{\alpha}(\pi_x^2 + \pi_y^2),$$

$$\sum_{i,j=1}^{2} \left(\epsilon_{ij} Q_i(t) \Pi_j(t) \right) = x\,\pi_y - y\,\pi_x. \tag{3.9}$$

The changes introduced by the NC variables can be evinced by setting $\pi_x = \pi_y = \sqrt{\alpha\hbar/2\beta}$, and $x = y = \sqrt{\beta\hbar/2\alpha}$, so that the associated x and y translational energy contributions evolve as

$$E_i = \alpha\beta \left(\frac{\alpha}{\beta} Q_i(t)^2 + \frac{\beta}{\alpha} \Pi_i(t)^2 \right) = \frac{\hbar\Omega}{2} \left(1 - (-1)^i \sin(2\gamma t) \right), \tag{3.10}$$

with $i = 1, 2$, from which a typical low frequency γ-dependent beating behaviour is encountered [46]. Such a time-dependent periodic modification is a new feature of the NC harmonic oscillator ground state. The *stargen*functions for the Hamiltonian, Eq. (3.5), are obtained from the *stargen* value equation,

$$H_{HO}^W \star \rho_{n_1,n_2}^W(\mathbf{Q}, \mathbf{\Pi}) = E_{n_1,n_2}\, \rho_{n_1,n_2}^W(\mathbf{Q}, \mathbf{\Pi}), \tag{3.11}$$

where $W(\mathbf{Q}, \mathbf{\Pi})$ is the eigenstate associated Wigner function, from which one has [24],

$$\rho_{n_1,n_2}^W(\mathbf{Q}, \mathbf{\Pi}) = \frac{(-1)^{n_1+n_2}}{\pi^2\hbar^2} \exp\left[-\frac{1}{\hbar} \left(\frac{\alpha}{\beta}\mathbf{Q}^2 + \frac{\beta}{\alpha}\mathbf{\Pi}^2 \right) \right] L_{n_1}^0 \left(\Omega_+/\hbar \right) L_{n_2}^0 \left(\Omega_-/\hbar \right), \tag{3.12}$$

where L_n^0 are the associated Laguerre polynomials, n_1 and n_2 are non-negative integers, and

$$\Omega_\pm = \frac{\alpha}{\beta}\mathbf{Q}^2 + \frac{\beta}{\alpha}\mathbf{\Pi}^2 \mp 2 \sum_{i,j=1}^{2} (\epsilon_{ij}\Pi_i Q_j), \tag{3.13}$$

such that the energy spectrum is given by $E_{n_1,n_2} = \hbar\left[2\alpha\beta(n_1 + n_2 + 1) + \gamma(n_1 - n_2) \right]$.

It has been shown that the 2-dimensional harmonic oscillator on the NC plane approaches the classical limit, and exhibits well-marked quantum effects such as state swapping, quantum beating, and some extent of loss of quantum coherence. These properties are not due to extrinsic or artificial time-dependent effects, but due to the entanglement [34] induced by NC "x" and "y" Hilbert spaces mapped by the time-independent Hamiltonian. From Eq. (3.5) one sees that quantum states associated to x and y degrees of freedom are no longer independent. This differs from the standard quantum mechanical configuration, for which x and y modes are each of them associated to decoupled stationary behaviour [46]. In order to clarify the relation with the time crystal behaviour, one should get back to the Hamiltonian Eq. (3.1) and examine the contributions of $\{\hat{q}_1, \hat{p}_1\}$ and

$\{\hat{q}_2,\, \hat{p}_2\}$ to the energy and to the eigenstates. Indeed, identifying the associated energy of each i-sector $(i = 1,\, 2)$ as

$$\hat{\xi}_i = \frac{\hat{p}_i^2}{2m} + \tfrac{1}{2}m\omega^2\hat{q}_i^2, \tag{3.14}$$

from standard QM, one would have $\dot{\hat{\xi}}_i = i\hbar^{-1}\langle[H,\,\hat{\xi}_i]\rangle = 0$ (with $\langle\hat{\xi}_i\rangle \equiv \xi_i$). However, after recasting $\{\hat{q}_i,\, \hat{p}_i\}$ in terms of the SW map, Eqs. (3.2)-(3.3), with Ω, ω and γ constrained by Eq. (3.8), one finds an unexpected non-stationary behaviour for each of the energy contributions,

$$\xi_i(t) \;\; = \;\; \frac{\hbar\Omega}{2}\left\{1 - (-1)^i\left[\sqrt{1 - \tfrac{\omega^2}{\Omega^2}}\,\left(\cos(2\gamma t)\cos(2\Omega t) - \tfrac{\gamma}{\Omega}\sin(2\gamma t)\sin(2\Omega t)\right)\right.\right. \tag{3.15}$$

$$\left.\left. + \tfrac{\omega}{\Omega}\sqrt{1 - \tfrac{\gamma^2}{\Omega^2}}\,\sin(2\gamma t)\right]\right\},$$

from which arise the time crystal non-stationary behaviour driven by Ω and a beating behaviour driven by γ. This is depicted in Fig. 40; if either θ or η vanishes, one has $\Omega^2 = \omega^2 + \gamma^2$ and

$$\xi_i(t) = \frac{\hbar\Omega}{2}\left\{1 - (-1)^i\left[\tfrac{\gamma}{\Omega}\left(\cos(2\gamma t)\cos(2\Omega t) - \tfrac{\gamma}{\Omega}\sin(2\gamma t)\sin(2\Omega t)\right) + \left(1 - \tfrac{\gamma^2}{\Omega^2}\right)\sin(2\gamma t)\right]\right\}. \tag{3.16}$$

For the arbitrary choice of $\gamma/\Omega = 0.002$, it is shown in the smaller window of Fig. 40, the energy decoupled γ-frequency NC quantum beating (dashed lines) and the externally driven Ω-frequency time crystal behaviour (dotted lines) for $\gamma t \gtrsim 0$. In Fig. 41 the time derivative of the energy is depicted, from which the magnitude of the time crystal oscillating behaviour can be quantified. From Eq. (3.15), the externally driven oscillation amplitude, $\hbar\Omega/2$, is modulated by a factor γ/Ω.

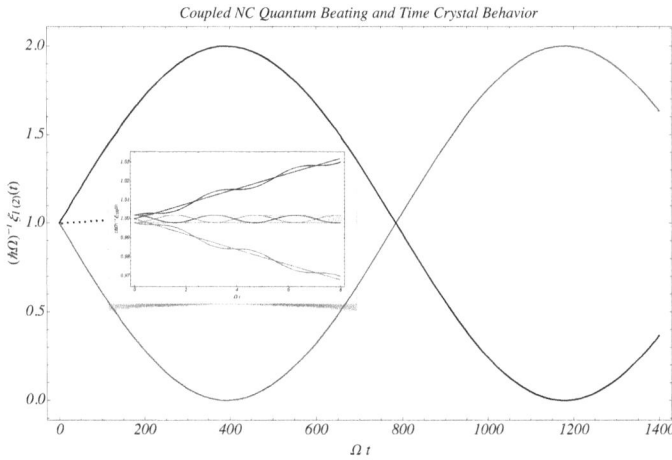

Figure 40: (Colour on line) Dimensionless NC associated energies , $(\hbar\Omega)^{-1}\xi_{1(2)}$ (black (blue) line) as function of Ωt, for $\gamma/\Omega = 0.002$. Decoupled γ-frequency NC quantum beating (dashed lines) and Ω-frequency time crystal behaviour (dotted lines) for $\gamma t \gtrsim 0$ are identified in the zoom in window. Figure from Ref. [1]

Given that the corrections due to the γ parameter are small, the beating oscillations are presumably difficult to measure. On the other hand, an effect is acessible for $\gamma \ll \Omega$ and $\gamma t \gtrsim 0$ implying that

$$\xi_i(t) \approx \frac{\hbar\Omega}{2}\left[1 - (-1)^i\tfrac{\gamma}{\Omega}\left(2\,\Omega t + \cos(2\Omega t)\right)\right], \tag{3.17}$$

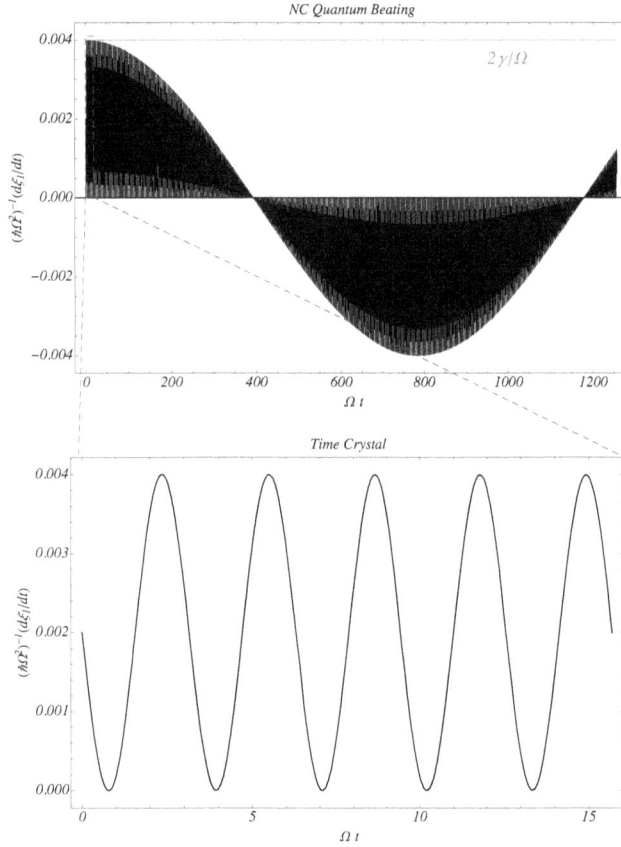

Figure 41: (Colour on line) Dimensionless time derivative, $(\hbar\Omega^2)^{-1}\dot{\xi}_1$ as function of Ωt, for $\gamma/\Omega = 0.002$. The NC beating behaviour is depicted by the first plot and the time crystal periodic behaviour, driven by the amplitude modulation, $\hbar\gamma\Omega(\equiv \gamma/\Omega \times \hbar\Omega^2)$ (red line) is depicted in zoom in plot. Figure from Ref. [1]

at first order in γ. In this case,

$$\dot{\xi}_i(t) \approx (-1)^{i+1}\hbar\gamma\Omega\left[1 - \sin(2\Omega t)\right], \tag{3.18}$$

a time crystal periodic behaviour arise with a measurable energy time derivative oscillation amplitude, $\hbar\gamma\Omega$, driven by both the NC parameter, γ, and the external oscillation frequency $\Omega \sim \omega$.

It should be added that the states of the 2-dimensional NC quantum harmonic oscillator here examined satisfy the no-cloning and no-deleting theorems without additional constraints [47], given that these theorems depend only on the unitarity of QM, which is shared by PSNCQM. From these results it follows that some specific features of the system studied here cannot be ruled out on the basis of the same no-go theorems that render the Wilczek's hypothesis untenable [16].

Let us close this section pointing out that, as discussed in Ref. [1] ,the effects presented above can also be inferred from the behaviour of the time derivative of the Wigner eigenfunctions for each i-sector of the 2-dimensional harmonic oscillator.

148

4 Discussion and Conclusions

Let us briefly discuss our results. First of all, it is natural to expect that for too small values of γ/Ω (say $\gamma/\Omega \ll 0.002$), time crystal and NC beating patterns are hard to measure. Nevertheless, it is interesting that 2-dimensional (or 3-dimensional) Bose-Einstein condensates with time crystal like behaviour have been detected through a resonance between two (or three) oscillating mirror atoms [48, 49]. Our results suggest that the observed behaviour is a natural explanation for the quasi-periodic eigenstates driven by ξ_1, ξ_2, and γ. In fact, for extended time intervals, in which the NC quantum beating takes place, the short time scale Ω-frequency periodic behaviour turns into a quasi-periodic one, due to the periodic corrections from γ-frequency. This suggests that a connection of our results with the spontaneous formation of time quasi-crystals from atoms bouncing between a pair of orthogonal mirror atoms [48, 50] is possible.

On general grounds, our results show that the non-stationary behaviour associated to time crystals, arises entirely from either position (q) or momentum (p) noncommutativity, i.e., from $[\hat{q}_i, \hat{q}_j] \neq i\theta\epsilon_{ij}$ and/or $[\hat{p}_i, \hat{p}_j] \neq i\eta\epsilon_{ij}$, with no need of an *ab initio* hypothesis of spontaneous breaking of time translation symmetry. Thus, we can conclude that the NC parameters naturally give origin to periodic oscillations that resemble time crystals. Conversely, besides accounting for the emergence of such unexpected properties, the measurable oscillation amplitude $\propto \hbar\gamma\Omega$ (γ/Ω driven by the NC parameters, γ, and the external oscillation frequency $\Omega \sim \omega$, can themselves be tested in order to set bounds to the NC parameters. Thus, we hope that our discussion in the context of the 2-dimensional noncommutative quantum harmonic oscillator might stimulate further attempts to experimentally test such a fascinating phenomena as time crystals. In fact, recent claims on the observation of continuous time crystal behaviour in quantum processors [51, 52], atom-cavity system excited by photon oscillations [53] and in an electron-nuclear spin system [54] show that there is a vivid interest in searchimg for concrete experimental evidence of time crystals in Nature.

As a final remark, we point out that some discrete or Floquet time crystals have been considered, for instance, in finite dimensional Hilbert spaces [55, 56], however these might not be related to the ones discussed here, which emerge from the continuous deformed algebra, Eq. (2.1).

Acknowledgments

The work of one of us (O.B.) is partially supported by FCT (Fundação para a Ciência e Tecnologia, Portugal) through the project CERN/FIS-PAR/0027/2021, with DOI identifier 10.54499/CERN/FI.

References

[1] A.E. Bernardini and O. Bertolami, Phys. Lett. B **835**, 137549 (2022).

[2] F. Wilczek, Phys. Rev. Lett. **109**, 160401 (2012).

[3] A. Shapere and F. Wilczek, Phys. Rev. Lett. **109**, 160402 (2012).

[4] K. Sacha and J. Zakrzewski, Rep. Prog. Phys. **81**, 016401 (2018).

[5] K. Sacha, Phys. Rev. A **91**, 033617 (2015).

[6] J. Smits, L. Liao, H. T. C. Stoof and P. van der Straten, Phys. Rev. Lett. **121**, 185301 (2018).

[7] V. Khemani, A. Lazarides, R. Moessner and S. L. Sondhi, Phys. Rev. Lett. **116**, 250401 (2016).

[8] D. V. Else, B. Bauer and C. Nayak, Phys. Rev. Lett. **117**, 090402 (2016).

[9] S. Pal, N. Nishad, T. S. Mahesh and G. J. Sreejith, Phys. Rev. Lett. **120**, 180602 (2018).

[10] J. Rovny, R. L. Blum and S. E. Barrett, Phys. Rev. Lett. **120**, 180603 (2018).

[11] A. Kyprianidis, F. Machado, W. Morong, P. Becker, K. S. Collins, D. V. Else, L. Feng, P. W. Hess, C. Nayak, G. Pagano, *et al.*, Science **372**, 1192 (2021).

[12] J. Randall, C. E. Bradley, F. V. van der Gronden, A. Galicia, M. H. Abobeih, M. Markham, D. J. Twitchen, F. Machado, N. Y. Yao and T. H. Taminiau, Science **374**, 1474 (2021).

[13] J. Zhang, P. W. Hess, A. Kyprianidis, P. Becker, A. Lee, J. Smith, G. Pagano, I.-D. Potirniche, A. C. Potter, A. Vishwanath, *et al.*, Nature **543**, 217 (2017).

[14] S. Choi, J. Choi, R. Landig, G. Kucsko, H. Zhou, J. Isoya, F. Jelezko, S. Onoda, H. Sumiya, V. Khemani, *et al.*, Nature **543**, 221 (2017).

[15] O. Bertolami, "The Physics of Time: Current Knowledge and Unanswered Challenges", Proceedings of the 13th Bial Symposium, "The mystery of time", Behind and Beyond the Brain, 6-9 April 2022, Porto, Portugal; https://arxiv.org/abs/2208.13737.

[16] H. Watanabe and M Oshikawa, Phys. Rev. Lett. **114**, 251603 (2015).

[17] H. S. Snyder, Phys. Rev. **71**, 38 (1947).

[18] A. Connes, M. R. Douglas and A. Schwarz, JHEP **02**, 003 (1998).

[19] M. R. Douglas and C. Hull, JHEP **02**, 008 (1998).

[20] N. Seiberg and E. Witten, JHEP **9909**, 032 (1999).

[21] M. R. Douglas and N. A. Nekrasov, Rev. Mod. Phys. **73**, 977 (2001).

[22] J. Gamboa, M. Loewe and J. C. Rojas, Phys. Rev. D **64**, 067901 (2001); J. Gamboa *et al.*, Mod. Phys. Lett. A **16**, 2075 (2001).

[23] O. Bertolami, J. G. Rosa, C. M. L. de Aragão, P. Castorina and D. Zappalà, Phys. Rev. D **72**, 025010 (2005).

[24] M. Rosenbaum, J. David Vergara and L. Roman Juarez, Phys. Lett. A **367**, 1 (2007); M. Rosenbaum and J. David Vergara, Gen. Rel. Grav. **38**, 607 (2006).

[25] Jing S-C, Tao L-P, Liu Q-Y and Ruan T-N, Commun. Theor. Phys. **45**, 249 (2006).

[26] C. Bastos, O. Bertolami, N. C. Dias and J. N. Prata, J. Math. Phys. **49**, 072101 (2008).

[27] N. C. Dias and J. N. Prata, Annals Phys. **324**, 73 (2009).

[28] C. Bastos and O. Bertolami, Phys. Lett. A **372**, 5556 (2008).

[29] H. Groenewold, Physica **12**, 405 (1946).

[30] J. Moyal, Proc. Camb. Phil. Soc. **45**, 99 (1949).

[31] E. Wigner, Phys. Rev. **40**, 749 (1932).

[32] C. Bastos, O. Bertolami, N. C. Dias and J. N. Prata, Int. J. Mod. Phys. A **24**, 2741-2752 (2009).

[33] C. Bastos, O. Bertolami, N. C. Dias and J. N. Prata, Phys. Rev. D **86**, 105030 (2012).

[34] C. Bastos, A. E. Bernardini, O. Bertolami, N. C. Dias and J. N. Prata, Phys. Rev. D **88**, 085013 (2013).

[35] C. Bastos, A. E. Bernardini, O. Bertolami, N. C. Dias and J. N. Prata, Phys. Rev. D **93**, 104055 (2016).

[36] C. Bastos, A. E. Bernardini, O. Bertolami, N. C. Dias and J. N. Prata, Phys. Rev. A **89**, 042112 (2014).

[37] C. Bastos, A. E. Bernardini, O. Bertolami, N. C. Dias and J. N. Prata, Phys. Rev. D **90**, 045023 (2014).

150

[38] C. Bastos, A. E. Bernardini, O. Bertolami, N. C. Dias and J. N. Prata, Phys. Rev. D **91**, 065036 (2015).

[39] C. Bastos, O. Bertolami, N. C. Dias and J. N. Prata, Phys. Rev. D **78**, 023516 (2008).

[40] C. Bastos, O. Bertolami, N. C. Dias and J. N. Prata, Phys. Rev. D **80**, 124038 (2009).

[41] C. Bastos, O. Bertolami, N. C. Dias and J. N. Prata, Phys. Rev. D **82**, 041502 (2010).

[42] C. Bastos, O. Bertolami, N. C. Dias and J. N. Prata, Phys. Rev. D **84**, 024005 (2011).

[43] C. Bastos, A. E. Bernardini, O. Bertolami, N. C. Dias and J. N. Prata, Class. Quant. Grav. **28**, 125007 (2011).

[44] O. Bertolami and P. Leal, Phys. Lett. B **750**, 6-11 (2015).

[45] C. Bastos, A. E. Bernardini, O. Bertolami, N. C. Dias and J. N. Prata, Phys. Rev.D **93**, 10, 104055 (2016).

[46] A. E. Bernardini and O. Bertolami, Phys. Rev. A **88** 012101 (2013).

[47] P. Leal, A. E. Bernardini and O. Bertolami, J. Phys. A **52**, 375302 (2019).

[48] P. Hannaford1 and K. Sacha, "A Decade of Time Crystals: Quo Vadis?", arXiv:2204.06381 [cond-mat.quant-gas](https://doi.org/10.48550/arXiv.2204.06381).

[49] K. Giergiel, A. Miroszewski and K. Sacha, Phys. Rev. Lett. **120**, 140401 (2018).

[50] K. Giergiel, A. Kurós and K. Sacha, Phys. Rev. B **99**, 220303 (2019).

[51] X. Mi, M. Ippoliti, C. Quintana, A. Greene, Z. Chen, J. Gross, F. Arute, K. Arya, J. Atalaya, R. Babbush, *et al.*, Nature **601**, 531-536 (2022).

[52] P. Frey and S. Rachel, https://doi.org/10.48550/arXiv.2105.06632.

[53] P. Kongkhambut, J. Skulte, L. Mathey, A. Hemmerich and H. Kessler, Science, **377**, 6606, 670-673 (2022).

[54] A. Greilich, N.Kopteva, A. Kamenskii, P. Sokolov, V. Korenev and M. Bayer, Nature Physics, (2024), https://doi.org/10.1038/s41567-023-02351-6.

[55] D. V. Else, B. Bauer and C. Nayak, Phys. Rev. Lett. **117**, 090402 (2016).

[56] T. Kuwahara, T. Mori, and K. Saito, Ann. Phys. **367**, 96 (2016).

151

Part IV

Ontology of Spacetime and Nature of Time

Eric Ling and Annachiara Piubello (Eds), SPACETIME 1908-2023. Selected peer-reviewed papers presented at the *Third Hermann Minkowski Meeting on the Foundations of Spacetime Physics*, 11-14 September 2023, Albena, Bulgaria (Minkowski Institute Press, Montreal 2024). ISBN 978-1-998902-25-5 (softcover), ISBN 978-1-998902-26-2 (ebook).

12   On the Nature of Black Holes

Vesselin Petkov

Abstract   Three years after Roger Penrose won the 2020 Nobel Prize in Physics "for the discovery that black hole formation is a robust prediction of the general theory of relativity" there was a dramatic development in black hole physics. On 1 December 2023 Roy Kerr (who discovered the Kerr geometry of a rotating black hole) posted a very important paper "Do Black Holes have Singularities?" (arXiv:2312.00841 [gr-qc]) where he demonstrated that "There is no proof that black holes contain singularities when they are generated by real physical bodies." What makes this development dramatic is that it, *de facto*, puts into question the very understanding that black hole formation is a prediction of general relativity (of the Schwarzschild solution of the Einstein equation), because the two defining features of black holes are a singularity and an event horizon (which is the accepted view). The problem with the singularity is the first problem that casts doubt on the physical existence of black holes. The second problem, which has not been noticed so far, is that general relativity predicts *both* (i) that black holes form for observers falling together with the gravitationally collapsing star and (ii) that black holes *never* form for distant observers (like all of us) because they require infinite time to form (obviously, infinite time means never).

Keywords: Black holes, assimptotic formation of black holes, Schwarzschild solution of the Einstein equation, singularities

1   Introduction

The question of the nature of black holes may go down in the history of physics as a truly unprecedented case. The two defining features of black holes – a singularity and an event horizon – both lead to serious problems:

- the prevailing view in the gravity community is that a singularity cannot exist in the physical world

- the very existence of an event horizon is problematic because

    - for an observer falling together with a gravitationally collapsing star, the star's "surface" will reach and pass the event horizon and a black hole will form for such an observer

    - for a distant observer, far away from the star, the star's "surface" will never reach the event horizon and a black hole will never form for such an observer.

Eric Ling and Annachiara Piubello (Eds), SPACETIME 1908-2023. Selected peer-reviewed papers presented at the *Third Hermann Minkowski Meeting on the Foundations of Spacetime Physics*, 11-14 September 2023, Albena, Bulgaria (Minkowski Institute Press, Montreal 2024). ISBN 978-1-998902-25-5 (softcover), ISBN 978-1-998902-26-2 (ebook).

In addition to these problems, there does not exist conclusive experimental evidence for the existence of black holes. The existence of super-compact stellar objects does appear to be an experimental fact, but it has not been proven that those objects are black holes. A recent review "Nomen non est omen: Why it is too soon to identify ultra-compact objects as black holes" of the experimental evidence for the existence of black holes comes to the conclusion, stated in the title, that that evidence is inconclusive [1]:

> Despite claims that the observed astrophysical black hole candidates are black holes, there is currently no evidence that any one of them possesses a horizon. Both genuine black holes and horizonless UCOs are compatible with observational data, but the observation of PBHs (those bounded by quasilocal and in principle observable horizons) requires never-before-seen exotic matter and appears to be incompatible with the predictions of semiclassical gravity.

Despite the lack of experimental confirmation and the two apparently insurmountable theoretical problems (stated above), physicists have been freely talking about black holes as if their existence had been unambiguously experimentally established and (i) as if the singularity problem did not exist and (ii) as if the same theory, which predicted the formation of a black hole for observers falling with a collapsing star, did not predict that the collapsing star will never become a black hole for distant observers (like all of us).

Fortunately, this unprecedented situation in spacetime physics seems to have started to change after 1 December 2023 when Roy Kerr showed that, contrary to the accepted *belief*, "There is no proof that black holes contain singularities when they are generated by real physical bodies" [2].

The two theoretical problems with black holes are discussed in Sections 2 and 3, respectively.

2   Do Black Holes have Singularities?

In 1916, shortly after Einstein published his general relativity [3], Schwarzschild published a solution of the Einstein equation [6] that was later understood, mostly due to the works of Finkelstein [7] and Kruskal [8], as describing a region of spacetime whose curvature is so great that nothing – even light – can escape from it. Later this spacetime region with extreme curvature was called a black hole.[1]

The first problem with black holes is the well-known fact that the Schwarzschild solution contains a singularity and many relativists feel uneasy about it, because the existence of a singularity in a physical theory is regarded as a clear sign that the theory breaks down in the circumstances where the singularity appears. In general relativity the situation becomes additionally complicated when trying to have a consistent understating of the term "singularity" because "the general covariance of relativity theory creates serious difficulties in formulating a suitable definition of a singularity in this theory" [9]. Einstein himself had been firmly against any attempts to regard singularities as existing in the physical world [10]:

---

[1] According to different accounts, the term "black hole" had been introduced in the sixties of the last century either by Wheeler or Dicke (Dicke compared that spacetime region to a prison in India called the Black Hole, because no one who entered it left it alive).

The essential result of this investigation is a clear understanding as to why the "Schwarzschild singularities" do not exist in physical reality. Although the theory given here treats only clusters whose particles move along circular paths it does not seem to be subject to reasonable doubt that more general cases will have analogous results. The "Schwarzschild singularity" does not appear for the reason that matter cannot be concentrated arbitrarily. And this is due to the fact that otherwise the constituting particles would reach the velocity of light. ... The problem quite naturally leads to the question, answered by this paper in the negative, as to whether physical models are capable of exhibiting such a singularity.

Regardless of the recent tendency of considering the existence of black holes as experimentally confirmed, the singularity problem has never stopped worrying deep-thinking relativists. The authors of a recent (2016) attempt to free general relativity of singularities stated their motivation clearly [11]:

We believe that no acceptable physical theory should have a singularity (!), not even a coordinate singularity of the type discussed above! The appearance of a singularity shows the limitations of the theory.

Even a more recent (2023) volume[2] "Regular Black Holes: Towards a New Paradigm of Gravitational Collapse" is devoted to the question of how spacetime singularities can be eliminated [12]:

One of the most outstanding and longstanding problems in General Relativity is the inevitability of spacetime singularities in physically relevant solutions of the Einstein Equations. At a spacetime singularity, predictability is lost and standard physics breaks down. It is widely believed that the problem of spacetime singularities can be solved within a theory of quantum gravity.

The expectation that a theory of quantum gravity should get rid of singularities is also clearly stated by Ashtekar, Olmedo and Singh in their contribution to this volume [13]:

There is general agreement in the gravity community that black hole singularities of classical general relativity (GR) offer excellent opportunities to probe physics beyond Einstein. However, as of now, there is no consensus on the fate of black hole singularities in full quantum gravity. Indeed, there is an ongoing debate even on a central question in the subject: Will singularities of classical GR be naturally resolved in full quantum gravity, or will they persist? As the very name of this Volume suggests, in many circles an affirmative answer is taken to be a necessary condition for the viability of a proposed quantum gravity theory.

The reason of why the issue of singularity has not been decisively confronted appears to be the universally accepted belief in the existence of a mathematical proof that the singularity is a rigorous prediction of general relativity. Even the Nobel Prize in Physics appears to have strengthened[3]

---

[2]Non-singular (or singularity-free) black holes are called regular black holes. So, the definition of regular black holes keeps only one of the defining features of a black hole (the event horizon) and abandons the other defining feature (the singularity).

[3]The prize strengthens the belief in the existence of a mathematical proof of singularities because the statement

that belief [15]:

> The Nobel Prize in Physics 2020 was divided, one half awarded to Roger Penrose "for the discovery that black hole formation is a robust prediction of the general theory of relativity"...

But on 1 December 2023 Roy Kerr showed[4] that no such proof existed [2]:

> The consensus view for sixty years has been that all black holes have singularities. There is no direct proof of this, only the papers by Penrose outlining a proof that all Einstein spaces containing a "trapped surface" automatically contain FALL's. This is almost certainly true, even if the proof is marginal. It was then decreed, without proof, that these must end in actual points where the metric is singular in some unspecified way. Nobody has constructed any reason, let alone proof for this. The singularity believers need to show why it is true, not just quote the Penrose assumption.

When discussing the collapse of a star and the moment of the formation of a black hole, Kerr cannot hide his amazement [2]:

> Why do so many believe that the star inside must become singular at this moment? Faith, not science! *Sixty years without a proof, but they believe!*[5]

3 Do Black Holes exist for Distant Observers?

This question is no less worrying than the singularity problem. The Schwarzschild solution implies that there exist two spacetime regions inside and outside the Schwarzschild sphere of radius $r = 2m$ (where m is the gravitating mass), which "do not join smoothly on the surface $r = 2m$."[6] The Schwarzschild solution also implies that there exist two "realities" – one for observers falling together

---

"the discovery that black hole formation is a robust prediction of the general theory of relativity" implies that the *singularity* is a robust prediction of the general theory of relativity since the singularity is a defining feature of a black hole and since Penrose's research, for which he was awarded the prize, makes it explicit that "the body contracts and continues to contract until a *physical singularity* is encountered at $r = 0$" [22] (Penrose's italics).

[4]Kerr's paper started with the note:

> *The word "singularity" will be used to mean a region or place where the metric or curvature tensor is either unbounded or not suitably differentiable. The existence of a FALL by itself is not an example of this.*

[5]Even when physicists do not believe that singularities exist in the physical world, they do seem to believe that general relativity rigorously predicts the existence of singularities [16]:

> Nobody really expects the centres of black holes to harbour true singularities. Instead, it is expected that, close to the classical singularity, quantum gravitational effects will occur that will prevent the divergences of classical general relativity.

[6]Papapetrou particularly emphasizes the serious anomaly on the Schwarzschild sphere, whose physical meaning, I think, has not been thoroughly examined [17]:

> But these geodesics are space-like for $r > 2m$ and time-like for $r < 2m$. The tangent vector of a geodesic undergoes parallel transport along the geodesic and consequently it cannot change from a time-like to a space-like vector. It follows that the two regions $r > 2m$ and $r < 2m$ do not join smoothly on the surface $r = 2m$.

with the constituents of the collapsing body (of mass m), and another for distant observers far away from the Schwarzschild sphere. This situation is perhaps most clearly formulated by Dirac [18]:

> We see that the Schwarzschild solution for empty space can be extended to the region $r < 2m$. But this region cannot communicate with the space for which $r > 2m$. Any signal, even a light signal, would take an infinite time to cross the boundary $r = 2m$, as we can easily check. Thus we cannot have direct observational knowledge of the region $r < 2m$. Such a region is called a black hole, because things may fall into it (taking an infinite time, by our clocks, to do so) but nothing can come out.

> The question arises whether such a region can actually exist. All we can say definitely is that the Einstein equations allow it. A massive stellar object may collapse to a very small radius and the gravitational forces then become so strong that no known physical forces can hold them in check and prevent further collapse. It would seem that it would have to collapse into a black hole. It would take an infinite time to do so by our clocks, but only a finite time relatively to the collapsing matter itself.

Rigorously and explicitly stated, the Schwarzschild solution of the Einstein equation in general relativity predicts *both* (i) black holes form for an observer falling with the collapsing body, and (ii) that they will never form for distant observers like us.[7]

I am not aware of any attempts to address[8] this paradoxical situation despite that it was mentioned as early as 1939 by Oppenheimer and Snyder[9] [21]:

> The total time of collapse for an observer comoving with the stellar matter is finite, and for this idealized case and typical stellar masses, of the order of a day; an external observer sees the star asymptotically shrinking to its gravitational radius.

---

[7]Sometimes I hear from colleagues that there is no such problem in the Kruskal coordinates. The general reply to such a statement is that different coordinates describe the *same* physical reality and therefore different coordinate descriptions do not present different realities; even if one assumed that they did (as in the case of the Schwarzschild and the Kruskal coordinates), the immediate question is "How do we know which is the *true* reality?" Also, both Dirac [18] and Penrose [22] wrote *after* the publication of Kruskal's results that it will take infinite time for a black hole to form for distant observers. A problem with the Kruskal time coordinate v (which replaces the Schwarzschild time coordinate t) is that "at space infinity v differs from t. This is inconvenient because v cannot be interpreted as time in Minkowski's sense" [19]. There seems to be another problem with Kruskal's coordinates: "the Kruskal coordinates implicitly involve *division by zero*, and which explains various oddities associated with them. We would also take note of the objections associated with the Kruskal coordinates by some other authors" [20].

[8]The only comments after my talk "On the asymptomatic formation of black holes" at the *Third Minkowski Meeting* (11-14 September 2023, Albena, Bulgaria) and during the panel discussion on the nature of black holes just stated that there was no contradiction between the two predictions of the Schwarzschild solution. This is self-evident when those predictions are regarded as mathematical results. But they become problematic when it is claimed that they apply to the physical world.

[9]It is interesting that Oppenheimer and Snyder themselves do not appear to believe that the concept of a black hole, particularly the singularity, could represent a physical object, because they explicitly admit that their calculations are based on the physically unrealistic assumption that the pressure is zero [21]:

> We have been unable to integrate these equations except when we place the pressure equal to zero... Physically such a singularity would mean that the expression used for the energy-momentum tensor does not take account of some essential physical fact which would really smooth the singularity out. Further, a star in its early stage of development would not possess a singular density or pressure; it is impossible for a singularity to develop in a finite time.

and in the same paper [21]:

> The star thus tends to close itself off from any communication with a distant observer; only its gravitational field persists. We shall see later that although it takes, from the point of view of a distant observer, an infinite time for this asymptotic isolation to be established, for an observer comoving with the stellar matter this time is finite and may be quite short.

Using the term "asymptotically"[10] is an attempt to suggest that despite that a collapsing body "would take an infinite time" to collapse into a black hole, somehow the black hole would form "asymptotically"!? It seems the fact that the term "asymptotically" is not used in the case of light – "Any signal, even a light signal, would take an infinite time to cross the boundary $r = 2m$" – can be only explained as employing double standards in physics. If double standards are not used, then, by the same argument (that black holes somehow form "asymptotically"), it follows that, if light also asymptotically approaches the event horizon (the Schwarzschild sphere), it will reach it and eventually escape (which is not the accepted understanding). In any case, if black holes do form asymptotically, then, by absolutely the same logic, light will asymptotically reach the event horizon and it will be as bright as a star.

In fact, the situation with understanding the nature of black holes is even worse, because there exists another instant of using double standards when interpreting the expression "take an infinite time" *differently* in different situations:

- From Dirac's quote above: "Any signal, even a light signal, would take an infinite time to cross the boundary $r = 2m$" – in this case "take an infinite time" is interpreted to mean "never."

- Again from Dirac's quote above: "It would seem that it would have to collapse into a black hole. It would take an infinite time to do so by our clocks" – in this case, inexplicably, "take an infinite time" is not taken to mean "never" as in the case of light; instead, *it is assumed without any justification that black holes exist for distant observers.*

That is why, it strikes me when physicists confidently talk about black holes perhaps without realizing[11] that they have been employing double double[12] standards.

The actual situation with the status of black holes in spacetime physics appears to be the following

---

[10]In addition, taking the term "asymptotically" seriously requires a number of explicit definitions such as "asymptotically existing" ("asymptotic existence"), "asymptotically non-existing," "asymptotically real" or even more confusing definitions such as "asymptotically alive" and "asymptotically dead"...

[11]I believe that the use of double standards has not been intentional, but rather following "an inner voice" which whispers that there is a contradiction (when the Schwarzschild solution is regarded as describing a real physical situation) that should be avoided or at least addressed.

[12]Unfortunately, double double standards have been indeed used. First, the "surface" of a collapsing star "would take an infinite time to cross the boundary $r = 2m$," but nevertheless it will asymptotically cross it (and a black hole will form!?), whereas "a light signal, would take an infinite time to cross the boundary $r = 2m$," but "asymptotically" is not used here and the light signal will never cross that boundary. Second, the statement "a light signal, would take an infinite time to cross the boundary $r = 2m$" is interpreted to mean that the light signal will *never* cross that boundary, because "infinite time" is correctly understood as "never;" however, the "surface" of a collapsing star "would take an infinite time to cross the boundary $r = 2m$," but nevertheless it will cross it (and a black hole will form!?), because in this case "infinite time" is inexplicably understood not to mean "never."

– as we are (and have always been) distant observers it will take an infinite time (for us) for black holes to form, which, in ordinary language, means that *black holes will never exist for us.* This is the situation without employing double standards, i.e., by explicitly examining *both* predictions of the *same* theory (general relativity) – the formation of black holes for observers falling with the constituents of a collapsing body and that they will never form for distant observers.

Conclusion

The Schwarzschild solution of the Einstein equation and its interpretation as a prediction of black holes constitutes an unprecedented situation in fundamental physics for two reasons:

- A singularity in a theory has been regarded as having physical meaning, that is, has been viewed as representing a real feature of the physical world. Fortunately, in December 2023 Roy Kerr demonstrated that "There is no proof that black holes contain singularities when they are generated by real physical bodies."

- The same theory – general relativity – implies that there exist two "realities" – one for observers falling with a collapsing body (for these observers the collapsing body becomes a black hole for a finite period of time) and another for distant observers like all of us (for these observers the body will take an infinite time to collapse and therefore a black hole will *never* form).

The reason for emphasizing the two problems with the concept of black holes is to draw the physicists' attention to them in the hope that more research would be done to find satisfactory explanations of these problems:

- To expand Kerr's analysis [2] and decisively demonstrate that a mathematical singularity should not have a counterpart in the physical world

- To clarify the physical meaning of the Schwarzschild solution – that black holes will form for some observers, but will never come into existence for others?

References

[1] S. Murk, Nomen non est omen: Why it is too soon to identify ultra-compact objects as black holes. *Int. J. Mod. Phys.* D **32**(14), 2342012 (2023)

[2] R. P. Kerr, Do Black Holes have Singularities? arXiv:2312.00841 [gr-qc], 1 December 2023.

[3] A. Einstein, Die Grundlage der allgemeinen Relativitätstheorie, *Annalen der Physik*, **49**, 1916. New publication of the original English translation ([4]) in [5].

[4] A. Einstein, The Foundation of the General Theory of Relativity, in: H. A. Lorentz, A. Einstein, H. Minkowski and H. Weyl, *The Principle of Relativity: A Collection of Original Memoirs on the Special and General Theory of Relativity*. With Notes by A. Sommerfeld. Translated by W. Perrett and G. B. Jeffery (Methuen and Company, Ltd., 1923; reprinted by Dover Publications Inc., 1952)

[5] *The Origin of Spacetime Physics*, 2nd ed. Foreword by A. Ashtekar. Edited by V. Petkov (Minkowski Institute Press, Montreal 2023)

[6] K. Schwarzschild, On the gravitational field of a mass point according to Einstein's theory. *Sitzungsber. K. Preuss. Akad. Wiss.* **1**, 189 (1916)

[7] D. Finkelstein, Past-Future Asymmetry of the Gravitational Field of a Point Particle. *Phys. Rev.* **110**, 965 (1958)

[8] M. D. Kruskal, Maximal Extension of Schwarzschild Metric. *Phys. Rev.* **119**, 1743 (1960)

[9] R. Geroch, What is a Singularity in General Relativity? *Annals of Physics* **48** (1968) pp. 526-540

[10] A. Einstein, On a Stationary System With Spherical Symmetry Consisting of Many Gravitating Masses, *The Annals of Mathematics*, Second Series, Vol. 40, No. 4 (Oct., 1939), pp. 922-936

[11] P. O. Hess, M. Schäfer, W. Greiner, *Pseudo-Complex General Relativity* (Springer, Heidelberg 2016)

[12] C. Bambi (ed.), *Regular Black Holes: Towards a New Paradigm of Gravitational Collapse* (Springer 2023), p. 5

[13] A. Ashtekar, J. Olmedo, and P. Singh, Regular Black Holes from Loop Quantum Gravity in [12], pp. 235-236

[14] R. Penrose, Gravitational collapse and space-time singularities, *Phys. Rev. Lett.*, **14**, 18 (1965)

[15] https://www.nobelprize.org/prizes/physics/2020/summary/

[16] M. P. Hobson, G. P. Efstathiou, A. N. Lasenby, *General Relativity: An Introduction for Physicists* (Cambridge University Press, Cambridge 2006), p. 271

[17] A. Papapetrou, *Lectures on General Relativity* (Reidel, Dordrecht 1974) pp. 85-86

[18] P. A. M. Dirac, *General theory of relativity* (Princeton University Press, Princeton 1996) pp. 35-36

[19] R. Utiyama, *The Theory of Relativity*, translated in Russian (Moscow, Atomizdat 1979), p. 189

[20] A. Mitra, Kruskal Coordinates and Mass of Schwarzschild Black Holes: No Finite Mass Black Hole at All, *International Journal of Astronomy and Astrophysics*, Vol.2 No.4 (2012), DOI:10.4236/ijaa.2012.24031 (https://www.scirp.org/html/8-4500105_26225.htm), p. 237

[21] J. R. Oppenheimer and H. Snyder, On Continued Gravitational Contraction *Phys. Rev.,* **56**, 455 (1939), p. 456

[22] R. Penrose, Gravitational collapse and space-time singularities, *Phys. Rev. Lett.*, **14**, 18 (1965)

13 SENTIENT OBSERVERS AND THE ONTOLOGY OF SPACETIME

OVIDIU CRISTINEL STOICA

Abstract I show that, by the same criteria that led to Galilean and Special Relativity and gauge symmetries, there is no way to identify a unique set of observables that give the structure of space or spacetime. In some sense, space is lost in the state space itself. Moreover, the relationship between the observables and the physical properties they represent becomes relative. But we can verify that they are not relative, and the spacetime structure is unique. I show that this implies that not all structures isomorphic with observers can be observers, contradicting Structural Realism and Physicalism. This indicates a strong connection between spacetime and the sentience of the observers, as anticipated by some early contributors to Special and General Relativity.

Keywords: Emergent spacetime; nature of spacetime; foundations of quantum mechanics; observers; structural realism; philosophy of mind; philosophical zombies; no-go theorem

1   Introduction

In a quantum world, space is identifiable with a structure associated with the position observables. However, due to the huge symmetry of the state space, this identification cannot be done uniquely based only on the structures and relations. Observers are needed to interpret which of the observables represent positions. But the same symmetry of the state space implies that, for any observer, there are infinitely many structures identical to the observer's structure, and all of them would identify as position observables different isomorphic structures. From the point of view of Structural Realism, there is no preferred choice, leading to a principle of "meta-relativity" at the level of the state space itself. Then what breaks this huge unitary symmetry, reducing it to the space or spacetime symmetries? I prove that only observer-like structures associated with a particular choice of spacetime can be observers, otherwise their brains wouldn't be able to contain reliable information about the external world. All other observer-like structures can only be philosophical zombies. This has the surprising implication that the ontology of spacetime coincides with that of consciousness, refuting those theories of mind in which consciousness is reducible to structural or relational aspects of the observer's brain.

In Section 2, I revisit the lessons learned from the Principle of Relativity. In Section 3 I show that these lessons don't extend to the state space, because there are many structures isomorphic with

Eric Ling and Annachiara Piubello (Eds), SPACETIME 1908-2023. Selected peer-reviewed papers presented at the *Third Hermann Minkowski Meeting on the Foundations of Spacetime Physics*, 11-14 September 2023, Albena, Bulgaria (Minkowski Institute Press, Montreal 2024). ISBN 978-1-998902-25-5 (softcover), ISBN 978-1-998902-26-2 (ebook).

spacetime, and there is no unambiguous way to deduce the physical meaning of the physical properties from the observables representing them. In Section 4 I show that to solve these ambiguities and give meaning of the observables, observers are required. But if we define observers by their structure and dynamics only, the ambiguity extends to the observers. In Section 5 I show that the fact that we can know the physical properties of the external world refutes Structural Realism. In Section 6 I explain the relation between sentient observers and spacetime. In Section 7 I explain why this refutes Physicalism, and how several thinkers, including some early contributors to Special and General Relativity, anticipated the relation between sentience and spacetime.

2 Lessons from the Principle of Relativity

We use coordinate systems to prove theorems in Euclidean geometry since antiquity [30]. This, along with the development of cartography and the lessons on perspective learned by artists, revealed to us that geometric properties are independent of coordinates. Coordinates are very useful conventions, but they are conventions. The usage of coordinate systems to express geometric properties should be done with discernment, to make sure we don't take the conventional, the relative, as an intrinsic truth.

Gradually, we arrived at the discovery of symmetry groups and invariance. It's not an overstatement to say that this revelation culminated with Felix Klein's *Erlangen Program* [25]. We finally came to the realization that geometry is the study of symmetry transformations. By choosing different symmetry groups and group representations, we can classify the previously known geometries and "predict" new ones, similar to how Mendeleev predicted new types of atoms.

Returning to Euclidean geometry, it applies to any structure that satisfies its axioms. It applies to both absolute and relative spaces. Geometry by itself is unable to distinguish between these two cases, because even if space were absolute, only its relational properties are captured by geometry. It could be the case that the ontology of space has a way to distinguish a special point, an origin, or a special direction, or three special orthogonal directions that would give a preferred basis. But Euclidean geometry is blind to these possible additional properties. In terms of forgetful functors between categories, the Euclidean space is an inner product vector space that forgot its origin, and also the space of triples from $\mathbb{R}^3$ that forgot the coordinate system, and so on. What do we learn from this?

Lesson 1. *Even if space were absolute, anisotropic, with a preferred origin, or any other preferred structure, if all we can measure are distances and angles, we can treat it as relative.*

Moving through space takes time, so in mechanics positions and angles can change with time. This requires an enlargement of the symmetry group used in Euclidean geometry, and also of the space on which it acts, leading to Galileo's group. Galileo's group is also captured by the Erlangen program, but now it acts on four dimensions, three of space and one of time.

Newton believed that space and time are absolute, whereas Leibniz believed that they are relative. Newton's motivation was to find something absolute in the world, something that testifies for the existence of God. Ironically, Leibniz's motivation is related to his theological views. But probably both agreed that Physics is about the truths that ignore the absolute space and time, whether or not they are absolute. In Newtonian Physics, the truth about the nature of space and time remains transcendent. So, there is still peace between Newton and Leibniz, in the common ground provided by Galileo's Principle of Relativity. From this, we learn:

Lesson 2. *Newtonian Mechanics is independent of whether space and time are relative or absolute, and of their ontology.*

But Special Relativity revealed that, if we ignore gravity, the right extension of the Euclidean group is not Galileo's group, but the Poincaré group. Space and time are not absolutely separated, the correct extended geometry is Minkowski's.

Minkowski's arguments that spacetime is relative, or better, relational, are right, of course. But this won't stop people who want them to be absolute for various reasons to continue to believe this. For example, someone who needs "spooky action at a distance" (at the ontological level and not just in correlations) in their preferred interpretation of Quantum Mechanics may say that there is an absolute foliation of spacetime. A preferred foliation is required for example by the *pilot-wave theory* [6] and the *objective collapse theories* [17]. It's just that this foliation doesn't transpire in classical experiments, and neither in the quantum ones for that matter, but it can always be claimed that it's there, in the ontology that gives us local beables [4, 29]. And when we're talking only about ontology, but nothing that can show up in experiments or be falsified, it's hard to argue. So we have to admit this as a possibility, no matter how dim it is:

Lesson 3. *Einstein's Special Relativity and the underlying Minkowski geometry are independent of whether space and time are absolute or relative.*

Of course, this independence can be understood simply as saying that space and time are relative, since nothing contradicts this. But they could be absolute, and the equations and our measurements could still be the same, conspiring somehow to conceal the absoluteness of space and time.

The unification of space and time resulted in more evidence for relativity: the unification of other apparently separate things, like the momentum 3-vector and energy as the components of a 4-vector (with mass as the invariant length of this 4-vector), the electric 3-vector field and the magnetic 3-pseudovector field as components of an antisymmetric 4-tensor, and so on.

Things don't stop here, since the same symmetry principles, promoted from global to local, shape other sectors of physics. Consider an extension of spacetime, whose extra dimensions are invisible not necessarily because they're curled in a too small compact circle or manifold, but because nothing changes in these directions. The internal geometry of these extra dimensions has its own symmetry group – the gauge group. Add a *connection* – a way to transport tangent vectors in a parallel way, so that we can define derivatives of the vector fields. The symmetries of such a geometry are local gauge transformations. The internal space is relative, so there is no absolute gauge (that is, no basis in the internal space), there are no absolute potentials. This didn't change with the discovery of the Aharonov-Bohm effect, despite a widespread view that it needs absolute potentials, because this effect is explained by connections or by the difference between potentials, and these are gauge invariant. But, again, we can't disprove those thinking that potentials are fundamental for whatever metaphysical reasons or simply by not taking the geometric description seriously. So, to indulge them too,

Lesson 4. *Gauge theory is the same whether or not there is an absolute but unobservable choice of the gauge.*

Let's find the common denominator of these lessons. We've seen, through the Erlangen Program and its local extension to gauge theory, that symmetry plays a central role in all of these lessons. Putting them all together,

Lesson 5. *The symmetries of the physical laws may be broken by the ontology, but if this doesn't transpire in the observations, it has no place in our descriptions of the laws.*

Of course, if the ontology breaks the symmetry of the laws in a way that is manifest in the experiments, it means we have to update the theory by replacing the laws with those satisfying the broken symmetry. Lesson 5 refers to the situation when no experiment contradicts the physical law.

Moreover, since the symmetry groups admit more representations, they can explain more things that we initially expected, in a unified way. The Poincaré symmetry allows us to classify particles, and their free evolution equations, by spin and mass [50, 3]. Other numbers associated with particles appear as the invariants of the gauge symmetries, giving us a more detailed classification of particles. This is how hadrons were classified, and new quarks and leptons were predicted, and later discovered [16, 20]. This leads us to another lesson:

Lesson 6. *Symmetry plays a central role in physics. If not a fundamental role, at least a very efficient systematizing one.*

These lessons add-up to a lesson in epistemic humility:

Lesson 7 (Structural Realism). *Experiments and inferences from experiments give us access to the relations, but not to the true nature of the relata.*

This may surprise those who expected that science gives us access to reality itself. But let's revise what measurements do: they compare sizes with some reference sizes. For example, they compare lengths or distances with some unit lengths, and this is true for any measurement unit. Experiments may also consist in counting, for example we can count how many particles of each type resulted from a collision. Or they may consist of detecting if an event happened or not. But in all cases, all we get are relations, not the nature of the relata. So the fair and epistemically modest position is that of *(Epistemic) Structural Realism* [27]: Science tells us about relations, but not about the nature of the relata.

And these relations are mathematizable, in fact mathematical structures are nothing but relations [19]. Counting particles gets us numbers. Ratios are numbers. We can also obtain non-numerical results, for example topological ones, but whatever relations there are, they turn out to be describable by mathematics. This makes mathematics very effective in physics, and from Lesson 6, this effectiveness is highly amplified by the power of symmetries.

From the raw data collected from experiments, we can guess more than just seems to be out there. But we can't read unequivocally what the universal laws are from these data, not even if there are such laws. Our theoretical models are a guesswork, we postulate general rules that fit the data, and then we make more experiments to falsify or corroborate these rules. And this seems to work well.

3 Do these lessons extend to the quantum world?

The lessons from the previous section should apply to the quantum world as well. Otherwise we could find quantum effects that break them. This not only didn't happen, but, as the classification of particles by the invariants of the representations of the symmetry groups showed, it powered Quantum Theory. In fact, the main ingredient of Quantum Field Theory (QFT) is the Poincaré symmetry, leading, together with the gauge symmetries, to the entire particle and interaction zoo [47].

One may think that, due to entanglement, relativity of simultaneity is violated, especially in the Einstein-Podolsky-Rosen experiment [15, 5, 4]. But in fact, no matter how entangled, the wavefunction is how it is due to the Poincaré symmetry itself [39], and can be understood as an object on space, not just on the configuration space [37, 42]. There is no "spooky action at a distance" [37, 42], unless it makes its way in, either by the wavefunction collapse, as in the objective collapse theory [17], or by non-local interactions, as in the pilot-wave theory [6]. But this may happen in a way that is not manifest in our experiments, and this is what Lesson 3 is about. However, these interpretations weren't so far able to explain simple quantum phenomena beyond non-relativistic quantum mechanics with a fixed number of particles, let alone to produce a quantum theory of fields, despite decades of attempts [46]. On the other hand, the Poincaré symmetry allowed the discovery of the first working formulations of relativistic quantum mechanics in a couple of years [12] and of QFT during the next couple of decades. In fact it enforced it upon us.

However, we will soon see that, despite Lesson 5 and other lessons, the very existence of space or spacetime requires a symmetry breaking of the very large symmetry group of the state space, a symmetry breaking that neither the quantum structures nor their dynamics provide it!

To see this, recall that a quantum world is described by a unit vector $|\psi(t)\rangle$ in a Hilbert space $\mathcal{H}$ (the state space). The state vector $|\psi(t)\rangle$ evolves unitarily, governed by a Schrödinger-type equation

$$|\psi(t)\rangle = e^{-\frac{i}{\hbar}\widehat{\mathbf{H}}t}|\psi(0)\rangle, \tag{3.1}$$

where the Hamiltonian operator $\widehat{\mathbf{H}}$ is a Hermitian operator.

The Hilbert space is endowed with a *tensor product structure* – a decomposition as a tensor product of the Hilbert spaces for the elementary particles. This tensor product structure is known to us from the empirical data, since it's determined by observables [51].

The observables represent physical properties, so they have associated physical meanings. The physical meaning of each observable follows from experiments. The observables are represented by Hermitian operators on $\mathcal{H}$. The observables pertaining to a subsystem are also represented by Hermitian operators on $\mathcal{H}$, with the additional constraint that they are independent of the states of other subsystems.

Among the observables, the position observables $\hat{x}$, $\hat{y}$, $\hat{z}$ are more fundamental, in the following sense. First, they are related to space. Second, the momenta $\hat{p}_x := -i\hbar\frac{\partial}{\partial x}$, $\hat{p}_y := -i\hbar\frac{\partial}{\partial y}$, $\hat{p}_z :=$

$-i\hbar\frac{\partial}{\partial z}$ are defined as partial derivatives with respect to the spectra of the positions. Third, all other observables (except for the spin observables and the observables corresponding to internal degrees of freedom, which are independent of positions and momenta) are functions of the position and the momentum observables.

And here we have a problem.

Problem 1. *There is no unique way to identify the subsystems and the observables – for example the position observables – only from the mathematical structure of Quantum Theory.*

This problem is not noticed in the literature, so I will have to explain it and convince the reader of its existence and importance.

Suppose we have a complete set of commuting observables. It can be chosen to contain the position observables of all particles. For n particles the position observables are

$$\widehat{\mathbf{q}} := (\hat{q}_1, \hat{q}_2, \ldots, \hat{q}_{3n}) = (\underbrace{\hat{x}_1, \hat{y}_1, \hat{z}_1}_{\text{particle 1}}, \underbrace{\hat{x}_2, \hat{y}_2, \hat{z}_2}_{\text{particle 2}}, \ldots, \underbrace{\hat{x}_n, \hat{y}_n, \hat{z}_n}_{\text{particle } n}). \tag{3.2}$$

Other observables, for spin and charges of the internal symmetries, are needed to obtain a CSCO, and may be included as additional observables. For simplicity I will ignore them, but we should remember that they exist and can be added at any time, as needed.

Let's collect the spectra of the operators $\hat{q}_j$ in a manifold $\mathcal{C}_{\widehat{\mathbf{q}}}$, named in the following position *parameter space*. Let $\boldsymbol{q} \in \mathcal{C}_{\widehat{\mathbf{q}}}$, where

$$\boldsymbol{q} = (q_1, q_2, \ldots, q_{3n}). \tag{3.3}$$

The spin and internal degrees of freedom can be included as additional parameters, or used to label different connected components of the parameter space $\mathcal{C}_{\widehat{\mathbf{q}}}$.

Then, the wavefunction of the n particles, with respect to the position basis, is obtained from the state vector $|\psi\rangle$ by

$$\psi(\boldsymbol{q}) = \langle \boldsymbol{q}, \psi \rangle. \tag{3.4}$$

The momentum observables are

$$\widehat{\mathbf{p}} := (\hat{p}_1, \hat{p}_2, \ldots, \hat{p}_{3n}). \tag{3.5}$$

The momentum parameter space $\mathcal{C}_{\widehat{\mathbf{p}}}$ is obtained by Fourier transforming the position space, and keeping the spin and internal degrees of freedom,

$$\boldsymbol{p} = (p_1, p_2, \ldots, p_{3n}). \tag{3.6}$$

All other observables are functions of the position and momentum operators, $\widehat{\mathbf{f}}(\widehat{\mathbf{q}}, \widehat{\mathbf{p}})$.

For a variable number of particles, as in the Fock representation in Quantum Field Theory, the manifolds $\mathcal{C}_{\widehat{\mathbf{q}}}$ and $\mathcal{C}_{\widehat{\mathbf{p}}}$ are the unions of the manifolds for fixed numbers of particles.

Evidently, the symmetries of space allow different choices of the axes and the origins determining the position observables. Changing the origin amounts to shifting the eigenvalues, and changing the axes amounts to replacing the position observables with linear combinations of the old position observables. Similar for the momentum observables, whose eigenvalues depend of the relative motion of the inertial reference frames. In the following we will assume that these equivalences go without saying.

Suppose we have identified what all observables mean physically. If we apply a unitary transformation $\widehat{\mathbf{S}}$ to the observables, we obtain another set of observables. If the unitary transformation doesn't correspond to a change of space coordinates, the resulting operators can't represent positions.

But any two observables $\widehat{\mathbf{S}}\hat{q}_j\widehat{\mathbf{S}}^\dagger$ and $\widehat{\mathbf{S}}\hat{p}_k\widehat{\mathbf{S}}^\dagger$ are in the same relation with one another as the original $\hat{q}_j$ and $\hat{p}_k$. And for any observable defined as a function $\widehat{\mathbf{f}}(\widehat{\mathbf{q}}, \widehat{\mathbf{p}})$, $\widehat{\mathbf{S}}\mathbf{f}(\widehat{\mathbf{q}}, \widehat{\mathbf{p}})\widehat{\mathbf{S}}^\dagger$ is in the same relation with $\widehat{\mathbf{S}}\widehat{\mathbf{q}}\widehat{\mathbf{S}}^\dagger$ and $\widehat{\mathbf{S}}\widehat{\mathbf{p}}\widehat{\mathbf{S}}^\dagger$ as the original $\widehat{\mathbf{f}}(\widehat{\mathbf{q}}, \widehat{\mathbf{p}})$ is with $\widehat{\mathbf{q}}$ and $\widehat{\mathbf{p}}$.

Then how can we distinguish, from the structural relations only, the observables $\widehat{\mathbf{q}}$, $\widehat{\mathbf{p}}$, and the other observables $\widehat{\mathbf{f}}(\widehat{\mathbf{q}}, \widehat{\mathbf{p}})$, from any other choice of observables $\widehat{\mathbf{S}}\widehat{\mathbf{q}}\widehat{\mathbf{S}}^\dagger$, $\widehat{\mathbf{S}}\widehat{\mathbf{p}}\widehat{\mathbf{S}}^\dagger$, and $\widehat{\mathbf{S}}\mathbf{f}(\widehat{\mathbf{q}}, \widehat{\mathbf{p}})\widehat{\mathbf{S}}^\dagger$? It is not sufficient to merely call $\widehat{\mathbf{q}}$ but not other $\widehat{\mathbf{S}}\widehat{\mathbf{q}}\widehat{\mathbf{S}}^\dagger$ positions, since the relational structure gives all of these choices equal footing. The operators $\widehat{\mathbf{S}}\widehat{\mathbf{q}}\widehat{\mathbf{S}}^\dagger$ and $\widehat{\mathbf{S}}\widehat{\mathbf{p}}\widehat{\mathbf{S}}^\dagger$ are functions of $\widehat{\mathbf{q}}$ and $\widehat{\mathbf{p}}$ as well, so they are as real as any operator $\widehat{\mathbf{f}}(\widehat{\mathbf{q}}, \widehat{\mathbf{p}})$. But what if the position observables $\widehat{\mathbf{q}}$ and the transformed ones $\widehat{\mathbf{S}}\widehat{\mathbf{q}}\widehat{\mathbf{S}}^\dagger$ have completely different nature, different ontology? Again, our lessons taught us that this is irrelevant to the observations. Only relations are supposed to matter.

We arrived at what seems to be a counterexample to the Lessons from Section 2, particularly Lesson 7 (Structural Realism):

Counterexample 1 (To Structural Realism). *Something has to break the symmetry of the quantum structure, resulting in a preferred choice of the position observables.*

But if not the structure, then what makes the position observables special compared to other observables? Even if the unitary transformation $\widehat{\mathbf{S}}$ commutes with the Hamiltonian, so that the dynamics has the same form as it does in the position basis $(|q\rangle)_{q\in\mathcal{C}_{\widetilde{q}}}$, we arrived at the following question:

Question 1. *What gives the position observables $\widehat{\mathbf{q}}$ their preferred role among all observables of the form $\widehat{\mathbf{S}}\widehat{\mathbf{q}}\widehat{\mathbf{S}}^\dagger$, even assuming that $\widehat{\mathbf{S}}\widehat{\mathbf{H}} = \widehat{\mathbf{H}}\widehat{\mathbf{S}}$?*

Some researchers believe that the space structure emerges uniquely from the Hamiltonian $\widehat{\mathbf{H}}$ in its basis-independent form, that is, from its spectrum (including multiplicities). This is supposed to proceed in steps: first obtain the tensor product structure that corresponds to regions of space, then use additional information from the state vector to obtain distances [9, 8]. The tensor product structure is supposed to emerge, at least for finite-dimensional Hilbert spaces, from the spectrum and a condition of k-locality [10]. But in [38] it was shown that for any structure that would lead to physically observable differences of the kind we observe, and for any invariant method to obtain it, there are infinitely many physically distinct solutions. The tensor product structure, regardless of the additional conditions, not only doesn't emerge uniquely, but the number of additional continuous parameters needed grows exponentially with the number of subsystems (or regions of space)

[43]. And even if we assume a tensor product structure as given, the description of the world is maximally ambiguous [41].

4 Observers and the meaning of the observables

So how can we answer Question 1? As I mentioned, we know the meaning of the observables from experiments. But to understand these meanings as extracted from experiments, we need to include into the equation the experimentalists, the observers. This is a different problem than the measurement problem, which is also sometimes understood as necessitating the inclusion of the observers. In the present case, the observers are needed to give physical meanings to the observables, and this would be necessary even in a classical world, where the symmetries are symplectomorphisms of the phase space (canonical transformations), instead of unitary symmetries.

If we take Lesson 7 into account, we will have to describe the observers only by their structure. And the entire structure is contained in the wavefunction ψ_O of the observer,

$$\psi_O(\boldsymbol{q}) = \langle \boldsymbol{q}, \psi_O \rangle. \tag{4.1}$$

Let us apply a unitary transformation $\widehat{\mathbf{S}}$. The position operators will transform into the operators $\widehat{\mathbf{q}}' := \widehat{\mathbf{S}}\widehat{\mathbf{q}}\widehat{\mathbf{S}}^\dagger$, having eigenvectors $|\boldsymbol{q}'\rangle = \widehat{\mathbf{S}}|\boldsymbol{q}\rangle$, for all $\boldsymbol{q} \in \mathcal{C}_{\widehat{\mathbf{q}}}$. The new basis is parametrized by another parameter space, $\boldsymbol{q}' \in \mathcal{C}_{\widehat{\mathbf{q}}'}$, consisting of the eigenvalues of $\widehat{\mathbf{q}}'$.

Then, in the new basis of eigenvectors of $\widehat{\mathbf{q}}'$, $(|\boldsymbol{q}'\rangle)_{\boldsymbol{q}' \in \mathcal{C}_{\widehat{\mathbf{q}}'}}$, the wavefunction has the form

$$\psi_O(\boldsymbol{q}') = \langle \boldsymbol{q}', \psi_O \rangle. \tag{4.2}$$

Therefore, in general, an observer that looks like an observer in the basis $(|\boldsymbol{q}\rangle)_{\boldsymbol{q} \in \mathcal{C}_{\widehat{\mathbf{q}}}}$, will, in general, not look like an observer in the new basis $(|\boldsymbol{q}'\rangle)_{\boldsymbol{q}' \in \mathcal{C}_{\widehat{\mathbf{q}}'}}$. But sometimes other systems that are not necessarily observers may look like observers in the new basis. Let's define such structures:

Definition 1. *An observer-like structure O' with respect to a basis $(|\boldsymbol{q}'\rangle)_{\boldsymbol{q}' \in \mathcal{C}_{\widehat{\mathbf{q}}'}}$ is a state that looks in the basis $(|\boldsymbol{q}'\rangle)_{\boldsymbol{q}' \in \mathcal{C}_{\widehat{\mathbf{q}}'}}$ just like an observer O in the position basis $(|\boldsymbol{q}\rangle)_{\boldsymbol{q} \in \mathcal{C}_{\widehat{\mathbf{q}}}}$,*

$$\langle \boldsymbol{q}', \psi_{O'} \rangle = \langle \boldsymbol{q}, \psi_O \rangle. \tag{4.3}$$

Let $\widehat{\mathbf{S}}$ be a unitary transformation that commutes with the Hamiltonian. Let O' be an observer-like structure in the basis $(|\boldsymbol{q}'\rangle)_{\boldsymbol{q}' \in \mathcal{C}_{\widehat{\mathbf{q}}'}}$. Then, Structural Realism implies the following

Observation 1 (Relativity of structure). *If an observer-like structure O' conducts experiments, the results appear in the basis $(|\boldsymbol{q}'\rangle)_{\boldsymbol{q}' \in \mathcal{C}_{\widehat{\mathbf{q}}'}}$ just like they would appear to an observer O in the position basis $(|\boldsymbol{q}\rangle)_{\boldsymbol{q} \in \mathcal{C}_{\widehat{\mathbf{q}}}}$. In particular, O' would interpret as the position basis $(|\boldsymbol{q}'\rangle)_{\boldsymbol{q}' \in \mathcal{C}_{\widehat{\mathbf{q}}'}}$, and not $(|\boldsymbol{q}\rangle)_{\boldsymbol{q} \in \mathcal{C}_{\widehat{\mathbf{q}}}}$.*

Observation 1 leads to the following consequence:

Observation 2 (Indeterminacy of spacetime). *If we assume Structural Realism, there are as many choices for the space structure (and consequently for spacetime) as there are unitary transformations that commute with the Hamiltonian and don't correspond to changes of the reference frame.*

Observation 2 shows that the answer to Question 1 can't be given by assuming Structural Realism. Moreover, all our Lessons from Section 2 seem to no longer apply. Or, if we insist that they apply, we have to admit a huge number of alternative ways to choose the spacetime structure!

5   The ambiguity of Structural Realism

One may think that, while different observer-like structures can in principle identify different position operators, this identification is somehow physically the same. That is, maybe these position observables represent the same space but in a different reference frame, or maybe they can be related by a gauge or another kind of symmetry transformation, so that the physical world looks the same to two observer-like structures. Now we will see that this is not the case, and the ambiguity from Observation 2 is maximal.

To see this, consider an observer O, and let E be her environment, the rest of the world. Let the total state of the world at the time t be $|\psi\rangle = |\psi_\omega\rangle \otimes |\psi_\varepsilon\rangle$ (what follows work for entangled states too). Suppose that there is a physical property A of the environment, having a definite value a. That is, the total state $|\psi\rangle$ corresponding observable $\hat{A}$ is an eigenstate of $\hat{A}$ with the eigenvalue a, $\hat{A}|\psi\rangle = a|\psi\rangle$. Suppose that the observer O knows this, that is, the value a is encoded in the observer's memory as the value of the property A of the environment. The property A can be any property of the environment, the position of an external object, its size, its color, or even a microscopic property like the spin of a Silver atom, or the fact that an atom is a Silver atom. This is how it looks on the parameter space $\mathcal{C}$ identified by the observer as her position configuration space.

Let $|\psi'\rangle = |\psi'_\omega\rangle \otimes |\psi'_\varepsilon\rangle$ be another state, so that $|\psi'_\omega\rangle = |\psi_\omega\rangle$, but for the environment $|\psi'_\varepsilon\rangle$, the property A has a different value $a' \neq a$. That is, $\hat{A}|\psi'\rangle = a'|\psi\rangle'$. Both states $|\psi\rangle$ and $|\psi'\rangle$ are possible states of the world, but the world is in the state $|\psi\rangle$, not in $|\psi'\rangle$. However, I will show that there is another parameter space $\mathcal{C}'$ on which the state $|\psi\rangle$ appears just like a state $|\psi'\rangle$ satisfying $\hat{A}|\psi'\rangle = a'|\psi\rangle'$ would appear on $\mathcal{C}$.

Let's consider first the particular case when for every eigenvalue a of $\hat{A}$, $-a$ is also an eigenvalue, and any two eigenspaces can be mapped one into the other by a unitary transformation. That is, if they are finite-dimensional, they all have the same dimension. In [41] it was shown that in this case there is a unitary transformation $\hat{\mathbf{S}}$ that commutes with $\hat{\mathbf{H}}$ and interchanges $|\psi_\omega\rangle \otimes |\psi_\varepsilon\rangle$ and $|\psi'_\omega\rangle \otimes |\psi'_\varepsilon\rangle$, for some $|\psi'_\omega\rangle \otimes |\psi'_\varepsilon\rangle$ as above. Then, the transformation $\hat{\mathbf{S}}$ also interchanges the parameter spaces $\mathcal{C}$ and $\mathcal{C}'$, so that $|\psi_\omega\rangle \otimes |\psi_\varepsilon\rangle$ appears on $\mathcal{C}'$ just like $|\psi'_\omega\rangle \otimes |\psi'_\varepsilon\rangle$ appears on $\mathcal{C}$. The proof was given for the *standard model of quantum measurements* (see for example [31], §2.2(b), and [7], §II.3.4), which can account for spin measurements with the Stern-Gerlach device, various photon counters and beam splitter experiments, photon polarization measurements *etc.* ([7], §VII).

This result can be extended easily to any Hermitian operator $\hat{\mathbf{A}}$. Let us extend it to the case when

$a' \neq -a$, but any two eigenspaces can still be related by unitary transformations. Then, we can find a basis in which $\hat{\mathbf{A}}$ is diagonal, interchange the diagonal elements so that a' is replaced by $-a$ but a remains untouched, and return to the original basis. We will get another operator $\widetilde{\hat{A}}$ having the same eigenvectors as $\hat{A}$, so that if we know that a vector $|\psi\rangle'$ is an eigenvector of $\widetilde{\hat{A}}$ with the eigenvalue $-a$, we can determine that $|\psi\rangle'$ is also an eigenvector of $\hat{A}$, but with eigenvalue a'.

Let us now consider a general Hermitian operator $\hat{A}$. In a basis in which $\hat{A}$ is diagonal, we can decompose each eigenspace as a direct sum of subspaces that can be all related by unitary transformations. Then, all vectors in each such subspace are eigenvectors with the same eigenvalue. This can be done so that the projector on each of these subspaces is diagonal in the basis in which $\hat{A}$ is diagonal. After that, we can change the diagonal elements of $\hat{A}$ so that they differ for different subspaces. After returning to the original basis, we get a Hermitian operator $\widetilde{\hat{A}}$, so that if we know that a vector $|\psi\rangle'$ is an eigenvector of $\widetilde{\hat{A}}$ with the eigenvalue $-a$, we can determine that $|\psi\rangle'$ is also an eigenvector of $\hat{A}$, but with eigenvalue a'.

We obtained that for every state $|\psi\rangle \in \mathcal{H}$ containing an observer O who knows the value a of a property A of the environment E, and for any other possible value a' of A, there is an observer-like structure O' isomorphic with O, on a different parameter space $\mathcal{C}'$, so that the memory of O' contains the information that the value of the property A of the environment of O' is a, but in fact it is a'. Note that the property A is represented by an operator $\hat{A}(\boldsymbol{q}, \boldsymbol{p})$ on $\mathcal{C}$, but on $\mathcal{C}'$ it is represented by $\hat{A}' := \hat{A}(\boldsymbol{q}', \boldsymbol{p}')$, where $\boldsymbol{q}$ represents the positions on $\mathcal{C}'$, and $\boldsymbol{q}$ the canonically conjugate momenta, as explained in Section 3. This is why the same vector $|\psi\rangle$ can be eigenvector of both $\hat{A}$ and $\hat{A}'$, but with different eigenvalues $a \neq a'$. Therefore, we can apply Lemma 1 from [41] to any observable $\hat{A}$.

This has the following consequence: if we choose randomly a parameter space $\mathcal{C}'$ so that

1. $|\psi\rangle$ appears on $\mathcal{C}'$ as containing an observer-like structure O' isomorphic with O,

2. O' (just like O) encodes in the structure of its brain the information that the property A of the environment of O' has a definite value a,

3. on $\mathcal{C}'$, the property A has a definite value,

then the value of the property A on $\mathcal{C}'$ can be any eigenvalue of A. Moreover, for any observer O on $\mathcal{C}$ as above, the parameter spaces $\mathcal{C}'$ on which there is an observer-like structure isomorphic with O as above, are uniformly distributed. So a random observer-like structure O' would know nothing about the value of the property A of its environment! And this applies to any property of the environment.

Let us summarize the result we obtained:

Theorem 1. *If observers were reducible to structures, any property of the external world would be unknown to them.* $\quad\quad\square$

But we do know many properties of the external world. Therefore,

Consequence 1. *There are observer-like structures that are not observers.*

Even if these observer-like structures have the same structure, and even if, in their proper position basis, the Hamiltonian has the same form as it does in the observer's position basis, not all these observer-like structures can actually be observers. If they could be, a random observer-like structure would not know the properties of its environment.

It also follows that

Corollary 1. *Structural realism is refuted.*

Proof. If structural realism were true, observers would be reducible to their structure and the dynamics. But this would contradict Consequence 1. □

6 Sentient observers and spacetime

Let's see if now we can answer Question 1. Theorem 1 shows that if observers were reducible to their structure, even if we take dynamics into account, they wouldn't know the value of any physical property. But we can know the values of the physical properties. This also means that we are able to give physical meaning to the operators representing observables. In particular, we give the physical meaning of positions to some of these operators, and not to others, even if the latter are unitarily equivalent with the former.

But can we identify the position observables uniquely? The answer is yes, because otherwise, if there were an ambiguity, we wouldn't be able to do this. In general, for any observables that can be measured, even in principle, we can assign a unique physical meaning. Any ambiguity would make it impossible for that observable to be measured. But are there properties that we can't measure? Obviously absolute positions and velocities, and absolute potentials. In general, we can't assign unambiguously physical meaning to the operators that are not observable. It follows that

Theorem 2. *The parameter space determined by the existence of the observers is unique up to spacetime and local gauge symmetries. This uniquely determines the meaning of the position operators $\widehat{\mathbf{q}}$ and of their canonical conjugates $\widehat{\mathbf{p}}$, and also of all other observables $\widehat{\mathbf{f}}(\widehat{\mathbf{q}}, \widehat{\mathbf{p}})$, up to spacetime and local gauge symmetries.* □

In particular, there is a relation between the existence of sentient observers and a preferred spacetime structure:

Corollary 2. *In a quantum theory where spacetime is classical, the existence of the observers uniquely determines a spacetime.*

Proof. Local gauge transformations act on a fiber bundle's fibers, and commute with the projector on the base manifold. Therefore, they don't change the position. Then, the position observables are determined up to space symmetries, or up to spacetime symmetries in relativistic formulations. □

This gives the following answer to Question 1. Recall that there is nothing in the structure or the dynamics that exhibits a preferred choice of the observables that identify space or spacetime as preferred structures of the theory. The needed breaking of the unitary symmetry of Quantum

173

Theory comes from the fact that only some of the observer-like structures corresponding to different choices of the parameter space can be sentient.

Note that this result applies to Classical Physics too, because a classical system can be described by a state vector, and Hamilton's equations can be replaced by a Schrödinger equation, as Koopman and von Neumann showed [26, 45]. The allowed state vectors, representing unsuperposed classical states, form a basis of the Hilbert space, and the momentum operators commute with the position operators, but these differences don't affect the discussion from this article and the proof of Theorem 1. The classical case can also be proved directly [40]. In particular it applies to Classical General Relativity, because it admits a classical Hamiltonian formulation [1].

For Quantum Gravity, this implies the existence of a preferred spacetime structure, but one having multiple instances, leading to a version of the *many-worlds interpretation* endowed with genuine probabilities ([42] ch. 8).

7   Physicalism vs. sentient observers

The definition of materialism changed over time. From a Newtonian mechanistic, billiard ball universe (which Newton himself didn't take as supporting materialism), the goal post seemed to move. There was a time when "magnetism" was considered by some as immaterial, being associated with hypnosis and telepathy, so others said this was impossible. Maybe this is why Maxwell tried to explain it by a gear-like mechanism in the atoms of aether, and aether theories continued to be mainstream until Special Relativity was discovered. Decades later, the content of the universe was still artificially divided in "matter" and "radiation" or "fields", until Quantum Field Theory managed to give a unified picture. Gradually, the term "materialism" was replaced by "physicalism" or "naturalism", as if it would be unphysical or even unnatural or supernatural for consciousness to be fundamental. There was a time, before the discovery of the Minkowski spacetime, when a fourth dimension was considered a spiritualist speculation.

With such a shift in the definition of physicalism, we may wonder if there is a way to test it. To be a falsifiable hypothesis and not an always changing, always evasive claim, it needs to be pinned in a clear definition. Let's try:

Definition 2. *Physicalism is the thesis that everything is made of stuff that doesn't have phenomenal powers. Consciousness is reducible to the structure and the dynamics of that stuff, which can be called "matter".*

The term "phenomenal" is understood in the sense of *phenomenology*, "the study of structures of consciousness as experienced from the first-person point of view" [36]. That is, according to Definition 2, there is no need for more than Structural Realism to account for consciousness. Not all who see themselves as physicalists bother to clarify their position, but this definition is consistent with those given by people who are more precise [44, 18], and it's captured in the idea that there are no *philosophical zombies* identical to us but lacking sentient experience [24, 23]. According to physicalism, either philosophical zombies don't exist, or we all are philosophical zombies [11].

The difference between observers and observer-like structures that can't be observers is precisely the fact that someone can be an observer, but there is no "what is like to be" an observer-like structure that is not an actual observer, even though, according to Theorem 1, such structures exist.

Maybe Newton's intuition (see Section 2) that connected space and time with the existence of something that transcends and grounds the material world was not completely vacuous after all.

After the discovery of General Relativity but even before the discovery of Quantum Mechanics, Bertrand Russell, one of the most lucid and thoughtful exponents of skepticism and atheism, stated that materialism was refuted [33], and concluded that there is a unique substance with sentient powers beyond both mind and matter [34, 35]. Similarly monistic conclusions were reached by James [21], Mach [28], Eddington [13], Whitehead [49], and Weyl [48]. Also see Auger [2]. Not to mention the founders of Quantum Theory.

In connection with relativity, Eddington writes ([14], p. 146-147)

> And yet, in regard to the nature of things, this knowledge is only an empty shell—a form of symbols. It is knowledge of structural form, not knowledge of content. All through the physical world runs that unknown content, which must surely be the stuff of our consciousness.

These proposals were motivated by the need to include sentient experience in a geometric or structural-realist world. The present article confirms their vision, by starting from the quantum world, and exploring how meaning is assigned to the quantum operators, in particular to identify the structure corresponding to spacetime. Theorem 1 led to the conclusion that consciousness is fundamental and accounts for the needed breaking of unitary symmetry.

Some of these proposals were motivated both by the inclusion of consciousness, and by the need to explain how our sense of time flowing, called by Einstein "a stubbornly persistent illusion", is compatible with the relativistic block universe. Weyl ties consciousness and its experience of flow of time in the relativistic block universe to consciousness [48]. Jeans elaborates, proposing that [22]:

> our consciousness is like that of a fly caught in a dusting-mop which is being drawn over the surface of the picture; the whole picture is there, but the fly can only experience the one instant of time with which it is in immediate contact [...]

A profound analysis of the sense of the experience of time in the block universe, in particular of Weyl's proposal, was done by Petkov [32].

This article doesn't clarify the relation between the sense of time flowing and the block universe. The flow of time seems, in relation to the block universe, similar to what the formulation of geometry in a particular basis is to their relational formulation based on invariants. A worm's eye view *vs.* God's eye view. Dual aspects of the same sentiential ontology?

The results from this article don't explain what consciousness is, or what its sentiential ontology is. It only shows that sentience is not reducible to structure and dynamics, and that it is required to ground all observable physical properties, endowing them with meaning.

References

[1] R. Arnowitt, S. Deser, and C. W. Misner. The dynamics of general relativity. In L. Witten, editor, *Gravitation: An Introduction to Current Research*, pages 227–264, New York, U.S.A., 1962. Wiley, New York.

[2] P. Auger. The methods and limits of scientific knowledge, Lecture delivered september 5, 1952. In V. Petkov, editor, *On fundamental physics and scientific knowledge*, pages 77–104. Minkowski Institute Press, 2021.

[3] V. Bargmann. Note on Wigner's theorem on symmetry operations. *J. Math. Phys.*, 5(7):862–868, 1964.

[4] J.S. Bell. *Speakable and unspeakable in quantum mechanics: Collected papers on quantum philosophy.* Cambridge University Press, 2004.

[5] D. Bohm. *The Paradox of Einstein, Rosen, and Podolsky*, pages 611–623. Prentice-Hall, Englewood Cliffs, 1951.

[6] D. Bohm. A suggested interpretation of quantum mechanics in terms of "hidden" variables, I & II. *Phys. Rev.*, 85(2):166–193, 1952.

[7] P. Busch, M. Grabowski, and P. Lahti. *Operational quantum physics.* Springer, Berlin, Heidelberg, 1995.

[8] S.M. Carroll. Reality as a vector in Hilbert space. Technical Report CALT-TH-2021-010, Cal-Tech, 2021.

[9] S.M. Carroll and A. Singh. Mad-dog Everettianism: Quantum Mechanics at its most minimal. In A. Aguirre, B. Foster, and Z. Merali, editors, *What is Fundamental?*, pages 95–104. Springer, 2019.

[10] J.S. Cotler, G.R. Penington, and D.H. Ranard. Locality from the spectrum. *Comm. Math. Phys.*, 368(3):1267–1296, 2019.

[11] D.C. Dennett. *Consciousness explained.* Penguin UK, 1993.

[12] Dirac, P.A.M. The Quantum Theory of the Electron. *Proc. R. Soc.*, A117, 1928.

[13] A. Eddington. *The nature of the physical world.* Dent, London, 1928.

[14] A.S. Eddington. *Space, Time and Gravitation.* Minkowski Institute Press, 2017.

[15] A. Einstein, B. Podolsky, and N. Rosen. Can quantum-mechanical description of physical reality be considered complete? *Phys. Rev.*, 47(10):777, 1935.

[16] H. Georgi. *Lie algebras in particle physics: From isospin to unified theories.* Benjamin/Cummings Pub. Co., Advanced Book Program (Reading, Mass.), 1982.

[17] G.C. Ghirardi, A. Rimini, and T. Weber. Unified dynamics of microscopic and macroscopic systems. *Phys. Rev. D*, 34(2):470–491, 1986.

[18] P. Goff. *Consciousness and fundamental reality.* Oxford University Press, 2017.

[19] G. Grätzer. *Universal algebra.* Springer Science & Business Media, 2008.

[20] M.J.D. Hamilton. *Mathematical Gauge Theory: With Applications to the Standard Model of Particle Physics.* Springer International Publishing, 2017.

[21] William James. A world of pure experience. *The Journal of Philosophy, Psychology and Scientific Methods*, 1(20):533–543, 1904.

[22] J. Jeans. *The Mysterious Universe.* Minkowski Institute Press, 2020.

[23] Robert Kirk. Zombies. In Edward N. Zalta and Uri Nodelman, editors, *The Stanford Encyclopedia of Philosophy.* Metaphysics Research Lab, Stanford University, Summer 2023 edition,

2023.

[24] Robert Kirk and Roger Squires. Zombies v. materialists. *Proc. Aristot. Soc., Supplementary Volumes*, 48:135–163, 1974.

[25] F. Klein. Vergleichende Betrachtungen über neuere geometrische Forschungen. *Math. Ann.*, 43(1):63–100, 1893.

[26] B.O. Koopman. Hamiltonian systems and transformation in Hilbert space. *Proc. Nat. Acad. Sci. U.S.A.*, 17(5):315, 1931.

[27] J. Ladyman. Structural realism. In Edward N. Zalta, editor, *The Stanford Encyclopedia of Philosophy*. Metaphysics Research Lab, Stanford University, spring 2020 edition, 2020.

[28] Ernst Mach. *The analysis of sensations, and the relation of the physical to the psychical*. Open Court Publishing Company, 1914.

[29] T. Maudlin. *Philosophy of physics: Quantum Theory*, volume 33 of *Princeton Foundations of Contemporary Philosophy*. Princeton University Press, Princeton, 2019.

[30] U. Merzbach and C. Boyer. *A history of mathematics*. John Wiley & Sons, Hoboken, New Jersey, 2011.

[31] P. Mittelstaedt. *The interpretation of quantum mechanics and the measurement process*. Cambridge University Press, Cambridge, UK, 2004.

[32] V. Petkov. *From Illusions to Reality: Time, Spacetime and the Nature of Reality*. Minkowski Institute Press, 2013.

[33] B. Russell. *Analysis of mind*. George Alien and Unwin Ltd; London; The Macmillan Company, New York, London, New York, 1921.

[34] B. Russell. *The analysis of matter*. London: Kegan Paul, 1927.

[35] B. Russell. *The problems of philosophy*. Minkowski Institute Press, 2021.

[36] David Woodruff Smith. Phenomenology. In Edward N. Zalta, editor, *The Stanford Encyclopedia of Philosophy*. Metaphysics Research Lab, Stanford University, Summer 2018 edition, 2018.

[37] O.C. Stoica. Representation of the wave function on the three-dimensional space. *Phys. Rev. A*, 100:042115, 10 2019.

[38] O.C. Stoica. 3d-space and the preferred basis cannot uniquely emerge from the quantum structure. *To appear in Advances in Theoretical and Mathematical Physics. Preprint arXiv:2102.08620*, 2021.

[39] O.C. Stoica. Why the wavefunction already is an object on space. *Preprint arXiv:2111.14604*, 2021.

[40] O.C. Stoica. Asking physics about physicalism, zombies, and consciousness. *philsci-archive:23108*, 2023.

[41] O.C. Stoica. The prince and the pauper. A quantum paradox of Hilbert-space fundamentalism. *Preprint arXiv:2310.15090*, 2023.

[42] O.C. Stoica. The relation between wavefunction and 3D space implies many worlds with local beables and probabilities. *Quantum Reports*, 5(1):102–115, 2023.

[43] O.C. Stoica. Does the Hamiltonian determine the tensor product structure and the 3d space? *Preprint arXiv:2401.01793*, 2024.

[44] Daniel Stoljar. Physicalism. In Edward N. Zalta and Uri Nodelman, editors, *The Stanford Encyclopedia of Philosophy*. Metaphysics Research Lab, Stanford University, Summer 2023 edition, 2023.

[45] J. v. Neumann. Zur Operatorenmethode in der klassischen Mechanik. *Ann. Math.*, pages 587–642, 1932.

[46] D. Wallace. The sky is blue, and other reasons quantum mechanics is not underdetermined by evidence. *Eur. J. Philos. Sci.*, 13(4):54, 2023.

[47] S. Weinberg. *The quantum theory of fields*. Cambridge University Press, Cabrige, UK, 2005.

[48] H. Weyl. *Space-Time-Matter*. Minkowski Institute Press, 2021.

[49] A.N. Whitehead. *Process and Reality: An Essay in Cosmology*. Free Press, New York, 1978.

[50] E.P. Wigner. *Group Theory and its Application to the Quantum Mechanics of Atomic Spectra*. Academic Press, New York, 1959.

[51] P. Zanardi, D.A. Lidar, and S. Lloyd. Quantum tensor product structures are observable induced. *Phys. Rev. Lett.*, 92(6):060402, 2004.

14   Did the Universe Construct Itself?

Stuart Kauffman and Stephen Guerin

Author Contributions:

Kauffman: conceptualization, formulation, analysis, writing initial and final draft
Guerin: computational modeling, visualizations, analytics and deployment

Abstract   Nonlocality is now established loophole free. Therefore, in a choice between locality – spacetime – as fundamental and nonlocality as fundamental, there is no *a priori* reason to choose locality. If we choose nonlocality, General Relativity, String Theory, Loop Quantum Gravity, the AdS/CFT duality, and the Holographic Principle are ruled out as fundamental, as all assume locality. AdS/CFT provides a mapping from entangled particles with no geometry (CFT) to a spacetime geometry (AdS). If we start with nonlocality, N coherent entangled particles from $SU(3) \times SU(2) \times U(1)$, then locality – spacetime – must somehow emerge from the behaviors of these particles. We report here that the particles of $SU(3) \times SU(2) \times U(1)$ are capable of collective autocatalysis. Via this autocatalysis, the universe can start with no matter and no spacetime and construct itself – Cosmogenesis. The autocatalytic process yields particles that break matter-antimatter symmetry – baryogenesis. Then, by entangling and actualizing, they yield a power law construction of classical spacetime. This is a candidate for the unknown physics of Inflation. It also proposes a mapping from entangled particles with no geometry to a spacetime geometry. It may become possible to explain our Laws and our values of the 25 constants: These Laws and values may maximize some measure of cosmogenesis.

Significance Statement

We report here that the particles of $SU(3) \times SU(2) \times U(1)$ are formally capable of collective autocatalysis that can drive Cosmogenesis. We propose that the universe started with no spacetime and no matter. The autocatalytic behavior of the particles yields more particles, breaks matter-antimatter symmetry – baryogenesis, and is also a new theory of cosmic inflation. This offers a possible explanation for our Laws and the 25 values of the Constants.

Keywords: SU(3) x SU(2) x U(1); the lambda-CDM theory; collectively autocatalytic sets; baryogenesis; inflation

1   Introduction

Of the three great mysteries, the origin and evolution of the universe, of life, and of mind, we understand the origin and evolution of the universe best. The fundamental theory of particle physics, SU(3) x SU(2) x U(1), is very well tested [1]. General Relativity is very well tested [2]. Our account of cosmology is its standard Lambda CDM model plus Inflation [3, 4, 5, 6, 7, 8]. With the few relevant equations [6, 7], it is now possible to predict statistical features of the Cosmic Microwave Background, the formation and distribution of galaxies, and the abundance of the first elements [6, 7].

Eric Ling and Annachiara Piubello (Eds), Spacetime 1908-2023. Selected peer-reviewed papers presented at the *Third Hermann Minkowski Meeting on the Foundations of Spacetime Physics*, 11-14 September 2023, Albena, Bulgaria (Minkowski Institute Press, Montreal 2024). ISBN 978-1-998902-25-5 (softcover), ISBN 978-1-998902-26-2 (ebook).

Despite these successes, much of the fundamental physics is unknown. We do not know how spacetime suddenly appeared [2, 6, 7]. We do not know why the universe has far more matter than antimatter, baryogenesis [6, 7]. We do not know how the universe started in such a low entropy state, perhaps as Boltzmann suppressed as $1/e^{10^{124}}$ [9]. We do not know the physics of Inflation [5, 6, 10], Dark Matter [7, 11], or Lambda – Dark Energy [6, 7, 12].

There are times in the evolution of science when a new conceptual framework may prove useful. The magnificent example is Copernicus, 1543 [13]. He proposes seven postulates, none of which has independent evidence. However, the seven together constitute an entirely new conceptual framework for The World. He writes the Pope to explain his audacity. The book is published as Copernicus lies on his deathbed.

The short-term success of Copernicus is to reduce the number of epicycles with some loss of predictive success compared to the Ptolemaic theory. The long-term success is transformative. Given the sun at the center of The World, might emanations from the Sun hold the planets in orbit? Then Kepler, Galileo, and Newton.

The difficulties we now have in our understanding of Cosmology might benefit from a new conceptual framework. That hope is the aim of this article.

We start with three claims. There is independent evidence for each. i. The quantum state corresponds to ontologically real potentia. This is Heisenberg's 1958 interpretation of quantum mechanics [14]. Thus, quantum mechanics can be interpreted in terms of potentia, neither true nor false. The variables of classical physics are True or False. ii. Nonlocality is firmly established and loophole free [15, 16, 17]. iii. The particles of SU(3) x SU(2) x U(1) are formally capable of collective autocatalysis [18]. We establish the truth of this claim in the present article.

Quantum Gravity If Nonlocality is Fundamental

The first two claims invite the following: If potentia are ontologically real, they may not exist in spacetime. Thus, we can conceive of something real that is not in spacetime. If nonlocality is real, then in a choice between locality versus nonlocality as fundamental, there is no *a priori* reason to choose locality. Cosmology has largely insisted on locality. If instead we choose nonlocality as fundamental, then locality – spacetime – cannot be fundamental. Should we choose nonlocality as fundamental, General Relativity cannot be fundamental, [2, 19], nor can String Theory [20], nor can Loop Quantum Gravity [21], be fundamental. These all start with locality.

The AdS/CFT duality [22], and Holographic Principle [23], map from entangled particles on a $D-1$ dimensional surface to a D dimensional spacetime. Nearby versus distant points on the $D-1$ dimensional surface map to points near and deeper into the D dimensional spacetime bulk [22, 23]. The AdS/CFT duality with the Holographic Principle is a famous way to map from a set of entangled particles with no geometry to a spacetime geometry. The AdS/CFT duality and the Holographic Principle depend upon locality. Both are ruled out if we start with nonlocality as fundamental.

There is an entirely independent way to map from entangled particles without a geometry to a spacetime geometry. If we choose nonlocality as fundamental, taken as $N = 2$ or more entangled coherent particles of SU(3) x SU(2) x U(1), then we must explain locality. Locality must somehow emerge from the behaviors of the N coherent entangled particles. But this flatly contradicts the foundation of General Relativity. General Relativity is local. There is no emergence of spacetime in General Relativity. General Relativity can be formulated in the absence of matter fields, so matter cannot be necessary for the very existence of spacetime. Yet if we start with nonlocality, spacetime cannot emerge without the matter comprised of the N coherent entangled particles.

The fundamental implication, if we start with nonlocality, is that matter somehow constructs spacetime. This implication is entirely new and surely not part of General Relativity.

One approach to taking nonlocality as fundamental is published [24]. This is a mapping from entangled coherent particles with no geometry to a classical spacetime geometry. The central steps are:

1. Start with N coherent particles entangled in some pattern. Between each pair of entangled particles, define the von Neumann entropy (VNE). Because VNE is sub-additive, it fits the Triangle Inequality and is a Norm, so can define a distance between each pair of entangled particles. From this there is a metric in Hilbert space.

2. Because Heisenberg demonstrates that quantum particles in Hilbert space can be interpreted as potentia, neither true nor false, and because all variables of classical physics are Boolean true or false, it becomes necessary to map from a metric in Hilbert space to true false variables. This is the "measurement problem", not solved by decoherence [25]. One choice is to use "actualization" to map from a metric among potentia in Hilbert space to actual events that will constitute classical spacetime. A means to construct a linear map of von Neumann distances in Hilbert space to real spacetime distances invokes "remember" where particles remember their former VNE distances to particles that have actualized. When the particles then actualize, they convert their former von Neumann distances to real distances on some length scale [24].

The resulting theory constructs an emergent and growing classical Minkowski [24], spacetime one element at a time by successive actualization events among four mutually entangled particles. At each actualization step a new tetrahedron arises adjacent to an old tetrahedron. That emergent classical spacetime has a metric so can have a Ricci Tensor. Because it arises from the quantum particles of SU(3) x S(2) x U(1) it can have a stress energy tensor [24].

In this view, quantum gravity does not somehow equate to the classical spacetime of General Relativity. Rather quantum gravity is to construct the classical spacetime in which General Relativity operates.

The particles of SU(3)xSU(2)xU(1) are formally capable of collective autocatalysis

We establish this third claim, fundamental to our efforts, in this article. The claim is quite astonishing. Again, like Copernicus, one is invited to wonder if a capacity for collective autocatalysis among the particles of the standard model of particle physics might allow the universe to construct itself autocatalytically. In the remainder of this article, we hope to demonstrate that the universe, based on such collective autocatalysis, might indeed have constructed itself.

2 Testing the Autocatalytic Hypothesis: Baryogenesis and Cosmogenesis

We currently have no pathway to derive Cosmogenesis, Baryogenesis, and Inflation from the Standard Model of Particle Physics [1]. In this Part II, we propose a new and specific theory, now modelled computationally by the PAM model, as a possible pathway from SU(3) × SU(2) × U(1) to Cosmogenesis, Baryogenesis, and Inflation. The Particle Apothecary Model (PAM) is at best a "toy model." PAM treats the particles of SU(3) × SU(2) × U(1) as classical variables, not quantum variables. Doing so allows very complex stochastic dynamical systems to be studied. But use of classical variables may be badly misleading. With this caveat, this model, PAM, finds a kinetic phase transition breaking matter-antimatter symmetry, hence baryogenesis.

The PAM computational model constitutes our methods. The running version is online, access is at the end of this article.

In turn, baryogenesis with an increasing number of particles and their interactions, including entanglement and actualization, then drives a consequent quantum construction of spacetime. The resulting steep power law construction of spacetime becomes a candidate for Inflation itself.

The PAM model transiently breaks conservation of matter and energy in a controlled way. The total matter and energy of the universe increases in this model throughout Inflation, then stops increasing and is conserved thereafter. Is this possible or ruled out? General Relativity has no global conservation of energy [2, 19, 26]. Further, the proposal that Dark Energy is constant per unit volume of space implies that as Dark Energy drives an accelerating expansion of the universe, the total energy of the universe is, in fact, increasing. On these bases, we take a transient non-conservation of matter plus energy as a proposal.

2.1 Collectively Autocatalytic Motifs

Famously, the Standard Model of particle physics is formulated in three interwoven mathematical Groups, SU(3) × SU(2) × U(1). Because these are mathematical groups, all the particles transform directly or indirectly into one another and only into one another, see Figure 1 and Figure 2 below.

The set of particle transformations in SU(3) × SU(2) × U(1), taken as classical variables, can, in fact, function as a "collectively autocatalytic set," see below, as often considered with respect to the origin of life.

Nghe and colleagues have shown that there are only five collectively autocatalytic Motifs [18], Figure 3. Within a subset of transitions among only 13 particles of the Standard Model, Table 1: [Down quarks, Down antiquarks, Up quarks, Up antiquarks, electrons, positrons, neutrinos, antineutrinos, muons, antimuons, muon neutrinos, antimuon neutrinos, photons] one finds a very large number, four hundred and eighty-six, Nghe Collectively Autocatalytic Motifs. Of these 192 are Nghe type II motifs, 294 are Nghe Type III motifs, Figures 2 and 3 and Table 2.

Nghe motifs range from type I to type V [18], Figure 1. Higher numbered motifs have higher survival probability.

Collective autocatalysis need not include "catalysts" but can refer only to the structure of the set of transformations. Consider classical chemistry. Let substances A + B undergo a two-substrate one-product reaction to form C. A + B → 2C. Let C + D → 2E. Let E + F → 2A. The set of three reactions has a cycle A → C → E → A. If A, C, and E are present in the system, then given exogenous input of B, D, and F in an open thermodynamic system, A, C, and E will accumulate in concentration. There is no "catalyst." The cyclic structure of the reactions constitutes the "catalyst."

For the particles of $SU(3) \times SU(2) \times U(1)$ to function autocatalytically, the universe may be closed, but the initial state of the universe is not at equilibrium and transient excesses of some particles can occur as a changing equilibrium is gradually approached while the temperature of the universe falls.

Ongoing collective autocatalysis requires exogenous input. In the PAM model, we propose a controlled way to do so below.

Figure 1 below includes the reversible transformations among the 9 particles: Down quarks, Down antiquarks, Up quarks, Up antiquarks, electrons, positrons, neutrinos, antineutrinos, and photons used in PAM. The other bosons are assumed. The PAM model does not yet include muons, antimuons, muon neutrinos, or antimuon neutrinos in the study of the branching stochastic processes among the particles. This work is in progress, see Figure 2. Their role in forming Nghe motifs is now included, Table 2.

3   Cosmogenesis

In the remainder of this article, we turn to initial considerations of the potential implications for Cosmogenesis of the fact that the woven group structure of particle physics can function as an autocatalytic system.

It is well known that there is no established pathway from the Standard Model of physics to Cosmogenesis, Baryogenesis, and Inflation [1, 2, 3, 4]. Our familiar theory of cosmogenesis starts with an existing universe near the Initial Singularity at extreme temperature and density, and a rapid power law or exponential Inflation driven by an unknown mechanism, the Inflaton field, from about 10^{-37} seconds to about 10^{-32} seconds for a 10^{27}-fold expansion of spacetime [4, 5, 6].

The fact that particle physics can possibly function as an autocatalytic system invites a radically different approach: The universe starts with nothing other than the quantum vacuum equipped with the standard model of particle physics. There is no matter, and no spacetime. The reversible time of quantum mechanics is present. Then, via collective autocatalysis, the universe is to construct itself.

There are two independent reasons to consider a theory in which the universe starts with nothing but the quantum vacuum:

3.1   The evidence for the Big Bang is twofold: The galaxies are receding according to the Hubble Law and the Cosmic Microwave Background.

Both strongly suggest the universe was very small long ago. However, we also typically start at or near an "Initial Singularity" which is naturally suggested by General Relativity. Yet we have no detailed evidence for such an Initial Singularity, and it is of some interest that we posit such a singularity exactly where General Relativity fails. Thus, it is of interest to consider possible theories that do not start with an initial singularity.

3.2 As noted, one reasonable approach to quantum gravity takes nonlocality as fundamental.

If we start with nonlocality, an absence of spacetime, as fundamental, we need not explain nonlocality, but must explain locality. Consider $N > 1$ entangled coherent quantum particles. These are nonlocal. Then for locality – spacetime – to emerge, something about these N-coherent quantum particles and their behaviors must be relevant. But this does flatly contradict the assumptions of General Relativity [2, 19, 24]:

1. General Relativity is the definition of local.

2. General Relativity can be formulated without matter, hence without N-entangled particles. Thus, matter can have nothing to do with the very existence of spacetime. But if we start with nonlocality, spacetime will not emerge without matter.

3. There is no "emergence" of spacetime in General Relativity.

4. There is no *a priori* reason not to take nonlocality as fundamental.

5. If we start with nonlocality, spacetime is not fundamental.

6. Then any initial state of the universe cannot yet have spacetime.

If starting with nonlocality as fundamental is starting without spacetime, this suggests considering starting the universe itself with no spacetime, no particles, merely the quantum vacuum. But then if entangled, coherent quantum particles are somehow to construct spacetime, from whence come the quantum particles? However, if the particles of $SU(3) \times SU(2) \times U(1)$ can act autocatalytically, might this create the requisite quantum particles and, thus, a quantum creation of spacetime as a new candidate for Cosmogenesis? This Section answers, YES.

The hypothesis that the universe starts with no matter and no spacetime has important advantages:

1. The hypothesis provides a new account of the Arrow of Time. The Arrow of Time requires the Past Hypothesis. The Past Hypothesis itself requires that the entropy of the initial state of the universe be the reciprocal of the current estimated complexity of the universe, $e^{10^{124}}$. [27]. Penrose points out how extremely improbable such a state with a very low but positive initial entropy is [27]. If the universe starts with no matter and no spacetime, its entropy is 0. The Arrow of Time emerges automatically.

2. There is no initial singularity. This obviates concern about why black holes were not formed [28].

3. The puzzle of the low gravitational entropy of the universe [29] is automatically explained.

4. If the universe starts with no matter and no spacetime, this is a unique initial state of the universe. The laws themselves do not specify any initial state at all.

If the universe is to start with no matter, it cannot initially be a Hot Big Bang. Moreover, a universe cannot come to exist without matter, in contradiction to major models of Infinite Inflation [10, 30].

4 PAM: The Particle Apothecary Model

The Particle Apothecary Model (PAM) uses a subset of 9 of the total standard model: [up quarks, up antiquarks, down quarks, down antiquarks, electrons, positrons, neutrinos, antineutrinos, photons]. PAM is a "toy model" because it treats particles as classical objects. It uses a modified "Gillespie algorithm" [31] implemented in Netlogo [51] as a stochastic particle interaction model to study the branching processes in which these 9 classical variable particles undergo the 14 transformations from the first 7 bi-directional equations given in Figure 1 and Table 1. Transition rates are tunable in each direction. Access to the PAM code and running model online is available at https://particleapothecary.org.

4.1 Methods

Consider a set of particles $\mathcal{P}$ and a set of reversible transformations $\mathcal{R}$ between these particles:

$$\mathcal{P} = \{u, \bar{u}, d, \bar{d}, e^-, e^+, \nu_e, \bar{\nu}_e, \gamma\}$$

$$\mathcal{R} = \{u + \bar{u} \rightleftharpoons \gamma, \ldots\}$$

where u and $\bar{u}$ represent up quarks and anti-up quarks, d and $\bar{d}$ represent down quarks and anti-down quarks, e^- and e^+ represent electrons and positrons, ν_e and $\bar{\nu}_e$ represent electron neutrinos and antineutrinos, and γ represents photons.

For each transformation $j \in \mathcal{R}$, let c_j denote the associated probability for the transformation to occur, and let X_i denote the count of each particle $i \in \mathcal{P}$.

#	Transformation
1	$u + \bar{u} \rightleftharpoons \gamma$
2	$d + \bar{d} \rightleftharpoons \gamma$
3	$u + d \rightleftharpoons W^+$
4	$\bar{u} + \bar{d} \rightleftharpoons W^-$
5	$e^- + e^+ \rightleftharpoons \gamma$
6	$\nu_e + \bar{\nu}_e \rightleftharpoons \gamma$
7	$e^- \rightleftharpoons W^- + \nu_e$
8	$e^+ \rightleftharpoons W^+ + \bar{\nu}_e$
9	$\nu_e \rightleftharpoons W^+ + e^-$
10	$\bar{\nu}_e \rightleftharpoons W^- + e^+$
11	$\gamma + \gamma \rightleftharpoons Z^0$
12	$\nu_e + \bar{\nu}_e \rightleftharpoons Z^0$
13	$u + \bar{u} \rightleftharpoons Z^0$
14	$d + \bar{d} \rightleftharpoons Z^0$

Table 14.1: Particle Transformations in the PAM Model

- u = up quark
- $\bar{u}$ = up antiquark
- d = down quark
- $\bar{d}$ = down antiquark
- e^- = electron
- e^+ = positron

- ν_e = electron neutrino
- $\bar{\nu}_e$ = electron antineutrino
- γ = photon
- W^+ = W boson (positive)
- W^- = W boson (negative)
- Z^0 = Z boson

Tuning these rates indirectly tunes the relevant constants of the Standard Model. This allows study of the consequences of the specific values of the constants for the NGHE autocatalytic motifs among these 9 variables, Figures 1 and 2, for Baryogenesis and the construction of spacetime it drives. This means we may be able to answer, "Why our values of the constants?" Optimal values may optimize some measure of Cosmogenesis. This may be testable.

4.2 Simulations and results

A reasonable hypothesis for the vacuum to be an open source of energy can be based on the standard view that a single quark-antiquark pair borrows energy from the vacuum, transiently emerge from nothing out of the vacuum, then returns the borrowed energy and vanishes within $\Delta E \times \Delta T \geq \hbar/2$.

We therefore propose a working hypothesis: If two or more quark-antiquark pairs transiently emerge from nothing out of the vacuum, and if two or more quarks, or if two or more antiquarks "interact," for example entangle, this delays their return of the borrowed energy. Alternately stated, Delay extends the Lifetime of particles.

In [32] S. Patra and Kauffman propose a possible mechanism for Delay. We propose that the classical world emerges as a symmetry breaking among an initial set of 2^n bases to "choose one basis." An emerging basis shared among the N entangled particles can decay "slowly." As quark-antiquark pairs emerge and pairs of quarks or pairs of antiquarks entangle, an emergence of a basis shared among quarks or among the antiquarks, while still present, could delay return of the borrowed energy until the basis decays.

The delay assumption immediately implies the possibility of a phase transition. Consider a two-dimensional parameter space of: i. the delay, D. ii. the frequency, f, with which each variable interacts. Let "f" increase linearly with the number of variables that are available to interact. Then each particle undergoes interactions more frequently as the total number of particles increases. Generically as ever more quarks and antiquark pairs borrow energy from the vacuum, emerge and quarks "interact" or antiquarks "interact" with one another, interacting quarks or interacting antiquarks can mutually persistently delay return of the borrowed energy. Thus, there must be a second order phase transition in the two dimensional "delay" × "f" parameter space when the matter, or the antimatter, particles are so abundant and interact so rapidly that eventually they just persistently "steal" the borrowed energy. The further the system is beyond the phase transition in this two-dimensional parameter space, the more rapidly they just steal the energy. The phase transition forms what we will call a Kinetically Stable Nucleus of fermions and photons that persistently delays return of the borrowed energy. The particles in the Nucleus itself may change. The PAM model exhibits this in the baryogenesis it exhibits.

In short, the very collectively autocatalytic behavior of particle physics can supply the ever-increasing numbers of particles that entangle and persistently delay one another's return of the borrowed energy. The system of particles emerging from the vacuum steals the energy. The autocatalytic system of produced particles continues to interact, transform, and annihilate in a stochastic branching autocatalytic process as defined by the Standard Model.

The present theory requires free quarks and antiquarks, such as those in a quark gluon soup before hadronization.

Figures 4a and 4b demonstrate this second order phase transition. This numerical study using the Gillespie algorithm was carried out for a specific set of parameter values of the PAM model. One axis of the figure is labeled "lifetime extension". Increasing lifetime extension is identical to increasing delay in return of the energy borrowed from the vacuum. The second axis is labeled "probability of interaction". Each probability value is linearly proportional to the total number of particles with which a particle can interact, "f". For each pair of parameter values, PAM was run for 500-time steps and for 50 repetitions with different random seeds. The total number of quarks plus antiquarks that were created in 500-time steps were recorded for each run. The results reported for each pair of parameter values are the means of those 50 repetitions.

The results in 4a and 4b clearly demonstrate the second order phase transition. The dark purple region in figure 4a in the vicinity of both axes corresponds to a formation of less than 5 quarks in 500-time steps. Figure 4b shows that in most of this region near both axes, the average number of quarks formed is less than 1 but greater than 0.

Well beyond the second order phase transition the formation of quarks or of antiquarks is rapid. This is the basis for baryogenesis, discussed next. In addition, the rapid formation of quarks, or of antiquarks well beyond the phase transition will become the basis for the missing physics of Inflation, also discussed below. Slightly beyond the phase transition the rate of formation of quarks or antiquarks is slow.

4.3 Spontaneous baryogenesis

The Particle Apothecary Model is entirely symmetric with respect to matter and antimatter. Yet the dynamical stochastic processes investigated by PAM kinetically break matter-antimatter symmetry. The present theory includes consideration of Up quarks and Up antiquarks as well as Down quarks and Down antiquarks. Figures 5 a, b, c, d, show the resulting ratio of quarks/antiquarks for four conditions: i. Tuning the probability of Up quark or Up antiquark versus Down quark or Down antiquark emerging from the vacuum from .06 to 0.5. ii. Tuning whether the rate of emergence of quark-antiquark pairs is or is not proportional to the current number of particles. In all four conditions the initial symmetry between matter, quarks, and antimatter, antiquarks, is maintained.

The results are striking. Either quarks strongly win, or antiquarks strongly win. The stochastic dynamics of the nine particles breaks matter-antimatter symmetry. This is baryogenesis with respect to quarks and antiquarks.

Importantly, symmetry is not broken with respect to electrons versus positrons and neutrinos versus antineutrinos, Figures 6a and 6b.

Our current universe is dominated by quarks, not antiquarks and by electrons not positrons. The present theory seems unable to fully account for this asymmetry. However, the theory predicts that if Up quarks and Down quarks win over Up antiquarks and Down antiquarks, then when hadrons are formed, they will be neutrons and protons, not antineutrons and antiprotons. It is therefore of interest that "positron capture" by neutrons eliminates positrons and yields protons plus antineutrinos [33]. If this process is sufficiently abundant, it is a candidate to remove positrons from the early universe after hadrons form. In short, it seems worth considering that later processes as hadronization occurs, given a predominance of quarks not antiquarks, may further break matter antimatter symmetry.

Thus, a kinetic matter antimatter symmetry breaking, hence baryogenesis, naturally emerges in PAM. We have not been able to account for baryogenesis. The present theory offers a way.

Because the universe here starts with no matter and no spacetime, it is vastly out of equilibrium. Hence this process fulfills Sakharov's criteria for baryogenesis [34]. Such a symmetry breaking could not be seen were the universe to start with very many particles as in the standard Big Bang model. Only tiny fluctuations would be seen, as is normally assumed to account for the prevalence of matter over antimatter at one part in 5×10^7 [35].

Di Biagio and Rovelli have introduced the idea of mutually "stable facts" among a set of entangled particles not all of which are entangled, and they have noted that such stable facts imply breaking the symmetry of reversible time [36]. The autocatalytic behaviors in the PAM theory with its mutually induced "delay" among a set of quarks or a set of antiquarks constitutes just such mutually stable facts. The delayed quarks or delayed antiquarks are stable facts with respect to one another as they mutually entangle and interact and further delay one another, even as not all pairs are entangled at any moment. In yielding baryogenesis, the system presumably breaks CPT symmetry. A quantum arrow of time emerges [24].

4.4 The emergence of spacetime and inflation

Several models exist that attempt to relate multiparticle entanglement to spacetime. Carroll's Bulk Entanglement Gravity, BEG, is a well-known example [37]. BEG maps from a stable mutual information among a set of entangled quantum particles in Hilbert Space to classical spacetime via the Radon transform [37]. More recently, Singh and Doré [38] have proposed a model with a fixed number of quantum particles. An increasing entanglement among these drives Inflation. Inflation stops in this model because the total number of quantum particles is finite and fixed. This theory does not propose a mechanism for the existence of the initial fixed set of quantum particles [38]. Raamsdonk suggests that decreasing entanglement is associated with regions of spacetime pulling apart [39].

PAM yields baryogenesis with an increasing number of particles. Like the theory of Singh and Doré above with a fixed number of particles [38], in PAM the increasing number of entanglements among the increasingly numerous multiparticle system could be taken to drive Inflation. However, a mapping from Hilbert space to classical spacetime remains uncertain.

PAM proposes that the universe starts with pure potentia, here identified as the quantum vacuum. As noted, quantum superpositions do not obey the law of the excluded middle. With Heisenberg [14], we identify superpositions with ontologically real Possibles, *Res potentia* [14, 24]. Actualization then converts potentia to ontologically real Actuals, *Res extensa*, whose variables do obey the Law of the Excluded Middle, hence are true or false [24]. The variables of classical physics, including General Relativity, are all Actuals, either true or false. Decoherence alone is not sufficient to yield specific true false specific outcomes [25]. It becomes natural to propose that sequential actualization of the quantum potentia of the quantum vacuum "constructs" classical spacetime as the relations among the "true" actualized events. This is sketched above and discussed in detail in the companion to this paper, where the growing classical three-dimensional Minkowski spacetime has a metric, Ricci Tensor and Stress Energy tensor [24].

In PAM, each "interaction, entanglement, and actualization event" among the particles constructs a "unit volume of spacetime." PAM yields a steep power law construction of spacetime, Figures 6a, 6b and 7. This is an obvious candidate for Inflation itself [4, 5, 6, 7]. For example, PAM easily yields a power law creation of space with a slope equal to 4.0 or greater, Figure 8. If Inflation is to occur between 10^{-37} seconds and 10^{-32} seconds, 5 orders of magnitude, a power law creation of spacetime with slope 4.0 yields a 10^{20} expansion of the universe. A typical guess at the expansion during Inflation is 10^{27}. Were the duration of Inflation to occur from 10^{-38} seconds to 10^{-31} seconds, the universe would expand 10^{28}-fold.

5 Further Implications for Cosmogenesis

5.1 Inflation ends naturally

Further aspects of this broad new theory of cosmogenesis suggested by PAM are not yet numerically or formally studied; however, they seem plausible, interesting, and require further analysis. Propose that a metric exists, see [24], and propose merely that the probability of interaction among particles falls off monotonically with distance between particles. Then if the Universe expands very rapidly via Inflation, particle density should fall. This is seen in PAM. If so, the frequency, f, of interactions

189

per particle must decline, moving the system in the two-dimensional parameter space of "delay" and "f" to ever smaller values of f. This is modeled in Figures 4a and 4b by a decreasing probability of interaction. This moves the system parallel to the probability of interaction axis. As this occurs, the system must pass from above to below the phase transition. In short, there must be a slowing then cessation of the capacity of particles to steal the borrowed energy. This cessation is a second-order transition from above where increasing distance between particles reduces the number of interactions per unit time for each particle so much that formation of kinetically stable nuclei is no longer possible. The universe no longer breaks matter-energy conservation.

A natural stopping of inflation needs careful study. If confirmed, this theory of Inflation is unlike most models of Inflation that yield infinite inflation and a multiverse of non-interacting pocket universes [10]. In these theories of Infinite Inflation, universes with no matter can exist. This is ruled out on the present theory which posits that nonlocality must be taken as fundamental. Thus, any universe must have matter to come to exist.

5.2 A possible union with General Relativity

This attempt based on PAM and taking nonlocality as fundamental as a theory of quantum gravity that constructs spacetime is not General Relativity. There is no construction of spacetime in General Relativity. General Relativity is local. Because General Relativity can be formulated without matter fields, General Relativity is incapable of addressing the formation of matter itself. Nor can General Relativity propose a role for matter in constructing spacetime.

Matter and energy are present from the first moment in this model of Cosmogenesis with its baryogenesis and construction of spacetime. Fermions form, transform, vanish, and exchange their bosons. Spacetime is the emergent relational metric among these events.

The ambition is to use a more mature version of the ideas discussed to modify General Relativity, perhaps a bit like the Chadwick *et al.* model where matter curves and also creates spacetime [40]. On such a view, the approach to quantum gravity constructing spacetime sketched above does not constitute General Relativity but unites with it in some new way.

The conceptual ingredients to do so may not be too far away. The forming spacetime with fermions and bosons emerging from the vacuum and transforming autocatalytically, has matter and energy within an emerging spacetime that has a metric. Thus, there is some stress energy tensor and a Ricci curvature tensor. It may not be too far-fetched to hope for union with General Relativity modified by a scalar field with amplitudes for a local construction of spacetime.

Here quantum gravity does not constitute General Relativity, but quantum actualization constructs the spacetime in which classical physics General Relativity operates [24]. In such a version of Quantum Gravity and General Relativity, matter constructs and curves spacetime. Curved spacetime tells matter how to move.

6 Discussion and Further Work

This article is one of a set of three articles:

The first article, "Quantum Gravity If Nonlocality Is Fundamental" [24], is based on two established facts:

1. Quantum Mechanics allows interpretation of the "quantum state" as "potentia," neither true nor false [14].

2. Nonlocality is now established [15, 16, 17].

Based on these, starting with nonlocality implies that matter somehow constructs spacetime [24]. The resulting quantum construction yields the sequential construction of a successive classical three-dimensional Minkowski space like slices that then constitutes a growing four-dimensional spacetime. The theory may be testable using the Casimir effect [24].

The present second article builds on the first and establishes a third claim: The particles of SU(3) × SU(2) × U(1) are formally capable of collective autocatalysis by which ever-new particles can be "stolen" from an "exogenous source" – the quantum vacuum. In short, by adding the "delay" hypothesis, the universe can start with no matter and no spacetime. Then, via the collectively autocatalysis and delay with its second-order phase transition, particles can be constructed and stochastically break matter-antimatter symmetry yielding baryogenesis. The creation, entanglement, and actualization of particles then drives a rapid power law construction of spacetime that is a candidate for the unknown physics of Inflation. The present theory offers a theory of Cosmogenesis, Baryogenesis, and Inflation.

One test of this theory of Cosmogenesis is to attempt to account for our values of the 25 Constants because this combination of values, better than other combinations, maximizes some measure of the efficiency of Cosmogenesis. Such a theory would require some form of Cosmic Natural Selection as first proposed by Smolin [41]. The current PAM computational model can be used to begin such efforts.

A second test of this theory of Cosmogenesis attempts to account for "Why our Laws?" SU(3) × SU(2) × U(1). C. Furey has located these among the Octonions [42]. If Cosmogenesis does indeed depend upon the capacity of our laws to support collective autocatalysis, then SU(3) × SU(2) × U(1) should be richer in Nghe Motifs than sub-adjacent groups among the Octonions. We can test this now. Confirmation could be quite startling.

A third test is experimental. The unexpected discovery that the particles of SU(3) × SU(2) × U(1) are capable of collective autocatalysis may well be open to direct experimental test, for example at CERN. The basic experiment is to partition the set of particles into at least two subsets, Set [A] and Set [B]. Use the set [A] injected as input and test if some or all members of Set [B] are produced. Reciprocally, use some or all of Set [B] injected as input and test if some or all of Set [A] is produced. If, in the most definitive outcome, the union of [A] and [B] is all or almost all of the particles, and injection of [A] produces all of [B], while injection of [B] produces all of [A], we would have strong evidence that the particles are collectively autocatalytic.

If we confirm that the particles of SU(3) × SU(2) × U(1) are indeed collectively autocatalytic, it would become difficult not to explore the potential role of such autocatalysis in cosmogenesis.

At present, our attempts to account for our values of the constants and our Laws rely on the Anthropic Principle [43], whose testability is widely doubted [44], and which most easily relies on the essentially untestable postulate of a multiverse [6, 7]. Our Laws and Constants may, instead,

191

have been selected for ease of efficient Cosmogenesis.

The third article in this series is online [45]. This paper also proposes that spacetime is constructed by matter in each locale proportional to its 4th root, $M^{1/4}$. It is obvious that a construction of spacetime by matter is already a form of Dark Energy [6, 7, 45]. If the universe starts, as often assumed, with a high density of matter at high temperature, that dense matter, by constructing spacetime, becomes a candidate for the unknown physics of Inflation [45]. It is more surprising that the same proposed construction of spacetime by matter, a force that expands spacetime, can be a candidate for Dark Matter and can explain MOND, an alternative to Dark Matter [46, 47, 49]. This third article proposes a direct test of the hypothesis that matter constructs spacetime. It must predict that galaxies that have existed longer must have constructed more spacetime so must rotate faster than galaxies that have lived less long. This can now be tested using the scatter in the data for the Baryonic Tully Fisher Relation [46, 47, 49]. Preliminary data now support this prediction [45]. Were this prediction strongly confirmed, we would conclude that matter does construct or expand spacetime [49]. If so, General Relativity would have to be modified [2, 19, 24, 45].

Taken together, the first two articles may be an example of G. Ellis' proposed "The Evolving Block Universe" [50]. Taken together, the three articles propose new theories of Cosmogenesis, Baryogenesis, Inflation, Dark Matter, and Dark Energy. Some aspects of theory are now testable.

7 Conclusions

Copernicus in 1543 created a new heliocentric view of the world. He based his work on seven postulates, none of which had independent support. Yet, these seven provided a new conceptual framework for astronomy. More, with the hint of something radiating from the sun that held the planets in orbit, the new vision led to Newton.

The profound success of General Relativity has led almost all work on cosmogenesis to start with locality and spacetime as fundamental. The universe somehow suddenly appears, and somehow has very low entropy. Somehow "exponential" – whose physics is unknown – Inflation happens. Then with Cold Dark Matter, whose physics is unknown, and Dark Energy, whose physics is unknown, the Lambda CDM model is now the Standard Model of Cosmogenesis. Fortunately, a 1 in 50,000,000 excess of matter over antimatter leads to a matter-dominated universe. And ours is a Fortunate Universe [6], whose 25 constants are exquisitely tuned such that life can exist.

We have attempted a mini-Copernicus. We start with three claims:

1. Quantum particles in superposition can be interpreted as potentia. Potentia are ontologically real and could be outside of spacetime. Then we can at least conceive of something real that is not in spacetime.

2. Nonlocality is firmly established. There are no *a priori* grounds to choose locality over nonlocality as fundamental. Why not start with nonlocality and see what can be done? Somehow spacetime must emerge from and be constructed from entangled coherent particles chosen among those of SU(3) × SU(2) × U(1).

3. Upon examination, the particles of SU(3) × SU(2) × U(1) are formally capable of collective autocatalysis. We establish this third claim in the present article. Our analysis can be

extended to all the particles and transformations among SU(3) × SU(2) × U(1). We can establish the total number of Nghe motifs and the number of Type 1, Type 2, Type 3, Type 4, and Type 5 Nghe motifs in our standard model and compare this objective measure of collective autocatalysis to other candidate laws among the octonions. Are our Laws better at collective autocatalysis than other candidate laws? We may even be able to experimentally test collective autocatalysis among the particles of SU(3) × SU(2) × U(1), stunning if confirmed.

Like the imagined rays from the sun holding the planets in orbit, might these three claims conspire to allow the universe to construct itself from no spacetime and no matter? We employ one further postulate: A "delay" in the return of borrowed energy back to the vacuum upon interaction of quarks or of antiquarks.

The postulates suffice. We can conceive of the universe starting from no matter and no spacetime and constructing itself. A construction of spacetime promises possible answers to: How did spacetime appear? If the universe starts from no spacetime and no matter, its entropy is 0, answering the struggle over the low entropy of the initial state needed for the Past Hypothesis. The universe starts from a unique initial state. Baryogenesis is natural, not an *ad hoc* 1 in 50 million-excess of matter over antimatter. A construction of spacetime by matter can be the unknown physics of Inflation.

The present article provides a testable new framework for cosmogenesis. The limitation of PAM to classical variables is severe. We hope our efforts prove useful.

Online access to the current Particle Apothecary Model is available at particleapothecary.org.

Acknowledgments

The Authors have the right to publish this material and acknowledge with gratitude conversations with Marina Cortês, George Ellis, Andrea Roli, Sauro Succi, and Pablo Tello. Warmest thanks to Philippe Nghe for his work on autocatalytic motifs and help identifying them in SU(3) × SU(2) × U(1). And warmest thanks to Stefan Bornholdt for his ever highly skeptical advice.

References

[1] R. Mann, *An Introduction to Particle Physics and The Standard Model*. Taylor and Francis, Boca Raton, FL, (2010).

[2] B. Harrison, S. Kent, K. Thorne, M. Wakano, J. A. Wheeler, *Gravitational Theory and Gravitational Collapse*. University of Chicago Press, (1965).

[3] N. Deruelle, *Relativity in Modern Physics*. Oxford University Press, (2018). https://doi.org/10.1093/oso/9780198786399.003.0059

[4] S. Tsujikawa, *Introductory review of cosmic inflation*. arXiv:hep-ph/0304257 (2003).

[5] P.J. Steinhardt, *The inflation debate: Is the theory at the heart of modern cosmology deeply flawed?* Sci. Am. 304(4), 18–25, (2011).

[6] G. Lewis, L. Barnes, *A Fortunate Universe*. Cambridge University Press, (2016).

[7] A. Liddle, *An Introduction to Modern Cosmology*. Third Edition. Wiley, Sussex, UK, (2015).

[8] B. Casaba, *A small review of the big picture*.
arxiv:1411.3398v1 (2014).

[9] M. Cortês, S. Kauffman, A. Liddle, L. Smolin, *Biocosmology: towards the birth of a new science*.
arXiv:2204.09378 [astro-ph.CO], (2022).

[10] A. Vilenkin, *Birth of inflationary universes*. Phys. Rev. D, 27(12), 2848–2855 (1983). doi:10.1103/PhysRevD.27.2848

[11] G. Bertone, D. Hooper, J. Silk, *Particle dark matter: evidence, candidates and constraints*. Physics Reports, 405(5-6), 279–390 (2005). arXiv:hep-ph/0404175

[12] P.J.E. Peebles, B. Ratra, *The cosmological constant and dark energy*. Rev. Mod. Phys, 75(2), 559–606 (2003). doi:10.1103/RevModPhys.75.559

[13] Wikipedia. *Nicolaus Copernicus*. Available at: https://en.wikipedia.org/wiki/Nicolaus_Copernicus

[14] W. Heisenberg, *Philosophy and Physics: The Revolution in Modern Science*. Penguin Books, (1958).

[15] A. Aspect, J. Dalibard, G. Roger, *Experimental test of Bell's inequalities using time-varying analyzers*. Phys. Rev. Lett, 49(25), 1804–1807 (1982). doi:10.1103/PhysRevLett.49.1804

[16] B. Hensen et al., *Loophole-free Bell inequality violation using electron spins separated by 1.3 kilometres*. Nature, 526(7575), 682–686 (2015). doi:10.1038/nature15759

[17] L.K. Shalm et al., *Strong loophole-free test of local realism*. Phys. Rev. Lett, 115(25), 250402 (2015). doi:10.1103/PhysRevLett.115.250402

[18] A. Blockhuis, D. Lacoste, P. Nghe, *Universal motifs and the diversity of autocatalytic systems*. PNAS, 117, 2523 (2020).

[19] S. Hawking, G. F. R. Ellis, *The Large Scale Structure of Space-Time*. Cambridge University Press, (1973). ISBN 0-521-09906-4.

[20] E. Witten, *String theory dynamics in various dimensions*. Nuclear Physics B, 443(1), 85–126 (1995). doi:10.1016/0550-3213(95)00158-O

[21] C. Rovelli, *Loop Quantum Gravity*. Living Reviews in Relativity, 11(1). doi:10.12942/lrr-2008-5 (2008).

[22] J. Maldacena, *The large N limit of superconformal field theories and supergravity*. Advances in Theoretical and Mathematical Physics, 2(2), 231–252 (1998).

[23] L. Susskind, *The World as a Hologram*. Journal of Mathematical Physics, 36(11), 6377 - 6396 (1995). doi:10.1063/1.531249

[24] S. Kauffman, *On quantum gravity if nonlocality is fundamental*. Entropy, 24, 554 (2022). doi:10.3390/e24040554

[25] M. Schlosshauer, *Decoherence, the measurement problem, and interpretations of quantum mechanics*. Rev. Mod. Phys, 76(4), 1267–1305 (2005).

[26] R. Feynman, *The Feynman Lectures on Physics Vol I*. Addison Wesley, (1970). ISBN 978-0-201-02115-8

[27] R. Penrose, *The Road to Reality: A Complete Guide to the Laws of the Universe*. Vintage, (2007).

[28] S. W. Hawking, *The Singularities of Gravitational Collapse and Cosmology*. Proc. R. Soc. A, 314, 1519 (1970).

[29] A.V. Nesteruk, *Inflation with low gravitational entropy at the singularity.* Europhys. Lett, 36, 233 (1996).

[30] A.D. Linde, *Eternally existing self-reproducing chaotic inflationary universe.* Phys. Lett. B, 175, 395-400 (1986). doi:10.1016/0370-2693(86)90611-8

[31] D.T. Gillespie, *Exact stochastic simulation of coupled chemical reactions.* Jour. Phys. Chem, 81, 2340 (1977).

[32] S. Kauffman, S. Patra, *A testable theory for the emergence of the classical world.* Entropy, 24 (2022). doi:10.3390/e24060844

[33] M. Khankhasyev, C. Scarlett, *Positron on neutron capture reaction, radiative corrections and neutron EDM.* arXiv:1305.6642 (2013).

[34] D. Sakharov, *Violation of CP invariance, C asymmetry, and baryon asymmetry of the universe.* JETP Lett. USSR, 5, 24 (1967).

[35] D.P. Kirilova, Y.V. Valchanov, *Early universe baryogenesis.* Proc. IV. Serbian-Bulgarian Astronomical Conference, M.S. Dimitrijevic et al. eds, 5, 209 (2005).

[36] A. Di Biagio, C. Rovelli, *Stable facts, relative facts.* Found. Phys, 51, 30 (2021).

[37] C. Cao, S.M. Carroll, S. Michalakis, *Space from Hilbert Space: recovering geometry from bulk entanglement.* Phys. Rev. D, 95, 024031 (2017).

[38] A. Singh, O. Doré, *Does quantum physics lead to cosmological inflation?* arXiv:2109.03049 (2021).

[39] M. Van Raamsdonk, *Building up spacetime with quantum entanglement.* arXiv:1005.3035, 42, 2323 (2010).

[40] E. Chadwick, T. Hodgkinson, G. McDonald, *A gravitational development supporting MOND.* Phys. Rev. D, 88, 024036 (2013). doi:10.1103/PhysRevD.88.024036

[41] L. Smolin, *Life of the Cosmos.* Oxford University Press, (1997).

[42] C. Furey, *SU(3) u SU(2) u U(1) (u U(1)) as a symmetry of division algebraic ladder operators.* Eur. Phys. J. C, 5, 375–414 (2013).

[43] J. Barrow, F. Tipler, *The Anthropic Cosmological Principle.* Oxford University Press, (1986).

[44] L. Smolin, *Scientific alternative to the anthropic principle.* In Universe or Multiverse, Cambridge University Press (2007), 20(4), 323–366.

[45] S. Kauffman, *Are Dark Matter Dark Energy and Inflation A Construction of Spacetime by Matter.*
doi:10.31219/osf.io/5chn4 (2024).

[46] S. McGaugh, *The Baryonic Tully–Fisher Relation of Gas-Rich Galaxies as a Test of ΛCDM and MOND.* Astrophysical Journal, 143(2), 40 (2012).
arXiv:1107.2934.
doi:10.1088/0004-6256/143/2/40. S2CID 38472632.

[47] E. Papastergis, E.A.K. Adams, M. van der Hulst, *An accurate measurement of the baryonic Tully-Fisher Relation with heavily gas dominated ALAFLA galaxies.* Astronomy and Astrophysics, vol 593 (2016).

[48] G. Mohan, U. Dev Goswamiy, *Galactic rotation curves of spiral galaxies and Dark Matter in f(R; T) gravity theory.*
arXiv:2211.02948 [gr-qc] (2022).

[49] S. Marongwe, S. Kauffman, *Dark Matter as a Ricci Soliton.*
arxiv.org/abs/0907.2492 (2024).

[50] G. F. R. Ellis, *The Evolving Block Universe and the Meshing Together of Time.* arxiv:1407.7243 (2014).

[51] U. Wilensky, *NetLogo.* http://ccl.northwestern.edu/netlogo/ (1999). Center for Connected Learning and Computer-Based Modeling, Northwestern University, Evanston, IL.

Table 14.2: Nghe Motifs Summary

#	Motif
486	core(s) found, including:
0	type I core(s)
192	type II core(s)
294	type III core(s)
0	type IV core(s)
0	type V core(s)

Table 14.3: Nghe Motifs: Type II and III

#Autocatalytic core number 53 of type 2

External set = { e^+, d, $\bar{d}$, $\bar{\nu}$, μ }

$2\gamma \rightleftharpoons u + \bar{u}$

$\bar{\mu} \rightleftharpoons \bar{u}$

$\bar{\mu} \rightleftharpoons u$

$e^- \rightleftharpoons u$

$2\gamma \rightleftharpoons e^-$

as part of the reactions:

$2\gamma \rightleftharpoons u + \bar{u}$

$\mu + \bar{\mu} + \bar{d} \rightleftharpoons \bar{u}$

$\mu + \bar{\mu} + d \rightleftharpoons u$

$e^- + \bar{\nu} + d \rightleftharpoons u$

$2\gamma \rightleftharpoons e^+ + e^-$

#Autocatalytic core number 80 of type 2

External set = { e^+, $\bar{d}$, $\bar{\nu}$, ν_μ, $\bar{\nu}_\mu$ }

$2\gamma \rightleftharpoons \mu + \bar{\mu}$

$\bar{\mu} \rightleftharpoons \nu$

$\nu \rightleftharpoons \bar{u}$

$\mu + \bar{\mu} \rightleftharpoons \bar{u}$

$\mu \rightleftharpoons e^-$

$2\gamma \rightleftharpoons e^-$

as part of the reactions:

$2\gamma \rightleftharpoons \mu + \bar{\mu}$

$\bar{\mu} \rightleftharpoons e^+ + \nu + \bar{\nu}_\mu$

$e^+ + \nu + \bar{d} \rightleftharpoons \bar{u}$

$\mu + \bar{\mu} + \bar{d} \rightleftharpoons \bar{u}$

$\mu \rightleftharpoons e^- + \bar{\nu} + \nu_\mu$

$2\gamma \rightleftharpoons e^+ + e^-$

#Autocatalytic core number 151 of type 2

External set = { d, $\bar{d}$, $\bar{\nu}$, ν, μ, $\bar{\nu}_\mu$ }

$2\gamma \rightleftharpoons u + \bar{u}$

$e^- \rightleftharpoons \bar{u}$

$e^- \rightleftharpoons u$

$e^+ \rightleftharpoons u$

$\bar{\mu} \rightleftharpoons e^+$

$2\gamma \rightleftharpoons \bar{\mu}$

as part of the reactions:

$2\gamma \rightleftharpoons u + \bar{u}$

$e^- + \bar{\nu} + \bar{d} \rightleftharpoons \bar{u}$

$e^- + \bar{\nu} + d \rightleftharpoons u$

$e^+ + \nu + d \rightleftharpoons u$

$\bar{\mu} \rightleftharpoons e^+ + \nu + \bar{\nu}_\mu$

$2\gamma \rightleftharpoons \mu + \bar{\mu}$

#Autocatalytic core number 233 of type 3

External set = { $\bar{d}, \bar{\nu}, \nu, \bar{\mu}, \nu_\mu$ }

$2\gamma \rightleftharpoons e^+ + e^-$

$e^+ \rightleftharpoons \bar{u}$

$2\gamma \rightleftharpoons \mu$

$\mu \rightleftharpoons \bar{u}$

$\mu \rightleftharpoons e^-$

as part of the reactions:

$2\gamma \rightleftharpoons e^+ + e^-$

$e^+ + \nu + \bar{d} \rightleftharpoons \bar{u}$

$2\gamma \rightleftharpoons \mu + \bar{\mu}$

$\mu + \bar{\mu} + \bar{d} \rightleftharpoons \bar{u}$

$\mu \rightleftharpoons e^- + \bar{\nu} + \nu_\mu$

#Autocatalytic core number 261 of type 3

External set = { $e^-, d, \bar{u}, \nu, \nu_\mu, \bar{\nu}_\mu$ }

$2\gamma \rightleftharpoons e^+$

$2\gamma \rightleftharpoons u$

$\bar{\nu} \rightleftharpoons u$

$2\gamma \rightleftharpoons \mu + \bar{\mu}$

$\mu \rightleftharpoons \bar{\nu}$

$\bar{\mu} \rightleftharpoons e^+$

as part of the reactions:

$2\gamma \rightleftharpoons e^+ + e^-$

$2\gamma \rightleftharpoons u + \bar{u}$

$e^- + \bar{\nu} + d \rightleftharpoons u$

$2\gamma \rightleftharpoons \mu + \bar{\mu}$

$\mu \rightleftharpoons e^- + \bar{\nu} + \nu_\mu$

$\bar{\mu} \rightleftharpoons e^+ + \nu + \bar{\nu}_\mu$

#Autocatalytic core number 324 of type 3

External set = { $e^+, d, \bar{d}, \mu, \bar{\nu}_\mu$ }

$2\gamma \rightleftharpoons e^-$

$2\gamma \rightleftharpoons u + \bar{u}$

$\nu \rightleftharpoons u$

$\bar{\mu} \rightleftharpoons e^-$

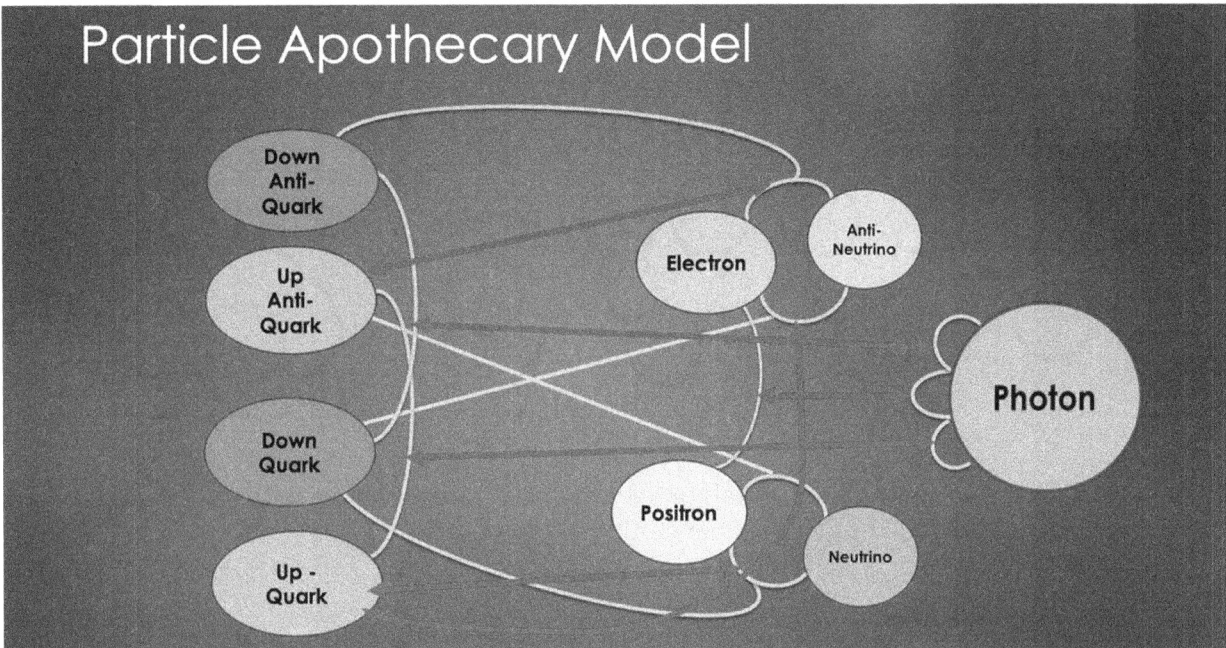

Figure 42: Transformations among the variables in the PAM Model. (See Transformations 1 – 14 in Table 1 below). Red lines with arrowheads at each end are transformations among the variables connected by yellow lines. Yellow lines connect variables that are inputs and outputs of transformations.

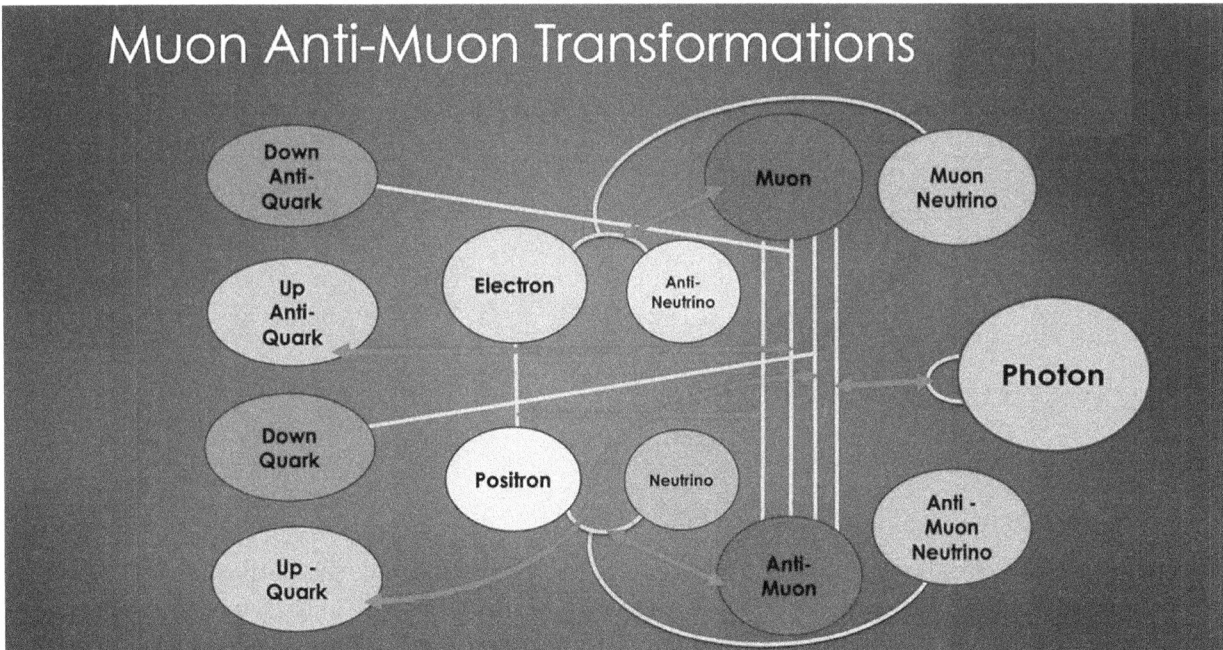

Figure 43: Transformations of muons, antimuons, muon neutrinos and antimuon neutrinos. Purple lines with arrow heads at each end are transformations among variables connected by yellow lines. Yellow lines connect variables that are inputs and outputs of transformations.

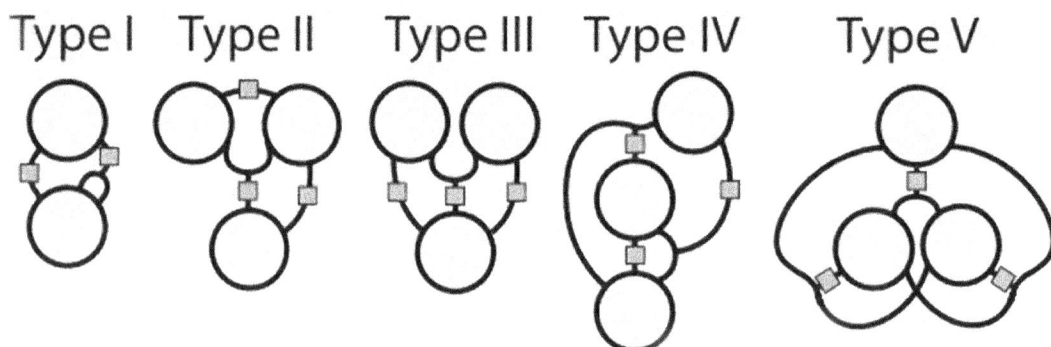

Figure 44: The five collectively autocatalytic motifs. The orange squares indicate locations where more reactions can be added as long as the motif type is preserved, (14).

Figure 4a: One axis, labeled Lifetime Extension is identical to Delay. The other axis, Probability of Interaction ranges from 1 to 1.0. Each probability of interaction is linearly proportional to the number of variables in the system. The figure clearly shows the second order phase transition. The dark purple zone along both axes and extending into the two-parameter space corresponds to a formation of 5 or fewer quarks or antiquarks in 500-time steps. Further beyond the phase transition, the total number of quarks or of antiquarks created in 500-time steps increases.

Mean Total Quarks

Probability Interaction (y-axis)

Particle Lifetime Extension (x-axis)

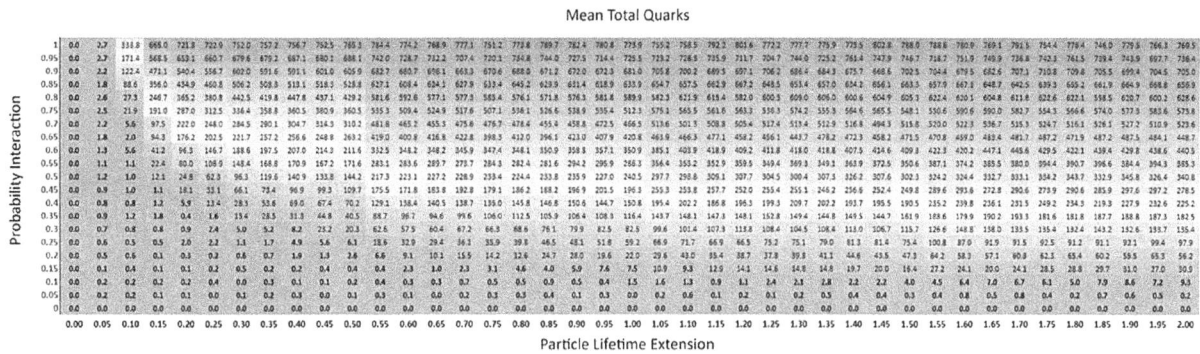

Figure 4b: The axes are identical to Figure 4a. Figure 4b shows the mean number of quarks plus antiquarks created in 500-time steps for all pairs of parameter values. Figure 4b shows the second order phase transition discussed. Decreasing probability of interaction can be used to model increasing spatial distance between variables during and after Inflation Note that even with a very small probability of interaction a slow rate of creation of quarks or of antiquarks continues.

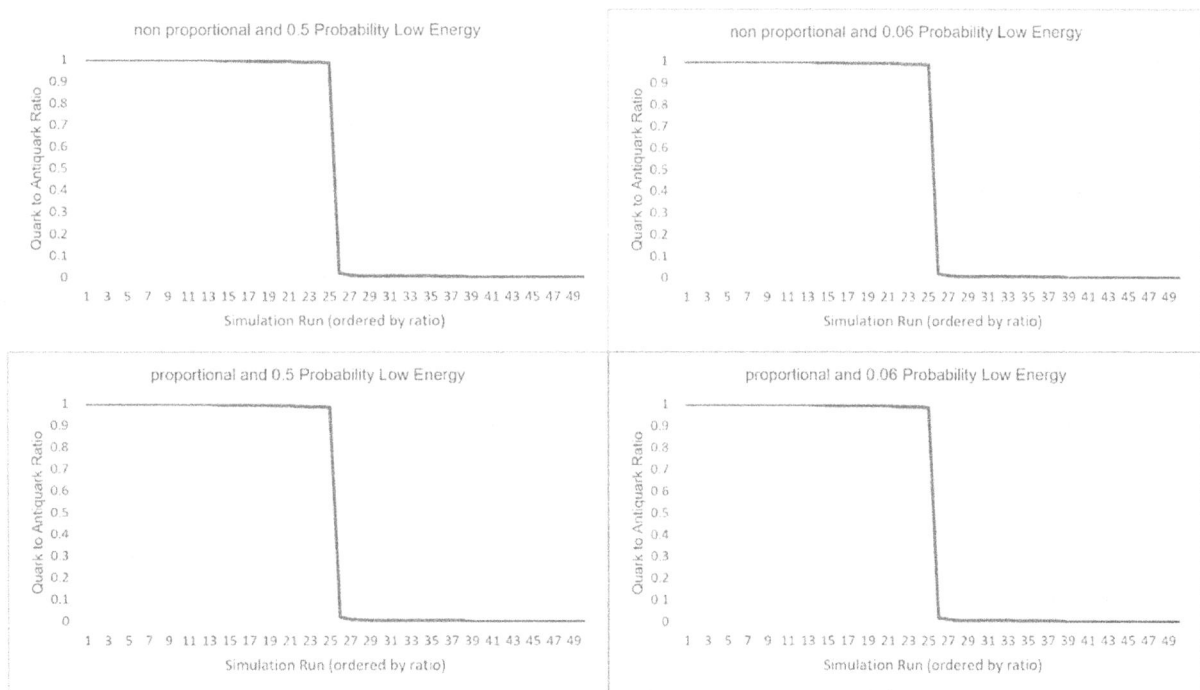

non proportional and 0.5 Probability Low Energy

Quark to Antiquark Ratio

Simulation Run (ordered by ratio)

non proportional and 0.06 Probability Low Energy

Quark to Antiquark Ratio

Simulation Run (ordered by ratio)

proportional and 0.5 Probability Low Energy

Quark to Antiquark Ratio

Simulation Run (ordered by ratio)

proportional and 0.06 Probability Low Energy

Quark to Antiquark Ratio

Simulation Run (ordered by ratio)

Figure 5: Panels a, b, c, and d, show 50 independent runs using different random "seeds" from the same PAM parameter settings. The data in each panel plot the "quark / antiquark ratio." The results are striking. For half of the "runs," the quark / antiquark ratio is 1.0 or slightly less, for the other half the ratio of quarks/antiquarks is 0.0 or slightly greater. The stochastic kinetic processes in PAM break matter antimatter symmetry and yields Baryogenesis with respect to quarks and antiquarks.

201

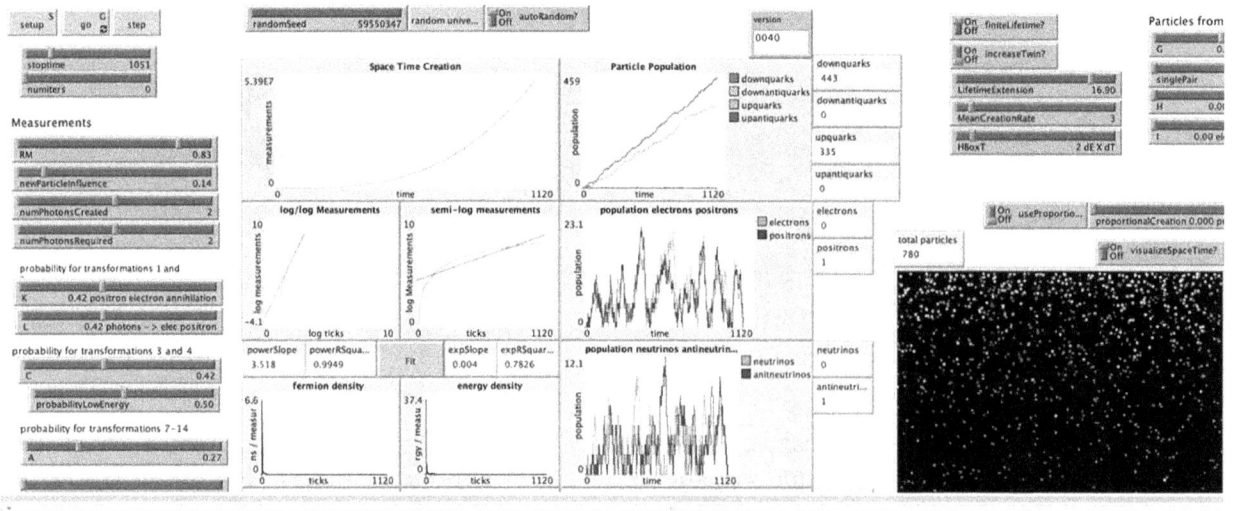

Figure 6a: PAM screen shot, power-law creation of spacetime and baryogenesis. PAM parameters make the rate of quark antiquark pair formation fully independent of the existing number of fermions and leptons. Power law creation of spacetime and baryogenesis with respect to quarks versus antiquarks. Above, upper right, quarks win. There is no baryogenesis with respect to electrons versus positrons or with respect to neutrinos versus antineutrinos, middle and lower right. Panel on the lower right shows the diversity of particles created in the system. The lower left panel shows the brief spike in matter density.

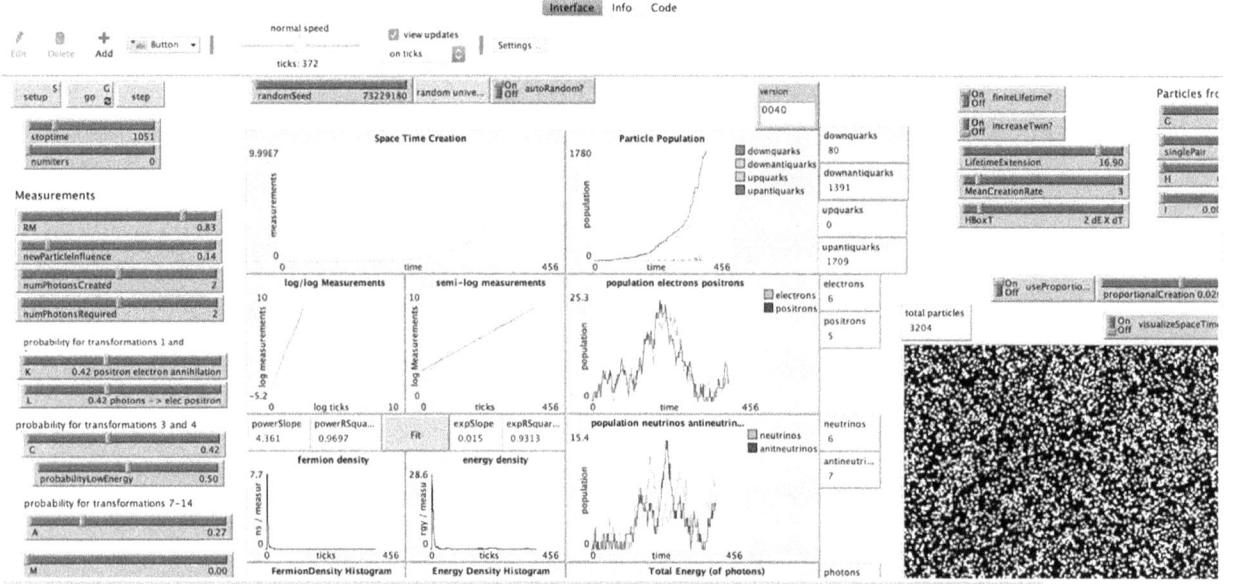

Figure 6b: PAM parameters make the rate of quark antiquark pair formation weakly dependent on the existing number of fermions and leptons. Power law creation of spacetime and baryogenesis with respect to quarks versus antiquarks. Above, upper right, antiquarks win. There is no baryogenesis with respect to electrons versus positrons or with respect to neutrinos versus antineutrinos, middle and lower right. Panel on the lower right shows the diversity of particles created in the system. The lower left panel shows the brief spike in matter density.

Space Time Creation

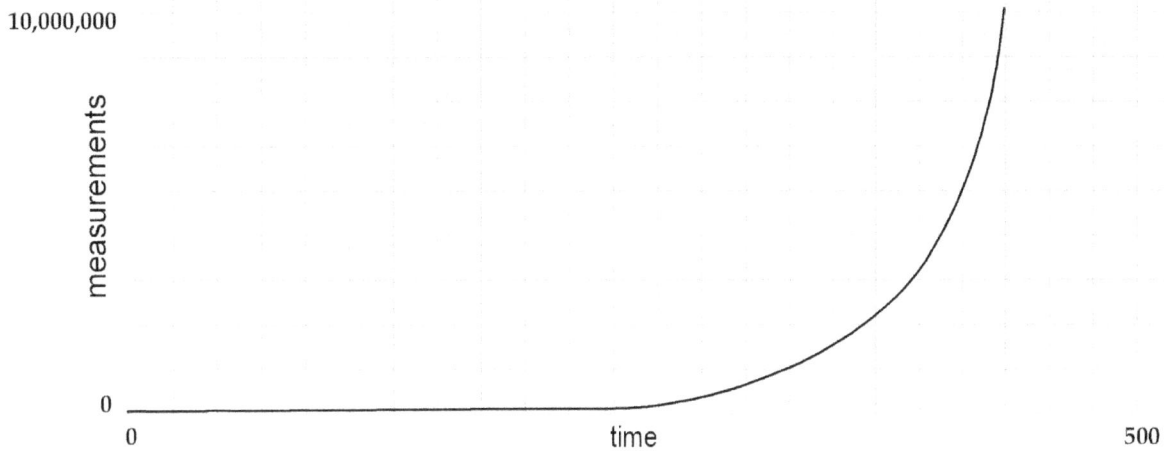

Figure 7: Spacetime Creation. Power law slope 4.643. PAM parameters set to make the rate of quark antiquark pair formation weakly proportional to the existing number of fermions and leptons. For each of the 100 runs from the same PAM parameter settings but from different random seeds, the kinetics of spacetime formation was analyzed both for the capacity to be fit by a power law and an exponential. In all PAM settings explored all curves are clearly best fit by a power law. The slopes vary, see Figure 8.

Prob Low Energy	mean slope of powerlaw	StdDev of slope	Max of slope	r^2 of powerlaw	r^2 of exponential
non proportional	3.909	0.612	5.743	0.988	0.856
0.06	3.844	0.494	4.940	0.989	0.853
0.5	3.975	0.710	5.743	0.987	0.858
proportional	4.571	0.790	6.862	0.983	0.915
0.06	4.843	0.885	6.862	0.984	0.922
0.5	4.300	0.572	5.794	0.983	0.908

Figure 8: Two conditions are varied independently: "i. The terms non-proportional" versus "proportional" refer to the independence or dependence of the process on the existing number of particles. ii. The terms 0.06 and 0.5 refer to the abundance of Up quarks or Up antiquarks relative to the abundance of Down quarks or Down antiquarks emerging from the vacuum. The ratio of Up to Down is higher for 0.06 than 0.5. The Figure shows the mean and standard deviation of the slope of the power laws for the creation of spacetime, the maximum slope seen, and the relative accuracy fitting of the observed curve of the creation of spacetime as a power law versus an exponential. In all cases the observed curves are better fit by a power law.

15 Modeling the Progression of Time as an Autocatalytic Reaction

Bruce M. Boman

Abstract   The progression of time was modeled as an autocatalytic reaction to understand the perpetual and asymmetric dynamics of time. Accordingly, the flow of time was modeled as the progression of events. Model design was based on a kinetic reaction and the direction of the reaction was assumed to progress from future to present to past events. Model rate equations were used to investigate how the progression of events might be explained by changes in energy and entropy. The results indicate that the asymmetric direction of event progression is due to a superposition between time future events and time now events. Modeling the progression of events as an autocatalytic reaction in which time past events act to catalyze their own production showed that this autocatalysis mechanism accelerates the reaction rate. That the autocatalytic production of time past events increases the overall rate of progression of events provides a mechanism that could explain why the progression of time events appears perpetual. The ability of time past events to autocatalyze their own production could lead to a self-sustaining progression of events when time past events at one point of reference serve as time future events at other points of reference. In this way, the Universe as a whole would dynamically be self-sustaining.

Keywords: Autocatalytic reaction, asymmetric direction of event progression, self-sustaining progression of events

1   Introduction

The goal of my study is to understand the perpetual and asymmetric dynamics of time. When I talk to others about the perpetual nature of time, the usual response is that we know from physics that perpetual motion is impossible. In physical chemistry, I learned the reason that perpetual motion is impossible is because it violates the first and second laws of thermodynamics [1]. In engineering, it is stated that you can't get work from nothing. Accordingly, while thinking about the perpetual nature of time, one needs to consider that energy cannot be created or destroyed (although it can be transferred). Also, consider that the entropy of isolated systems left to spontaneous evolution cannot decrease. So, if a physical process is irreversible, the entropy of the system and the environment must increase and the final entropy must be greater than or equal to the initial entropy. These laws clearly pertain to an understanding of time which appears to be not only perpetual, but also as a spontaneous and irreversible process.

In trying to reflect on why time seems perpetual but that perpetual motion is impossible, I con-

Eric Ling and Annachiara Piubello (Eds), Spacetime 1908-2023. Selected peer-reviewed papers presented at the *Third Hermann Minkowski Meeting on the Foundations of Spacetime Physics*, 11-14 September 2023, Albena, Bulgaria (Minkowski Institute Press, Montreal 2024). ISBN 978-1-998902-25-5 (softcover), ISBN 978-1-998902-26-2 (ebook).

sidered the meaning of motion. Motion is described as the act, process, or instance of change in position of an object with respect to its surroundings in a given interval of time [2]. In juxtaposition, I view time as being a measure of change in the position of an object with respect to its surroundings. This is consistent with Frank Wilczek's description of time in his book "Fundamentals", as "a quantity, usually written as t, which appears in our fundamental description of change that takes place in our physical world" [3]. In this view, it is the change that is perpetually taking place in our physical world that we describe as the flow of time. The change that is taking place is defined here as the happening of events.

My current thinking is that time is not some "thing or entity" that flows, rather time should be thought of as a reaction that progresses because time is an essential factor that serves as the basis for happening of events and processes. It is also the primary variable in measuring the rate of a reaction. Accordingly, my approach is to understand time in relation to the progression of events as a kinetic reaction. This line of thinking also provides a mechanism to explore why time progresses perpetually. To achieve this, I applied knowledge and concepts based on mechanisms for molecular reactions from chemistry to study the nature of time in terms of reaction kinetics. In this quest, the progression of events that are coming to be and then passing away is modeled like the generation of products from reactants in a chemical reaction. Thus, in contrast to the notion in physics that time flows from *past to present to future*, when viewed like a kinetic reaction, the direction of time progresses from *future to present to past*. Moreover, similar to the progression of a chemical reaction in which the reactants interact and go on to form products, the interaction of "future time" events can be modeled as the becoming of "now time" events that will then transition to "past time" events [4-7].

1.1   Time is a Function of Change in Space

In my model, spacetime is considered to be a dynamic system that is described in terms of changes in time that happen as a function of changes in space. This is consonant with the real world in which most, if not all, measurements of time are based on changes in space. Even the sense of time in our daily lives depends on spatial changes because it is intimately connected to spinning of Earth around its axis or the rotation of our planet around the sun. The standard way of quantifying time depends on the measurement of an interval of time which is quantified as a change in time relative to change in space. In fact, clocks are simply measuring devices that measure spacetime intervals as a regular succession of events. Atomic clocks measure the frequency of vibration within atoms. The base unit of time in the International System of Units (SI) is the second, defined based on the cesium standard or about 9 billion oscillations of the cesium atom. The shortest possible time interval that can theoretically be measured is Planck time, which is the time it would take a photon travelling at the speed of light to cross a distance equal to one Planck length [8, 9]. One unit of Planck time is approximately 5.39×10^{-44} seconds. Thus, even at the quantum level, the measurement of time is a function of change in space. In this line of thinking, it can be deduced that events are quantized.

1.2   Spacetime is the Happening of Events

The flow of time is usually described as the continued sequence of events that succeed one another [10]. In a different view, describing the flow of time in terms of events which occur in irreversible succession implies that it is actually occurrence of events that is fundamental. Indeed, in his recent book, "The Order of Time", Carlo Rovelli states "Ours is a world of events not things" [11]. The Merriam-Webster Dictionary defines event as "Something that happens" [12]. In this sense, it is not really time that flows, it is the happening of events [13]. From this perspective, events don't happen in spacetime, spacetime is the happening of events. Even as humans, we are not separate from spacetime, and the course of our lives and actions are made of events including any observations that we make about other events. Thus, in this study, spacetime is viewed as the happening of events, and the happening of events is produced by the interaction of events.

1.3 The Happening of Events Occurs via Progression of a Kinetic Reaction

Accordingly, my current thinking is that the happening of events occurs because events are components of a kinetic reaction. By viewing the happening of events as a kinetic reaction that dynamically progresses, it gives us the ability to apply concepts based on mechanisms in chemistry to explain events in terms of reaction kinetics. Analogous to molecules in a kinetic reaction, spacetime is assumed to be a system comprised of events which are in continuous motion and constant interaction. Therefore, the progression of events is modeled like a reaction that involves progression of reactants to form products. In my model, an event is defined as any change in time relative to a change in space. Thus, the progression of events can be modeled according to the mechanisms described by transition state theory (TST) that explains the kinetics of molecular reactions [14-21].

Transition State Theory

Transition State Theory (TST) provides a mechanism-based approach to understand the dynamics of time and space (in terms of reaction rates) from a reaction kinetics perspective. Based on reasoning from TST, modeling that time progresses because of the interaction of time events at a given point of reference, lets us understand how the progression of time events and their asymmetric direction occur as a spontaneous reaction [22]. This theory has been used for nearly 100 years to understand mechanisms of kinetic reactions. In TST, the interaction of the reactants leads to formation of an instantaneously activated complex during a transition state in the progression to products. TST also assumes that a quasi-equilibrium exists between the reactants and the activated complex in the transition state. The E_a is a threshold energy that the reactant(s) must acquire before reaching the transition state. Once in the transition state, the reaction can go in the forward direction towards product(s), or in the opposite direction towards reactant(s). The reaction products are then irreversibly formed from the transition state. Thus, the level of E_a that a reaction acquires to reach the transition state controls rate of progression to reaction products. By applying TST to explain progression of events in spacetime, we can begin to understand the thermodynamics in the transition of future time events to past time events.

TST also provides a mechanism-based approach to understand why the progression of time events might be perpetual. In modeling the progression of time as an autocatalytic reaction, the model mechanism involves the interaction of past time events with time future events, which catalyzes the production of past time events. That is, past time events catalyze their own production. In chemistry, a reaction is autocatalytic if one of the reaction products is also a catalyst for the same reaction. In this view, the autocatalytic production of past time events may provide a mechanism that explains why the progression of time is perpetual and self-sustaining.

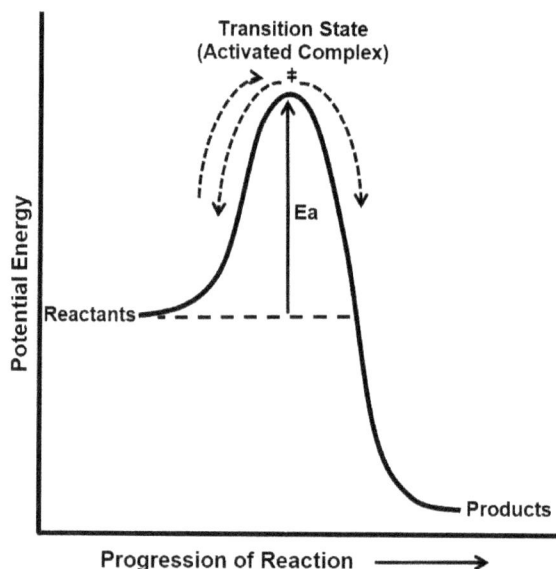

Figure 1. **Potential energy diagram for the progression of a chemical reaction.** The progression of a chemical reaction is explained by TST in terms of activation energy (E_a). The progress of the reaction is from reactants to transition state to products. The highest position for (E_a) required to initiate the reaction occurs in the transition state. The energy difference between the reactants and transition state equals the activation energy of the reaction. This energy is directly related to the rate of the reaction, which is determined by the shape of the energy curve connecting the reactants and products. The transition state (activated complex) is an instantaneous point in the conversion of reactants to products. In TST, it is assumed that an equilibrium exists between the reactants and the transition state activated complex.

2 Model Design

Accordingly, a mathematical model was created for the kinetics of spacetime events whereby an activated transition state occurs as a now time event in the progression of future time events to past time events. In this way, the present (time now) is considered to be an instantaneous transition state in the progression of future to past. Additionally, based on TST, an equilibrium exists between future time (t_F) events and now time (t_N) events that establishes a superposition-like state whereby time fluctuates forward and backward between future time and now time. From the transition state, the now time (t_N) events can (based on the activation energy) irreversibly progress to past time (t_P) events. Past time (t_P) events are different from future time (t_F) and now time (t_N) events because past time (t_P) events are immutable. Still, time past (t_P) events can interact with time future (t_F) events in an autocatalytic reaction mechanism in order to model the perpetual dynamics of time event progression.

2.1 Definition of an Event

In my model, spacetime is not considered to consist of fixed spatial or temporal dimensions, but simply as a dynamic variable (an event) based on changes in time as a function of changes in space (a spacetime vector or tensor). As noted above, this is consonant with the real world as our physical

208

measurement of time is based on changes in space. Thus, an event is defined as any change that takes place involving both space and time. And, the progression of spacetime is measured as a sequence of interacting events in motion at a given reference point where time changes (dt) as a function of changes in space (ds). In addition, time and space are variables expressed in terms of units of time and space.

Accordingly, in the model, the proposed quantitative definition of an event is as follows:

$\text{Event} = \frac{Change\ in\ time}{Change\ in\ space} = \frac{dt}{ds} = \frac{t_2 - t_1}{s_2 - s_1}$, where $t = $ time and $s = $ space

In scalar terms, an event is a change in time relative to a change in space which occurs at all given points of reference in the Universe.

2.2   Model Assumptions

The model assumptions on which it is based, are discussed below.

1. Time past (t_P), time now (t_N), and time future (t_F) events are distinct from one another.

2. Time now (t_N) is an instantaneous transition state between t_F and t_P in the progression of t_F to t_P Namely, time future (t_F) progresses to time now (t_N) which progresses to time past (t_P).

3. Time now (t_N) is also considered to occur at a point of reference anywhere in space where events progress in one direction and time progresses from future to past. In this view, some events, termed time future (t_F), are happening prior to (precede) time now (t_N) and are coming to be in time now. Other events, termed time past (t_P), take place later (downstream) than at the given point of reference of time now (t_N).

4. Time past (t_P) arises from t_N as an irreversible reaction from t_N to t_P.

5. A quasi-equilibrium exists between t_F and t_N events. The equilibrium exists because not every interaction of t_F events results in the formation of the transition state. The outcome depends on thermodynamics (energy and entropy) of the interacting t_F events. Once interacting t_F events reach a transition state, a t_N event is formed. At that instant it immediately progresses to t_P events and does not collapse back to the t_F events. Based on the equilibrium between t_F and t_N events, the progression of spacetime events occurs according to thermodynamics (changes in energy and entropy) of the system.

6. Time past (t_P) can also interact with time future (t_F) events to simulate an autocatalytic reaction mechanism.

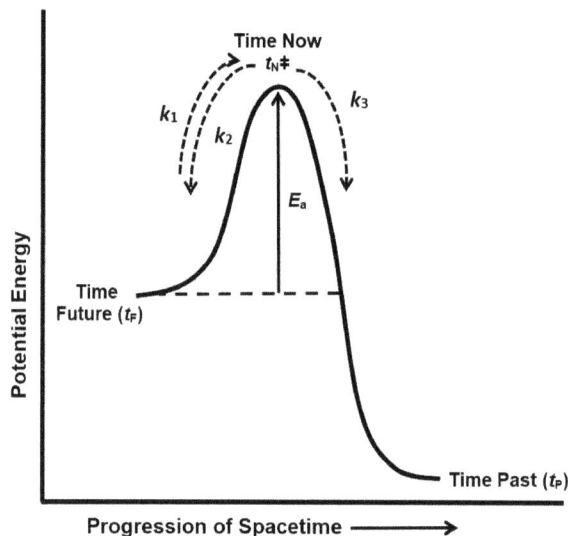

Figure 2. Potential energy diagram for the progression of time events. In the model: 1). Time events progress as a process or a reaction from $t_F \rightarrow t_N \rightarrow t_P$; 2). Time now ($t_N$) is an instantaneous transition state between t_F and t_P; 3). A reversible reaction exists between time future (t_F) and time now (t_N) events; 4). The reaction from t_N to t_P is irreversible.

3   Results

The expressions for the rate equations in different Models were used to study how the progression of time events might provide a mechanism that explains the perpetual and asymmetric dynamic nature of time. Different scenarios are presented based on whether sufficient energy of activation (E_a) is available to transition from a time now (t_N) event to a time past (t_P) event. That is, a time future (t_F) event by itself or by interacting with another a time future (t_F) event or a time past (t_P) event can surmount the energy barrier in order for the reaction to progress to a time past (t_P) event. Four different versions of my kinetic model were created to investigate the dynamics of the progression of time events. In each model, it is mathematically determined if the system depicts a steady state and whether an equilibrium can exist between time future (t_F) events and time now (t_P) events.

MODEL 1. First order time future (t_F) event reaction

MODEL 2. Reaction between two time future (t_F) events.

MODEL 3. Reaction of time future (t_F) event with time past (t_P) event.

MODEL 4. Reaction of two time future (t_F) events with a time past (t_P) event.

MODEL 1.

First Order Time Future (t_F) Event Reaction

Overall Reaction

$$t_F \rightarrow t_P$$

Reaction with a time now (t_N) intermediate

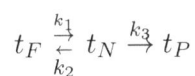

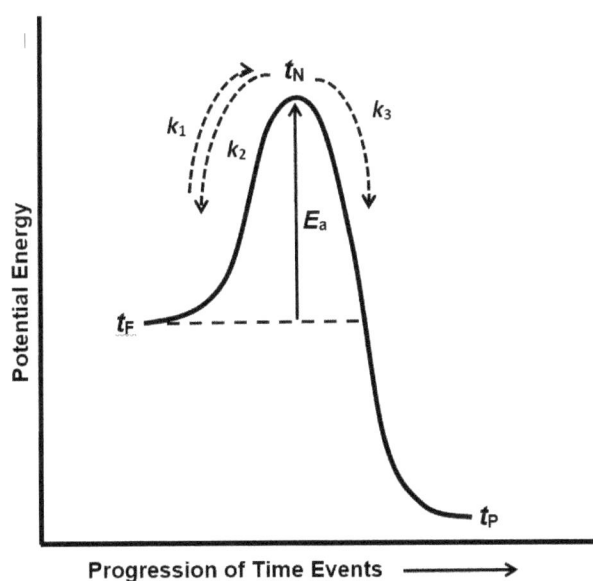

$$t_F \underset{k_2}{\overset{k_1}{\rightleftarrows}} t_N \overset{k_3}{\rightarrow} t_P$$

Figure 2. Potential energy diagram for the time future (t_F) first order reaction.

Note that, E_a = energy of activation and the rate constants have units of s^{-1}

Individual Reaction Steps

$$t_F \overset{k_1}{\rightarrow} t_N \qquad\qquad \text{rxn. 1.1}$$

$$t_N \overset{k_2}{\rightarrow} t_F \qquad\qquad \text{rxn. 1.2}$$

$$t_N \overset{k_3}{\rightarrow} t_P \qquad\qquad \text{rxn. 1.3}$$

Rate Equations

$$\frac{dt_F}{ds} = k_2 t_N - k_1 t_F \qquad\qquad \text{eq. 1.1}$$

$$\frac{dt_N}{ds} = k_1 t_F - k_2 t_N - k_3 t_N \qquad\qquad \text{eq. 1.2}$$

$$\frac{dt_P}{ds} = k_3 t_N$$

eq. 1.3

$$\frac{d(t_F + t_N)}{ds} = -k_3 t_N$$

eq. 1.4

$$\frac{d(t_F + t_N + t_P)}{ds} = 0$$

eq. 1.5

Two Scenarios are presented below to illustrate how the energy of activation (E_a) in the time future (t_F) event reaction affects the progression of time events

Scenario 1. If E_a of t_F is not equal to or greater than the minimum required for the transition to take place ($k_3 = 0$), this leads to a reversible reaction between the forward reaction ($k_1 t_F$) and the reverse reaction ($k_2 t_N$).

In this case, $\frac{dt_F}{ds} = k_2 t_N - k_1 t_F$ (eq. 1.1) and $k_3 = 0$

Then, $\frac{dt_N}{ds} = k_1 t_F - k_2 t_N - k_3 t_N$ (eq. 1.2) becomes $\frac{dt_N}{ds} = k_1 t_F - k_2 t_N$

eq. 1.6

So, $\frac{d(t_F + t_N)}{ds} = 0$, $\frac{dt_P}{ds} = 0$, and $\frac{d(t_F + t_N + t_P)}{ds} = 0$ from eqs. 1.3-1.6

Thus, in Scenario 1, if $k_3 = 0$, an equilibrium exists between t_F and t_N events and the overall system depicts a steady state, so the system doesn't expand.

Scenario 2. The E_a in t_F event is equal to or greater than the minimum that is required for a transition to occur, i.e. $k_3 \neq 0$.

An equilibrium between t_F and t_N events exists if $\frac{d(t_F + t_N)}{ds} = 0$

But, since $\frac{d(t_F + t_N)}{ds} = -k_3 t_N$ (eq. 1.4), an equilibrium and will not arise

And, since $\frac{d(t_F + t_N + t_P)}{ds} = 0$ (eq. 1.5), the overall system depicts a steady state, but the system doesn't expand

From eq. 1.3, the progression from $t_F \rightarrow t_P$ is first order with regard to t_N

Note $\frac{d(t_F + t_N + t_P)}{ds} = 0$, so that $t_F + t_N + t_P = t_F(0) + t_N(0) + t_P(0) = constant$

In Scenario 1. $k_3 = 0$, then from eqs. 1.3 & 1.5, $t_P(s) = t_P(0) = constant$

and $\frac{d(t_F + t_N)}{ds} = 0$, i.e. $t_F + t_N = T = t_F(0) + t_N(0) = constant$

It follows that $\frac{dt_F}{ds} + (k_1 + k_2) t_F = k_2 T$, and

$$t_F(s) = \frac{k_2 T}{(k_1 + k_2)} + \left(t_F(0) - \frac{k_2 T}{(k_1 + k_2)} \right) e^{-(k_1 + k_2)s}$$

eq. 1.7

$$t_N(s) = \frac{k_1 T}{(k_1 + k_2)} - \left(t_F(0) - \frac{k_2 T}{(k_1 + k_2)} \right) e^{-(k_1 + k_2)s}$$

eq. 1.8

From eqs. 1.7 and 1.8, we see that as $s \rightarrow \infty$,

$$t_F(s) \rightarrow \frac{k_2 T}{(k_1 + k_2)} \text{ and } t_N(s) \rightarrow \frac{k_1 T}{(k_1 + k_2)}$$

eq. 1.9

as $s \rightarrow -\infty$, $t_F(s) \approx \left(\frac{k_1 t_F(0) - k_2 t_N(0)}{(k_1 + k_2)} \right) e^{-(k_1 + k_2)s}$ and $t_N(s) \approx -t_F(s)$

eq. 1.10

In Scenario 2, $k_3 \neq 0$, The system (1.1)-(1.3) is linear, and (1.1), (1.2) decouple from (1.3).

The general solution provides expressions for eigenvalues ($\lambda_{1,2}$) as follows:

$$\lambda_{1,2} = \frac{-\sigma \pm \sqrt{\sigma^2 - 4k_1 k_3}}{2}. \tag{eq. 1.11}$$

Since $\sigma^2 - 4k_1 k_3 = k_2^3 + (k_1 - k_3)^2 + 2k_2(k_1 + k_3) > 0$, The eigenvalues are negative and distinct with two linearly independent eigenvectors $v_{1,2}$. Therefore, $t = c_1 e^{\lambda_1 s} v_1 + c_2 e^{\lambda_1 s} v_2$

eq. 1.12

and from eq. 1.5, $t_P(s) = t_F(0) + t_N(0) + t_P(0) - t_F(s) - t_N(s)$

So that as $s \to 0$, $t_F(s) \to 0$, $t_N(0) \to 0$ and $t_P(s) \to t_F(0) + t_N(0) + t_P(0)$

MODEL 2.

Reaction Between Two Time Future (t_F) Events

Overall Reaction

$$t_F + t_F \to t_P + t_P$$

Reaction with a time now (t_N) intermediate

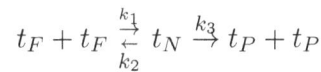

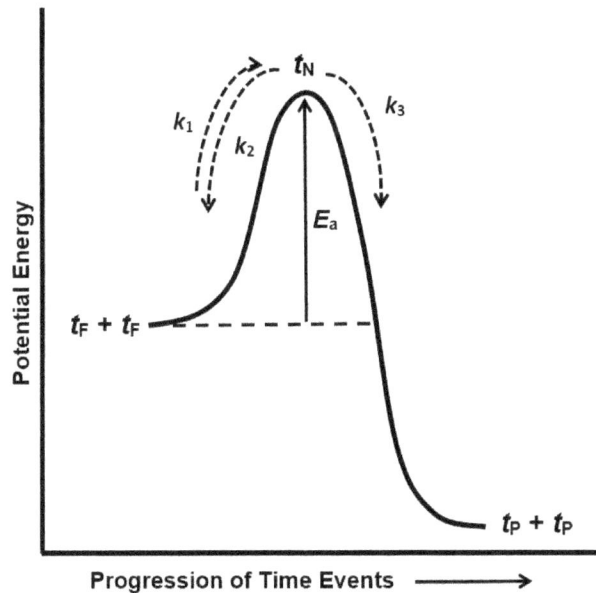

$$t_F + t_F \underset{k_2}{\overset{k_1}{\rightleftarrows}} t_N \overset{k_3}{\to} t_P + t_P$$

Figure 3. Potential energy diagram for the interaction between two time future (t_F) events. Note that, for second order reactions, the rate constants have units of s^{-1} t^{-1}.

Individual Reaction Steps

$$t_F + t_F \xrightarrow{k_1} t_N \qquad\qquad\qquad\qquad \text{rxn. 2.1}$$

$$t_N \xrightarrow{k_2} t_F + t_F \qquad\qquad\qquad\qquad \text{rxn. 2.2}$$

$$t_N \xrightarrow{k_3} t_P + t_P \qquad\qquad\qquad\qquad \text{rxn. 2.3}$$

Rate Equations

$$\frac{dt_F}{ds} = 2k_2 t_N - k_1 t_F^2 \qquad\qquad\qquad\qquad \text{eq. 2.1}$$

$$\frac{dt_N}{ds} = k_1 t_F^2 - k_2 t_N - k_3 t_N \qquad\qquad\qquad\qquad \text{eq. 2.2}$$

$$\frac{dt_P}{ds} = 2k_3 t_N \qquad\qquad\qquad\qquad \text{eq. 2.3}$$

$$\frac{d(t_F + t_N)}{ds} = k_2 t_N - k_3 t_N \qquad\qquad\qquad\qquad \text{eq. 2.4}$$

$$\frac{d(t_F + t_N + t_P)}{ds} = k_2 t_N + k_3 t_N \qquad\qquad\qquad\qquad \text{eq. 2.5}$$

Three Scenarios are presented below to illustrate how the energy of activation (E_a) in the interaction between time future (t_F) events affects the progression of time events

Scenario 1. If E_a of interaction ($t_F + t_F$) is not equal to or greater than the minimum required for the transition to take place ($k_3 = 0$), this leads to a reversible reaction between the forward reaction ($k_1 t_F^2$) and reverse reaction ($2k_2 t_N$).

In this case, $\frac{dt_F}{ds} = 2k_2 t_N - k_1 t_F^2$ (eq. 2.1) and $k_3 = 0$

Then, $\frac{dt_N}{ds} = k_1 t_F^2 - k_2 t_N - k_3 t_N$ (eq. 2.2) becomes $\frac{dt_N}{ds} = k_1 t_F^2 - k_2 t_N$ \qquad eq. 2.6

So, $\frac{d(t_F + t_N)}{ds} = k_2 t_N$, $\frac{dt_P}{ds} = 0$, and $\frac{d(t_F + t_N + t_P)}{ds} = k_2 t_N$ from eqs. 2.3-2.6

Thus, in Scenario 1, where $k_3 = 0$, an equilibrium does not exist between t_F and t_N events and the overall system does not depict a steady state.

Scenario 2. The E_a in the interaction between two t_F events is equal to or greater than the minimum that is required for a transition to occur, i.e. $k_3 \neq 0$.

An equilibrium between t_F and t_N events exists if $\frac{d(t_F + t_N)}{ds} = 0$

Since $\frac{d(t_F + t_N)}{ds} = k_2 t_N - k_3 t_N$ (eq. 2.4), an equilibrium arises if $k_2 = k_3$

And, eq. 2.2 becomes $\frac{dt_N}{ds} = k_1 t_F^2 - 2k_2 t_N$ \qquad\qquad\qquad eq. 2.7

Consequently, $\frac{d(t_F + t_N)}{ds} = 0$ since $\frac{dt_F}{ds} = -\frac{dt_N}{ds}$ from eqs. 2.1-2.7

Moreover, when $k_2 = k_3$, eq. 2.3 becomes $\frac{dt_P}{ds} = 2k_2 t_N$ \qquad\qquad eq. 2.8

214

And, eq. 2.5 becomes $\frac{d(t_F+t_N+t_P)}{ds} = 2k_2t_N$ eq. 2.9

Thus, in Scenario 2, an equilibrium can occur between t_F and t_N events, but a steady state will only exist if t_N equals zero, otherwise the system expands.

Also, if $k_2 = k_3$, $k_1t_F^2$ can be greater, less, or equal to k_2t_N.

The next scenario will address the latter situation.

<u>Scenario 3.</u> An equilibrium will also be established between the t_F events and t_N event if $k_2 = k_3$ and $k_1t_F^2 = k_2t_N$. Hence,

When $\frac{dt_F}{ds} = 2k_2t_N - k_1t_F^2$ (eq. 2.1) and $k_1t_F^2 = k_2t_N$

then $\frac{dt_F}{ds} = k_2t_N$ eq. 2.10

When $\frac{dt_N}{ds} = k_1t_F^2 - k_2t_N - k_3t_N$ (eq. 2.2) and $k_1t_F^2 = k_2t_N$

then $\frac{dt_N}{ds} = -k_3t_N$ eq. 2.11

Thus, $\frac{d(t_F+t_N)}{ds} = 0$, when $k_2 = k_3$ and $k_1t_F^2 = k_2t_N$

Additionally, when $\frac{dt_F}{ds} = k_2t_N$ (eq. 2.10) and $k_1t_F^2 = k_2t_N$,

then $\frac{dt_F}{ds} = k_1t_F^2$ eq. 2.12

Also, when $\frac{dt_N}{ds} = -k_3t_N$ (eq. 2.11) and $k_1t_F^2 = k_2t_N = k_3t_N$

then $\frac{dt_N}{ds} = -k_1t_F^2$ eq. 2.13

Since, $k_3t_N = k_1t_F^2$ (eqs. 2.11 & 2.13) and $\frac{dt_P}{ds} = 2k_3t_N$ (eq. 2.3)

It follows, $\frac{dt_P}{ds} = 2k_1t_F^2$ eq. 2.14

So, the progression from $t_F \rightarrow t_P$ is first order with regard to t_N and second order with regard to t_F

Moreover, since $\frac{d(t_F+t_N+t_P)}{ds} = k_2t_N + k_3t_N$ (eq. 2.5), then

for the overall system $\frac{d(t_F+t_N+t_P)}{ds} = \frac{dt_P}{ds} = 2k_1t_F^2$ eq. 2.15

In Scenario 3, modeling reveals that interaction between two t_F events leads to expansion of the system since a steady state doesn't exist, but an equilibrium between t_F and t_N events can occur.

<u>MODEL 3.</u>

Reaction Between a Time Future (t_F) Event and a Time Past (t_P) Event

Overall Reaction

$$t_F + t_P \rightarrow t_P + t_P$$

Reaction with t_N intermediate

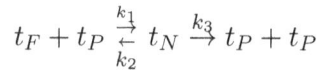

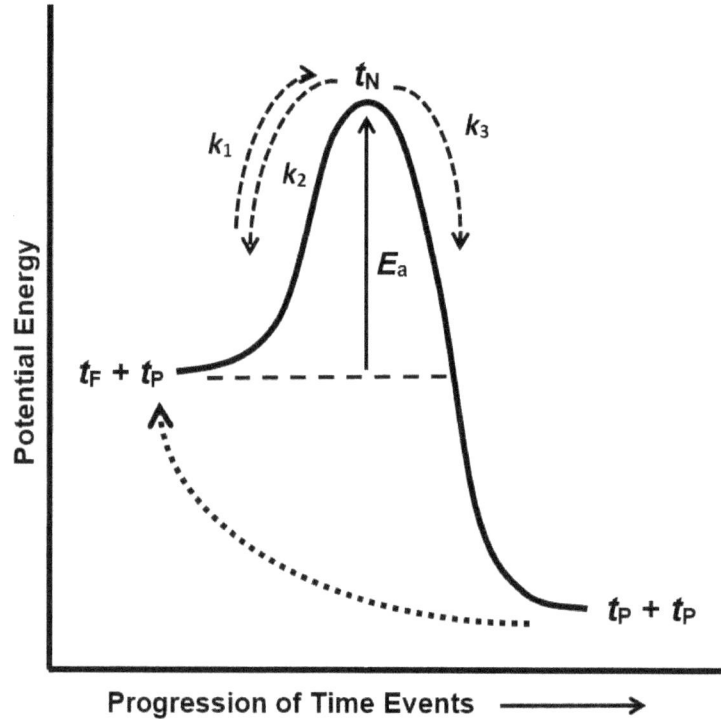

$$t_F + t_P \underset{k_2}{\overset{k_1}{\rightleftharpoons}} t_N \overset{k_3}{\rightarrow} t_P + t_P$$

Figure 4. Potential energy diagram for the interaction between a time future (t_F) event and a time past (t_P) event. Note that, for first order reactions, rate constants have units of s^{-1}; for second order reactions, the rate constant has units of s^{-1} t^{-1}.

Individual Reaction Steps

$$t_F + t_P \overset{k_1}{\rightarrow} t_N \qquad\qquad\qquad\qquad \text{rxn. 3.1}$$

$$t_N \overset{k_2}{\rightarrow} t_F + t_P \qquad\qquad\qquad\qquad \text{rxn. 3.2}$$

$$t_N \overset{k_3}{\rightarrow} t_P + t_P \qquad\qquad\qquad\qquad \text{rxn. 3.3}$$

Rate Equations

$$\frac{dt_F}{ds} = k_2 t_N - k_1 t_F t_P \qquad\qquad\qquad \text{eq. 3.1}$$

$$\frac{dt_N}{ds} = k_1 t_F t_P - k_2 t_N - k_3 t_N \qquad\qquad \text{eq. 3.2}$$

$$\frac{dt_P}{ds} = k_2 t_N + 2\,k_3 t_N - k_1 t_F t_P \qquad\qquad \text{eq. 3.3}$$

$$\frac{d(t_F + t_N)}{ds} = -\,k_3 t_N \qquad\qquad\qquad \text{eq. 3.4}$$

$$\frac{d(t_F + t_N + t_P)}{ds} = k_2 t_N + k_3 t_N - k_1 t_F t_P \qquad \text{eq. 3.5}$$

216

Three Scenarios are presented below to illustrate how the energy of activation (E_a) in the interaction between a time future (t_F) event and time past (t_P) event affects the progression of time events

Scenario 1. If E_a of interaction ($t_F + t_P$) is not equal to or greater than the minimum required for the transition to take place ($k_3 = 0$), this leads to a reversible reaction between the forward reaction ($k_1 t_F t_P$) and reverse reaction ($k_2 t_N$).

In this case, $\frac{dt_F}{ds} = k_2 t_N - k_1 t_F t_P$ (eq. 3.1) and $k_3 = 0$

Since $\frac{dt_N}{ds} = k_1 t_F t_P - k_2 t_N - k_3 t_N$ (eq. 3.2) and $k_3 = 0$

then $\frac{dt_N}{ds} = k_1 t_F t_P - k_2 t_N$ eq. 3.6

So, $\frac{d(t_F + t_N)}{ds} = 0$ eq. 3.7

Since $\frac{dt_P}{ds} = k_2 t_N + 2\,k_3 t_N - k_1 t_F t_P$, (eq. 3.3) and $k_3 = 0$

then $\frac{dt_P}{ds} = k_2 t_N - k_1 t_F t_P$, and $\frac{dt_P}{ds} = \frac{dt_F}{ds} = -\frac{dt_N}{ds}$ eq. 3.8

Also, $\frac{d(t_F + t_N + t_P)}{ds} = k_2 t_N - k_1 t_F t_P$ eq. 3.9

Thus, in Scenario 1 where $k_3 = 0$, an equilibrium exists between t_F and t_N events and the overall system depict a steady state if $k_2 t_N = k_1 t_F t_P$.

Scenario 2. The E_a in the interaction between the $t_F + t_P$ events is equal to or greater than the minimum that is required for a transition to occur, i.e. $k_3 \neq 0$.

An equilibrium between t_F and t_N events exists if $\frac{d(t_F + t_N)}{ds} = 0$

But, since $\frac{d(t_F + t_N)}{ds} = -k_3 t_N$ (eq. 3.4), an equilibrium won't arise between t_F and t_N events when $k_3 > 0$.

Scenario 3. Although an equilibrium won't occur between t_F and t_N events, a steady state will exist if $\frac{d(t_F + t_N + t_P)}{ds} = k_2 t_N + k_3 t_N - k_1 t_F t_P = 0$ (see eq. 3.5). Hence, when $\frac{dt_F}{ds} = k_2 t_N - k_1 t_F t_P$ (eq. 3.1) and $(k_2 + k_3) t_N = k_1 t_F t_P$,

then $\frac{dt_F}{ds} = -k_3 t_N$ eq. 3.10

When $\frac{dt_N}{ds} = k_1 t_F t_P - k_2 t_N - k_3 t_N$ (eq. 3.2) and $(k_2 + k_3) t_N = k_1 t_F t_P$,

then $\frac{dt_N}{ds} = 0$ eq. 3.11

When $\frac{dt_P}{ds} = k_2 t_N + 2\,k_3 t_N - k_1 t_F t_P$ (eq. 3.3) and $(k_2 + k_3) t_N = k_1 t_F t_P$,

then $\frac{dt_P}{ds} = k_3 t_N$ eq. 3.12

So, if $k_3 > 0$ and $(k_2 + k_3) t_N = k_1 t_F t_P$, $\frac{dt_P}{ds} = -\frac{dt_F}{ds}$, also $\frac{dt_N}{ds} = 0$ eq. 3.13

Thus, $\frac{d(t_F + t_N + t_P)}{ds} = 0$,   eq. 3.14

Thus, in Model 3, results shows that interaction between t_F and t_P events can establish a steady state, but an equilibrium between t_F and t_N events doesn't occur except when $k_3 = 0$.

In this steady state, k_2 can be greater or less than, or equal to k_3.

If the system is not in steady state, $k_1 t_F t_P$ can be greater or less than $(k_2 + k_3)t_N$, then it will expand or contract.

MODEL 4.

Reaction Between Two Time Future (t_F) Events and a Time Past (t_P) Event

Overall Reaction

$$t_F + t_F + t_P \rightarrow t_P + t_P + t_P$$

Reaction with t_N intermediate

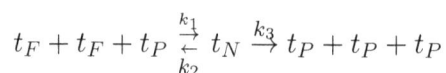

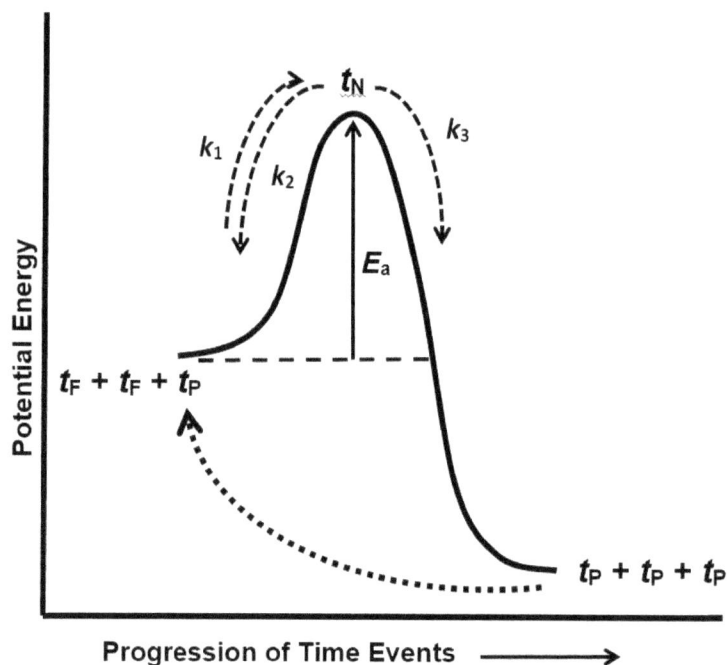

$$t_F + t_F + t_P \underset{k_2}{\overset{k_1}{\rightleftarrows}} t_N \overset{k_3}{\rightarrow} t_P + t_P + t_P$$

Figure 5. Potential energy diagram for the interaction between two time future (t_F) events and a time past (t_P) event. Note that, for first order reactions, the rate constants have units of s^{-1}; for second order reactions, the rate constant has units of $s^{-1} \; t^{-1}$.

Individual Reaction Steps

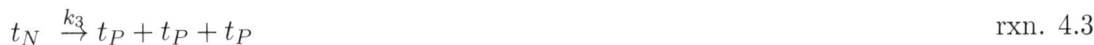

$$t_F + t_F + t_P \xrightarrow{k_1} t_N \qquad \text{rxn. 4.1}$$

$$t_N \xrightarrow{k_2} t_F + t_F + t_P \qquad \text{rxn. 4.2}$$

$$t_N \xrightarrow{k_3} t_P + t_P + t_P \qquad \text{rxn. 4.3}$$

Rate Equations

$$\frac{dt_F}{ds} = 2k_2 t_N - k_1 t_F^2 t_P \qquad \text{eq. 4.1}$$

$$\frac{dt_N}{ds} = k_1 t_F^2 t_P - k_2 t_N - k_3 t_N \qquad \text{eq. 4.2}$$

$$\frac{dt_P}{ds} = k_2 t_N + 3\, k_3 t_N - k_1 t_F^2 t_P \qquad \text{eq. 4.3}$$

$$\frac{d(t_F + t_N)}{ds} = k_2 t_N - k_3 t_N \qquad \text{eq. 4.4}$$

$$\frac{d(t_F + t_N + t_P)}{ds} = 2k_2 t_N + 2k_3 t_N - k_1 t_F^2 t_P \qquad \text{eq. 4.5}$$

Three Scenarios are presented below to illustrate how the energy of activation (E_a) in the interaction between time future (t_F) events and a time past (t_P) event affects the progression of time events

Scenario 1. If E_a of interaction ($t_F + t_F + t_P$) is not equal to or greater than the minimum required for the transition to take place ($k_3 = 0$), this leads to a reversible reaction between the forward reaction ($k_1 t_F^2 t_P$) and the reverse reaction ($2k_2 t_N$).

In this case, $\frac{dt_F}{ds} = 2k_2 t_N - k_1 t_F^2 t_P$ (eq. 4.1) and $k_3 = 0$

As, $\frac{dt_N}{ds} = k_1 t_F^2 t_P - k_2 t_N - k_3 t_N$ (eq. 4.2) and $k_3 = 0$,

then $\frac{dt_N}{ds} = k_1 t_F^2 t_P - k_2 t_N$ $\qquad$ eq. 4.6

So, eq. 4.4 becomes $\frac{d(t_F + t_N)}{ds} = k_2 t_N$ $\qquad$ eq. 4.7

Since, $\frac{dt_P}{ds} = k_2 t_N + 3\, k_3 t_N - k_1 t_F^2 t_P$, (eq. 4.3) and $k_3 = 0$

then $\frac{dt_P}{ds} = k_2 t_N - k_1 t_F^2 t_P$, and $\frac{dt_P}{ds} = -\frac{dt_N}{ds}$ $\qquad$ eq. 4.8

Since, $\frac{d(t_F + t_N + t_P)}{ds} = 2k_2 t_N + 2k_3 t_N - k_1 t_F^2 t_P$ (eq. 4.5)

then, $\frac{d(t_F + t_N + t_P)}{ds} = 2k_2 t_N - k_1 t_F^2 t_P$ $\qquad$ eq. 4.9

And, $\frac{d(t_F + t_N + t_P)}{ds} = \frac{dt_F}{ds}$ $\qquad$ eq. 4.10

Thus, in Scenario 1 where $k_3 = 0$, an equilibrium doesn't exist between t_F and t_N events, but the overall system depicts a steady state if $2k_2 t_N = k_1 t_F^2 t_P$.

Scenario 2. The E_a in the interaction between two t_F events and a time past (t_P) event is equal to or greater than the minimum that is required for a transition to occur, i.e.

$k_3 \neq 0$.

An equilibrium between t_F and t_N events exists if $\frac{d(t_F + t_N)}{ds} = 0$

Since $\frac{d(t_F + t_N)}{ds} = k_2 t_N - k_3 t_N$ (eq. 4.4), equilibrium arises if $k_2 = k_3$

When $k_2 = k_3$, then eq. 4.2 becomes $\frac{dt_N}{ds} = k_1 t_F^2 t_P - 2k_2 t_N$ eq. 4.11

Consequently, $\frac{d(t_F + t_N)}{ds} = 0$ since $\frac{dt_F}{ds} = -\frac{dt_N}{ds}$ from eqs. 4.1 & 4.11

Moreover, eq. 4.3 becomes $\frac{dt_P}{ds} = 4k_3 t_N - k_1 t_F^2 t_P$ eq. 4.12

And, eq. 4.5 becomes $\frac{d(t_F + t_N + t_P)}{ds} = 4k_3 t_N - k_1 t_F^2 t_P$ eq. 4.13

Thus, in Scenario 2, an equilibrium can occur between t_F and t_N events and a steady state will exist if $4k_3 t_N - k_1 t_F^2 t_P = 0$, otherwise the system expands.

Also, if $k_2 = k_3$, $k_1 t_F^2 t_P$ can be greater, less, or equal to $k_2 t_N$.

The next scenario will address the latter situation.

Scenario 3. An equilibrium will also be established between the t_F events and t_N event if $k_2 = k_3$ and $k_1 t_F^2 t_P = k_2 t_N$. Hence,

When, $\frac{dt_F}{ds} = 2k_2 t_N - k_1 t_F^2 t_P$ (eq. 4.1) and $k_1 t_F^2 t_P = k_2 t_N$

then $\frac{dt_F}{ds} = k_2 t_N$ eq. 4.14

When $\frac{dt_N}{ds} = k_1 t_F^2 t_P - k_2 t_N - k_3 t_N$ (eq. 4.2) and $k_1 t_F^2 t_P = k_2 t_N$

then $\frac{dt_N}{ds} = -k_2 t_N = -k_3 t_N$ eq. 4.15

Thus, $\frac{d(t_F + t_N)}{ds} = 0$, when $k_2 = k_3$ and $k_1 t_F^2 t_P = k_2 t_N$

Additionally, when $\frac{dt_F}{ds} = k_2 t_N$ (eq. 4.14) and $k_1 t_F^2 t_P = k_2 t_N$,

then $\frac{dt_F}{ds} = k_1 t_F^2 t_P$ eq. 4.16

Also, when $\frac{dt_N}{ds} = -k_3 t_N$ (eq. 4.15) and $k_1 t_F^2 t_P = k_2 t_N = k_3 t_N$

then $\frac{dt_N}{ds} = -k_1 t_F^2 t_P$ eq. 4.17

Since $k_2 t_N = k_3 t_N = k_1 t_F^2 t_P$ and $\frac{dt_P}{ds} = k_2 t_N + 3k_3 t_N - k_1 t_F^2 t_P$ (eq. 4.5)

It follows, $\frac{dt_P}{ds} = 3\ k_3 t_N$ *and also* $\frac{dt_P}{ds} = 3k_1 t_F^2 t_P$ eq. 4.18

So, the progression from $t_F \rightarrow t_P$ is first order with regard to t_N and second order with regard to t_F.

Moreover, since $\frac{d(t_F + t_N + t_P)}{ds} = 2k_2 t_N + 2k_3 t_N - k_1 t_F^2 t_P$ (eq. 4.5)

Then, for the overall system $\frac{d(t_F + t_N + t_P)}{ds} = \frac{dt_P}{ds} = 3k_1 t_F^2 t_P$ eq. 4.19

Thus, in Scenario 3, modeling reveals that interaction between two t_F events and a t_P event leads to expansion of the system since a steady state doesn't exist, but an equilibrium between t_F and t_N events can occur.

4 SUMMARY

Table 1. Modeling event interactions as not crossing the energy barrier (i.e. $k_3 = 0$)

Model	Reaction	Equilibrium between t_F and t_N events	Steady State
1	$t_F \overset{k_1}{\underset{k_2}{\rightleftarrows}} t_N \overset{k_3}{\rightarrow} t_P$	YES $\frac{d(t_F + t_N)}{ds} = 0$	YES $\frac{d(t_F + t_N + t_P)}{ds} = 0$
2	$t_F + t_F \overset{k_1}{\underset{k_2}{\rightleftarrows}} t_N \overset{k_3}{\rightarrow} t_P + t_P$	NO $\frac{d(t_F + t_N)}{ds} = k_2 t_N$	NO $\frac{d(t_F + t_N + t_P)}{ds} = k_2 t_N$
3	$t_F + t_P \overset{k_1}{\underset{k_2}{\rightleftarrows}} t_N \overset{k_3}{\rightarrow} t_P + t_P$	YES $\frac{d(t_F + t_N)}{ds} = 0$	YES , if $k_2 t_N = k_1 t_F t_P$
4	$t_F + t_F + t_P \overset{k_1}{\underset{k_2}{\rightleftarrows}} t_N \overset{k_3}{\rightarrow}$ $t_P + t_P + t_P$	NO $\frac{d(t_F + t_N)}{ds} = k_2 t_N$	YES, if $2 k_2 t_N = k_1 t_F^2 t_P$

Table 2. Modeling event interactions that can cross the energy barrier (i.e. $k_3 = 0$)

Model	Reaction	Equilibrium between reversible and irreversible reactions	Steady State
1	$t_F \overset{k_1}{\underset{k_2}{\rightleftarrows}} t_N \overset{k_3}{\rightarrow} t_P$	NO $\frac{d(t_F + t_N)}{ds} = -k_3 t_N$	YES $\frac{d(t_F + t_N + t_P)}{ds} = 0$
2	$t_F + t_F \overset{k_1}{\underset{k_2}{\rightleftarrows}} t_N \overset{k_3}{\rightarrow} t_P + t_P$	YES, if $k_2 = k_3$	NO $\frac{d(t_F + t_N + t_P)}{ds} = 2 k_1 t_F^2$
3	$t_F + t_P \overset{k_1}{\underset{k_2}{\rightleftarrows}} t_N \overset{k_3}{\rightarrow} t_P + t_P$	NO $\frac{d(t_F + t_N)}{ds} = -k_3 t_N$	YES, if $(k_2 + k_3) t_N = k_1 t_F t_P$
4	$t_F + t_F + t_P \overset{k_1}{\underset{k_2}{\rightleftarrows}} t_N \overset{k_3}{\rightarrow}$ $t_P + t_P + t_P$	YES, if $k_2 = k_3$	NO $\frac{d(t_F + t_N + t_P)}{ds} = 3 k_1 t_F^2 t_P$

5 DISCUSSION

Modeling the flow of time as a progression of events can illuminate thinking about time in new and different ways. For example, if the progression of time events involves an interaction of events, then one needs to consider how the nature of interaction will affect the reaction. In chemistry, we model reactions based on concentration of the components. However, in modeling the progression of time events, concentration doesn't make sense since it is not a molecule in a solution. Rather, I like to view the events as parcels of energy. That is, events are considered to be quantized (see Introduction) and the progression of events occurs as a reaction based on changes in energy.

To illustrate the changes in energy during the progression of time events, it is plotted as a potential energy diagram. Based on reasoning from TST, the kinetic energy in the interaction of time events must have reached a sufficient energy of activation (E_a) to pass the energy barrier during the transition state. The energy of t_N events at the transition state consists of high potential (available) energy. Since energy and entropy are interconnected, an increase in potential energy leads to a decrease in entropy. So, a t_N event will have high potential energy (i.e. available energy) and low entropy. Since energy is conserved [1, 23, 24], the equilibrium between between t_F and t_N will oscillate between kinetic energy and potential energy in association with increased (high) entropy and (low) decreased entropy, respectively [1, 24, 25]. Then, upon progression of a t_N event to t_P events, the potential energy decreases and kinetic energy increases. In parallel, the entropy increases, which obeys the 2nd Law of Thermodynamics [1, 23-25]. Another way to view the progression of a t_N event to t_P events is that sufficient low entropy at t_N must be attained to produce a net increase in entropy in formation of t_P events (i.e. a putative entropy of activation). In this view, it is both the low entropy and high potential energy created in the transition state (t_N event) that drives the spontaneous and irreversible generation of t_P events.

So, if the progression involves a first order-type reaction, then the energy level of only one t_F event determines the reaction rate [26]. This rate then depends on a t_F event having enough energy to pass the energy barrier to transition to a t_P event. On the other hand, if an interaction involves two t_F events, then the reaction rate will depend on the energy level of both t_F events. This second-order-type reaction progresses at a rate proportional to the square of the energy level of one t_F event as the two events have equivalent energy levels [27]. If the two events have different energy levels, then the rate will depend on the product of the energy levels. If the reaction involves the interaction of three events, it is a third-order-type reaction. This could involve the energy level of all three being equivalent, two being the same and one being different, or all three being dissimilar. The rate will be influenced by the energy of each event [28].

Notably, a second or third order reaction doesn't necessarily mean that it is faster than a first order reaction, just that the changes in the energy level in a first order reaction have less impact on the reaction rate if the change occurs within the same range of energy levels [29]. Thus, a higher order reaction might have a higher likelihood of having sufficient energy to pass the energy barrier to transition to a t_P event.

The other factor that can affect the rate of a reaction is the presence of a catalyst. A catalyst provides an alternative reaction mechanism having a lower energy of activation. So, the presence of a catalyst will increase the reaction by stabilizing the transition state [30]. Thus, a catalyst can significantly accelerate the rate of a second order reaction.

222

In an autocatalytic reaction, one of the reaction products acts as a catalyst for the same reaction [31]. In this case, the reaction product, by interacting with the other reactants, can act as an energizing factor that increases the energy level of the other reactants. As the products are not consumed, they can also catalyze many other reactions.

In my model, a t_P event was incorporated as a reactant to catalyze its own production in order to analyze the dynamics of an autocatalytic mechanism. The presence of a t_P event that catalyzes its own production is predicted to accelerate its own production. For example, comparing Models 2 and 4, the expression for rate of t_P event production was found to be $\frac{dt_P}{ds} = 3k_1 t_F^2 t_P$ for the autocatalyzed reaction as compared to $\frac{dt_P}{ds} = 2k_1 t_F^2$ for the uncatalyzed reaction.

Additionally, to study the effect of events having insufficient energy to pass the energy barrier, the k_3 rate constant was set to zero in each Model so that the transition from t_N to t_P was blocked (Table 1). This situation leads to a reversible reaction between the forward reaction and the reverse reaction from t_F to t_N, and vice versa. The results show that an equilibrium was established between t_F and t_N (i.e. $\frac{d(t_F + t_N)}{ds} = 0$) in Models 1 and 3, but not in Models 2 and 4. Modeling this theoretical inability to pass the energy barrier showed a variable effect on the overall system. In Models 1, 3, & 4, a steady state ($\frac{d(t_F + t_N + t_P)}{ds} = 0$) was determined to occur in the reversible reaction between t_F and t_N. In Models 1 and 3, a steady state is established because there is an equilibrium between between t_F and t_N. However, in Model 4, a steady state can be established even though there is not an equilibrium between t_F and t_N (i.e. $\frac{d(t_F + t_N)}{ds} \neq 0$). This is because t_P is also a reactant in modeling the autocatalytic process. In contrast, in Model 2, a steady state will not be established due to lack of an equilibrium between t_F and t_N ($\frac{d(t_F + t_N)}{ds} \neq 0$). Thus, the system is predicted to expand because t_F and t_N increase at a rate of $\frac{d(t_F + t_N)}{ds} = k_2 t_N$.

For Model 2, this result showing that the overall system expands when two t_F events interact and $k_3 = 0$, may have significance to physics. For example, the interaction of two t_F events in a reaction that doesn't progress from t_N to t_P may dynamically portray entanglement of two particles. That is, the dynamics of two interacting t_F events would simulate two entangled particles. So, from the model, it might be predicted that two particles when entangled will exist in space throughout the Universe until they are measured. This explanation from the model also has relevance when $k_3 > 0$ and a t_P event is incorporated into the reaction as a reactant. In this case, when a t_P event interacts with two t_F events and t_N can progress to t_P, the dynamics change such that an equilibrium occurs between t_F and t_N (i.e. $\frac{d(t_F + t_N)}{ds} = 0$) and the system is predicted to expand ($\frac{d(t_F + t_N + t_P)}{ds}$) at an even greater rate than when a t_P event is not a reactant ($3k_1 t_F^2 t_P$ versus $2k_1 t_F^2$). Thus, if the interaction of two interacting t_F events dynamically simulates two entangled particles and if their interaction with a t_P event portrays a measurement, then the model would predict "observation" of entangled particles contributes to expansion of the Universe.

One other prediction from the equilibrium between t_F and t_N events is that there is no change in time when the entangled particles are measured. That is, at the time of measurement, the particles are at exactly the same time as when they were entangled. If this is the case, there would be no change in time of the particles upon measurement. So the argument that information travels faster than the speed of light doesn't make sense as speed is a function of change in time. Thus, the particles can be at different places in space but they are at the same "when" in time. Consequently, perhaps in quantum mechanics, we should be considering the temporal changes that occur when we are thinking about spatial non-locality.

If an equilibrium exists between time future (t_F) events and time now (t_N) events, an event will exist in both the future and present. This equilibrium is composed of a reversible reaction – a forward reaction and a backward reaction, which simultaneously are constantly fluxuating back and forth in a wave-like dynamic. Thus, in the equilibrium, time events can exist in a superposition state whereby they are both t_F and t_N events. It can be envisioned that this superposition between t_F and t_N events pervades throughout the Universe much like gravity has an infinite range and light radiates everywhere. In this view, t_N events at each point of reference are linked to all possible t_F events throughout the Universe. This thinking seems to resonate with the interpretation many worlds interpretation (MWI) which envisions that our Universe is just one of numerous parallel worlds that branched off from each other. This difference, however, is that the MWI relates back to the big bang while the idea of a superposition between t_F and t_N events relates to the current state of the Universe. The latter idea seems to concur with the notion of free will (now events emerge from the interaction of future events), while in the MWI, the evolution of the Universe is deterministic. The concept of a superposition between t_F and t_N events might also relate to other superposition states in physics whereby a quantum system is able to be in multiple states at the same time.

This line of thinking also pertains to one of the goals of my study – To understand the asymmetric dynamics of time. If a superposition exists between t_F and t_N events that extends throughout the Universe, it would provide a state that is the same at every point of reference. This state is a condition where time does not change in the superposition between t_F and t_N events, it only changes at the transition state when a t_N event progresses to a t_P event. It doesn't mean t_F and t_N are zero or that time is the same everywhere, it means that time doesn't change in the equilibrium between t_F and t_N events ($\frac{d(t_F+t_N)}{ds} = 0$). If this condition exists, it establishes a "when" from which time could progress asymmetrically at all points of reference in the Universe. This idea also pertains to time reversal symmetry which holds that the laws of physics can't distinguish between forward and backward directions in time. It sounds counter intuitive, but the invariance in symmetry might account for the variance in asymmetry. In this view, maybe the ground state for progression of time events is based on a superposition involving a forward and backward direction in time. If so, then at each point of reference, the origin for asymmetric progression of time events might stem from this superposition.

In particle physics, the notion that actions and reactions propagate forward and backwards in time is not new. For example, in his Theory of Positrons [32], Richard Feynman gave his solution to the problem of the behavior of positrons and electrons as "negative energy states" that "appear in a form which may be pictured (as by Stückelberg) in space-time as waves traveling away from the external potential backwards in time. Experimentally, such a wave corresponds to a positron approaching the potential and annihilating the electron. A particle moving forward in time (electron) in a potential may be scattered forward in time (ordinary scattering) or backward (pair annihilation). When moving backward (positron) it may be scattered backward in time (positron scattering) or forward (pair production). For such a particle the amplitude for transition from an initial to a final state is analyzed to any order in the potential by considering it to undergo a sequence of such scatterings." Perhaps, modeling the actions and reactions of particles as events might give rise to new concepts about time.

At this juncture, I'd like to go full circle and return to another goal given in the Introduction – To understand the perpetual dynamics of time. My view on this issue is that since time appears to be perpetual then we might think of it as a perpetual progression of events. In my model, the autocatalysis by t_P events was incorporated as a mechanism to study the perpetual progression of events. Findings from the model indicate that this autocatalytic mechanism is predicted to accelerate the rate of the reaction. However, this was from the perspective of one point of reference and doesn't necessarily indicate that the dynamics in whole Universe as a system are perpetual. So, a better question might be: does autocatalysis that increases the progression of events lead to a system that is self-sustaining? For this to happen, it would require that t_P events at all points of reference throughout the Universe function as autocatalysts. In that line of thinking, the Universe would need to be comprised of t_P events which are created by other t_P events in the Universe, such that as a whole, the entire system is able to catalyze its own production.

Thus, the autocatalyzed production of events at one reference point would need to promote the autocatalyzed production of events at another reference point, which would drive the perpetual progression of events in the system. The way that this could happen is if t_P events at one point of reference exist as t_F events at another point of reference. For an example, the photons that are generated by nuclear reaction in the sun and released as sunlight might be thought of as t_P events from the reference point of the sun. However, these t_P events are t_F events from earth's point of reference during the 8 minutes it takes for photons to travel to our planet. When the sunlight reaches earth, the photons interact with other particles in t_N events to produce t_P events that then serve as t_F events with regard to other points of reference and so on. Thus, the ability of t_P events to serve as t_F events at other points of reference and the ability of t_P events to autocatalyze their own production would theoretically make the whole system self-sustaining. In this way the Universe as a whole would dynamically be self-sustaining.

Acknowledgements: I thank Drs. Gilberto Schleiniger for invaluable discussions and assistance with the mathematics,

Funding: This project was generously supported by CATX Inc.

References

[1] Sheehan WF. Physical Chemistry 2nd edition. Allyn & Bacon Inc, Boston. 1970.

[2] Wilczek F. Fundamentals Ten Keys to Reality. Penguin Press, 2021. p1-272.

[3] Boman BM. A Simple Mathematical Model for the Asymmetry of Time. Presented at the Fifth International Conference on the Nature and Ontology of Spacetime, Varna, Bulgaria, May 16, 2018.

[4] Boman BM. A Simple Mathematical Model for the Asymmetry of Time. arXiv:2102.08826v1, Feb 14, 2021.

[5] Boman BM. Modeling spacetime based on transition state theory. In: Slagter RJ, Keresztes Z (eds) Spacetime 1909 – 2019. Minkowski Institute Press, Montreal. 2020. ISBN 978-1-927763-54-4, pp179-206.

[6] Boman BM. Becoming as a Transition State in Spacetime. Presented at the Sixth International Conference on the Nature and Ontology of Spacetime, Varna, Bulgaria, September 12, 2022.

[7] Zwart PJ. The flow of time. Synthese 24:133 – 158. 1972.

[8] Beynon JH, Gilbert JR. Application of Transition State Theory to Unimolecular Reactions: An Introduction. John Wiley & Sons, New York. 1984.

[9] Glasstone S, Laidler KJ, Eyring H. The Theory of Rate Processes: The Kinetics of Chemical Reactions, Viscosity, Diffusion and Electrochemical Phenomena. McGraw-Hill Book Company Inc, New York & London. 1941.

[10] Moore JW, Pearson RG. Kinetics and Mechanism. John Wiley & Sons Inc, New York, NY, USA. 1981.

[11] Eyring H. The Activated Complex in Chemical Reactions. J Chem Phys 3:107-115. 1934.

[12] Evans MG, Polanyi M. Some applications of the transition state method to the calculation of reaction velocities, especially in solution. Trans Faraday Soc 31:875-94. 1935.

[13] Laidler K, King C. Theories of chemical reaction rates. R. E. Krieger Publishing Co., New York. 1979.

[14] Laidler KJ. Just what is a transition state? J Chem Educ 65:540-2. 1988.

[15] Laidler KJ. Theories of chemical reaction rates. R. E. Krieger Publishing Co., New York. 1979.

[16] Rovelli C. The Order of Time. Riverhead Books, New York. 2018.

[17] Feynman R. The Theory of Positrons. Physical Review. 76:749–759. 1949.

www.ingramcontent.com/pod-product-compliance
Lightning Source LLC
Chambersburg PA
CBHW051209200326
41519CB00025B/7058